The Determination of
Sulphur-containing Groups

THE ANALYSIS OF ORGANIC MATERIALS

An International Series of Monographs

edited by R. BELCHER and D. M. W. ANDERSON

The Determination of Sulphur-containing Groups

Volume 2. Analytical Methods for Thiol Groups

M. R. F. ASHWORTH

University of The Saarland
Saarbrücken, Germany

1976

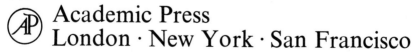

Academic Press
London · New York · San Francisco

A Subsidiary of Harcourt Brace Jovanovich, Publishers

ACADEMIC PRESS INC. (LONDON) LTD.
24/28 Oval Road
London NW1

United States Edition published by
ACADEMIC PRESS INC.
111 Fifth Avenue
New York, New York 10003

Copyright © 1976 by
ACADEMIC PRESS INC. (LONDON) LTD.

Library of Congress Catalog Card Number: 72-11465
ISBN: 0-12-065002-9

PRINTED IN GREAT BRITAIN BY
Page Bros (Norwich) Ltd, Norwich

PREFACE

This book aims at giving analytical methods for compounds in which the thiol group is carried primarily by a carbon atom joined either to hydrogen or to another carbon atom. This comprises alkane- and arenethiols and also includes compounds which contain other functional groups, such as mercaptocarboxylic acids, mercaptoamines, mercaptoalcohols, etc. Notable here are the amino acids cysteine and glutathione and also penicillamine. As a general rule, only those analytical procedures for amino acids are given in which the thiol group, or at least the sulphur atom, plays a part. To go beyond this to cover procedures for amino acids among which an occasional sulphur-containing representative chanced to be included, and which clearly depend on participation of groups other than the thiol group, would exceed the scope of the book and would in any case often duplicate material in other reviews or compilations, e.g. those devoted to the chromatography of amino acids.

Further, the applications to thiols in complex material such as peptides or proteins has been treated rather sparingly, since other factors such as structure or the presence of other groups are highly influential. Biochemical aspects have in general not been considered.

Xanthates, thioloacids, thionothioloacids, dithiocarbamates and other compound classes containing oxygen or nitrogen atoms do not fulfil the inclusion requirement but some exceptions have been made; an example is mercaptobenzothiazole, on which considerable work has been performed.

The accusation of inconsistency now trembling on the reader's lips may also be levelled at the author for his treatment of potential thiols. Their tautomeric system must inevitably include a heteroatom, thereby disqualifying them from inclusion. Although little mention is made of compounds such as thiourea or thiobarbituric acid, nevertheless the important mercaptopurines have been included.

In compilations on analytical methods, considerations of space force a decision to give either considerable experimental details of a limited number of methods, or brief descriptions of a larger number. The author has taken the latter way, hoping to show the reader the wide range of possibilities and to give him a greater chance of finding a procedure or principle suited to his own problem. In the author's opinion, it is usually best to consult the original literature for full experimental details of a procedure. Whole publications cannot be quoted in compilations and even the larger extract given may fail to contain a decisive clue or tip present, possibly, in the introduction, results or discussion.

The analytical information is subdivided into purely physical methods (including chromatographic procedures) and chemical methods. The latter, mainly classified according to the reaction undergone by the thiol group, comprise Chapters 1 to 10.

Some reagents participate in more than one type of reaction, e.g. iron(III) which may oxidize (Chapter 1), yield mercaptides (Chapter 2) or react in a hitherto unclarified way (Chapter 9). It is convenient to quote such reagents in only one place, and cross references are given from other chapters.

May, 1975   **M. R. F. Ashworth**

CONTENTS

1. OXIDATION

The oxidation of thiols is primarily a two-electron change, yielding disulphides:

$$2RSH \rightarrow RSSR + 2H^+ + 2e$$

Further oxidation to the sulphur(IV) and sulphur(VI) states, as in sulphinic, sulphonic or sulphuric acids, is possible with stronger oxidizing agents, but this is usually more difficult to control and quantitative procedures based on it are rare. Energetic total oxidation to sulphate, which belongs to elemental analysis, is unspecific for a particular sulphur-containing class; such procedures are therefore not given here.

About fifty oxidizing agents, arranged alphabetically, are described below.

1. AZO COMPOUNDS

Kosower *et al.* (1969) studied the oxidation of glutathione (in the test system of human erythrocytes) with an azo compound, methyl phenyl-diazenecarboxylate, and found that the reaction was rapid (complete in 1 to 2 min), unaffected by oxygen, and approximately stoichiometric:

$$2GSH + C_6H_5 - N = N - COOCH_3$$
$$\rightarrow GSSG + C_6H_5 - NH - NH - COOCH_3$$

There appears to be no example of the analytical application of this, although a spectrophotometric method, based on lowering of the absorbance due to the azo group, must be possible.

2. BROMATE

This relatively strong oxidizing agent has been used in direct and indirect titrimetric determinations of thiols. The stoichiometry depends on conditions, such as pH, $[Br^-]$, temperature and time of reaction. Amongst those who

1

Table I. Direct titrations with bromate

Sample	Conditions	End-point indication	References
Cysteine	In HCl or H_2SO_4, $+ Br^-$; reacts with 6Br atoms	Yellow, persisting for 5 min	Okuda (1924)
Cysteine	In acid, with or without Br^-	Potentiometric	Yamazaki (1930)
Thiophenol, thiocresols	In acetic acid, $+$ conc. HCl; 60–80°C	First yellow	Gregg and Blood (1951)
Mercaptobenzothiazole	In acid	Amperometric	Liberti (1951)
Mercaptoaneurin	$+$ conc. HCl	Potentiometric	Budešinský and Vaničková (1957)
Thiols	$+$ HCl $+$ KBr	First yellow	Willemart and Fabre (1958)
Higher thiols	$+$ acid $+$ KI	Yellow persisting for at least 30 s	Jaselskis (1959)
Mercaptoacetic acid	$+$ HCl	Decolouration of methyl orange	Nutiu and Bokenyi (1969)
Mercaptoacetic acid (with corresponding sulphide and disulphide)	In HCl	Decolouration of methyl orange	Nutiu (1974)

have investigated this point are Zöllner and Varga (1957) and Aravamudan and Rama Rao (1963). In direct titration, oxidation usually proceeds only to the disulphide; further oxidation may take place in indirect procedures, as, for example, the 12-electron oxidation of thiouracil derivatives noted by Wojahn and Wempe (1953). Some examples are given in Tables I and II.

3. BROMINE

This reagent is equivalent to bromate/bromide, which yields bromine in acid solution.

Bromine is a popular direct titration agent. Thus Tomíček and Valcha (1950) titrated benzyl mercaptan (as well as other compounds) in acetic acid containing sodium acetate, using a bromine reagent in acetic acid and potentiometric end-point indication. Hladký (1965) also titrated potentiometrically with a bromine reagent in a non-aqueous solvent, namely dimethylformamide, in an oxygen-free nitrogen atmosphere; he did not specify the thiols titrated.

Electrolytically generated bromine from acidified (sulphuric or hydro-chloric acid) bromide solutions is used in most procedures nowadays, end-point determination being generally potentiometric or biamperometric. Examples are the determination of thiols in gas streams (Austin *et al.* 1950); sodium thiopental (Kalinowski and Pietrowska, 1959); trace amounts (Podurovskaya *et al.*, 1966); malodorous sulphur compounds, *e.g.* methane-thiol, in air (Adams, 1969); cysteamine (Stoicescu and Beral, 1969); cysteine (Kreshkov and Oganesyan, 1971); thiols (ethanethiol) in natural gas (Beskova *et al.*, 1971); cysteine in mixtures with ascorbic acid, then combined with coulometric iodine titration of different stoichiometry for the two components (Kreshkov and Organesyan, 1973); cysteine/cystine mixtures, also employing parallel coulometric titration with iodine (Inkin and Kharlamov, 1973).

Saville (1956) and Bakes and Jeffery (1961) used bromine reagent to deter-mine thiols in a colorimetric procedure. The first stage is evidently oxidation to sulphonyl bromide, which then yields bromine cyanide with added cyanide ion (excess bromine is previously destroyed with phenol):

$$RSH + 3Br_2 + 2H_2O \rightarrow RSO_2Br + 5HBr$$

$$RSO_2Br + CN^- \rightarrow BrCN + RSO_2^-$$

A pyridine/benzidine reagent is then added and the bromine cyanide splits the pyridine ring to yield glutaconic dialdehyde:

Table II. Indirect titrations with bromate

Sample	Conditions	Final stage	References
Ag or Cd derivatives of glutathione, precipitated in its determination in biological material	+ HCl + excess reagent; 8 min reaction at room temp.	+ Na_2HPO_4 + I^- and titrated with $S_2O_3^{2-}$	Hartner and Schleiss (1936)
Cd derivative of glutathione from its determination in blood	+ HCl + Br^- + excess reagent; 8 min reaction (above method)	+ Na_2HPO_4 + I^- and titrated with $S_2O_3^{2-}$ to end-point starch.	Kul'berg and Presman (1940)
Thiouracils	In 10% NaOH, + BrO_3^-/Br^- excess + 25% HCl; 30–60 min	+ excess standard arsenite and back titrated with BrO_3^- to p-ethoxychrysoidine	Wojahn and Wempe (1953)
1-Methyl-2-mercaptoimidazole	+ BrO_3^-/Br^- + HCl; 15 min/room temp.	+ I^- and titrated with $S_2O_3^{2-}$	Varga and Zöllner (1955)
6-Mercaptopurine	In 6N H_2SO_4, + BrO_3^-; 15 min/room temp. in darkness	+ I^-, left for 5 min, then titrated with $S_2O_3^{2-}$	Jančík et al. (1957)

The dialdehyde forms a coloured condensation product with the diamine.

Bromine can be used as a test for certain sulphur compounds, usually under conditions which yield sulphate; this is then detected with barium ion as a white turbidity or precipitate. Bucher (1951) and Bucci and Cusmano (1962), for example, used this test for thiouracils. It has the disadvantage of being unspecific for thiols.

Bayfield et al. (1965) proposed a selective test for organic sulphur(II) compounds in which the sample, dried on Whatman No. 1 paper, was drawn through 3 % (v/v) aniline in petroleum ether (40°–60°C); the solvent was allowed to evaporate and the paper then exposed to bromine vapour. Blue or mauve spots were yielded within 30–60 s. The authors suggested a sulphenyl bromide as intermediate, acting as oxidizing agent in a way possibly analogous to hypochlorite in the colour reaction with aniline (which gives a similar colour). With thiols, the reaction scheme would be:

$$RSH \xrightarrow{Br} RSBr \xrightarrow{C_6H_5NH_2} RS-NH-C_6H_5 \text{ and oxidation products}$$

Sulphenamides were found to give a blue to violet colour directly with bromine. Bayfield and Cole (1969) later used the procedure for visualizing sulphur-containing amino acids in PC.

Bromine (and iodine) cyanides were used by Paul et al. (1974) to titrate thiols (oxidized to disulphates) with potentiometric or visual end-points. However, Srivastova and Bose (1974) reported that bromine cyanide and chloride were unsuitable for titrating 0·1 to 1 mmol amounts of thiols.

4. N-BROMOSUCCINIMIDE

Thibert and Sarwar (1969) titrated sulphur-containing amino acids (cysteine, cystine and methionine) directly with a 0·01 M N-bromosuccinimide reagent using Bordeaux Red (C.I. Acid Red 17) as indicator (colour change of red to yellow). They also used an excess of reagent, back-titrating iodometrically after 1 min.

They suggested the reaction equation:

$$6 \begin{array}{c} CH_2-CO \\ | \quad\quad\quad \rangle NBr \\ CH_2-CO \end{array} + 2HS-CH_2-CH(NH_2)COOH + 6H_2O$$

$$\rightarrow 6 \begin{array}{c} CH_2-CO \\ | \quad\quad\quad \rangle NH \\ CH_2-CO \end{array} + 6HBr + 2CO_2 + 2HO_3S-CH_2-CH_2-NH_2$$

(Cystine yielded the same product and methionine, the sulphoxide).

In a later publication, the same authors and Carroll (1969) titrated cysteine in the presence of cystine, using N-bromosuccinimide and a solution of phosphate buffer, pH 7, iodide and starch. As end-point, blue persisting for

30 s was taken. The sum of both amino acids was estimated by titration as above, in the absence of iodide and buffer.

Schneider *et al.* (1972) used the same methods for α-substituted DL-cysteines. Bachhawat *et al.* (1973) used an N-bromosuccinimide reagent to titrate thiols in a medium 4% in acetic acid and iodide, taking the blue colour with starch as end-point. Oxidation was, of course, to the corresponding disulphides.

5. CACOTHELINE

This reagent is a nitro derivative of brucine, from which it is prepared by treatment with 10% nitric acid at 60–70°C. It yields a blue colour with reducing agents, such as Sn(II), and Rosenthaler (1938) quotes this as a test for mercaptoacetic acid and, though less satisfactorily, for cysteine hydrochloride.

6. CERIUM(IV)

This powerful oxidizing agent has been used both qualitatively and quantitatively with thiols. Thus Trop *et al.* (1968) visualized cysteine, cysteamine and mercaptoacetic acid on paper chromatograms with an ammonium hexanitratocerate/dilute nitric acid reagent, which gave white spots on a yellow background. Grossert and Langler (1974) obtained similar results with thiophenols and other thiols on thin silica gel HF_{254} layers. Varga and Zöllner (1955) titrated 1-methyl-2-mercaptoimidazole(methimazole) with ceric sulphate at 0°C, using various indicators such as methyl red, methyl orange, thymol blue and *p*-ethoxychrysoidine. Piotrowska (1970) recently carried out direct titration of 6-mercaptopurine in 25% hydrochloric acid containing iodine monochloride and ferroin. She first added 0·05 N ceric sulphate until the solution was yellow–orange, then warmed to 50°C and finally titrated to the end-point of a blue colour.

Alexander *et al.* (1969) studied the oxidation of mercaptoacetic, thiolactic and thiomalic acids with cerium(IV). They diluted the sample with oxygen-free water and titrated with an ammonium disulphatocerate, $(NH_4)_2[Ce(SO_4)_2]$ generally about 100 times more concentrated than the titrand solution to limit dilution. They employed a thermometric procedure for the end-point indication. The formation of disulphide in the reqction products points to a 1:1 stoichiometry which they formulated:

$$Ce^{4+} + RSH \rightarrow Ce^{3+} + H^+ + RS^{\cdot} (\rightarrow RSSR)$$

Usually more cerium(IV) was consumed, evidently through reaction with the $RS^{\cdot}$ radical. In the presence of methyl acrylate, the expected 1:1 ratio was found.

7. CHLORAMINE T

Three direct titration methods for thiols using this reagent may be given: Mahadevappa (1965) titrated mercaptoacetic acid in the presence of potassium iodide/starch, taking pale blue as the end-point. Fecko and Zaborniak (1968) titrated methylthiouracil at pH 4·6, using an amperometric end-point with 2 platinum electrodes and 400 mV applied potential difference. This compound and others in pharmaceutical preparations were titrated also by Avakyants and Murtazaev (1969).

Aravamudan and Rama Rao (1963) studied the effect of pH, temperatures and reaction time on the reaction of thiomalic acid with various oxidizing agents, including chloramine T.

8. CHLORINE

Only two references to the use of chlorine in analytical work on thiols can be given. Kalinowski *et al.* (1957) titrated methyl thiouracil coulometrically with chlorine generated from 10% hydrochloric acid. Prochukhan *et al.* (1969) determined tertiary thiols through oxidation with chlorine water ultimately to sulphate which was estimated gravimetrically as barium salt. Primary and secondary thiols evidently were not oxidized to sulphate, but tertiary disulphides did yield it.

9. CHLORINE CYANIDE

Aldridge (1948) used this reagent to determine 1,2-dithiols, with which it evidently forms an unidentified product which yields thiocyanate ion in alkaline solution. He treated the sample with the reagent for 5 min, then added sodium hydroxide; after 10 min the solution was acidified, then aerated for 30 min with moist air and treated with bromine water. Excess of this last reagent was removed with arsenite, and the resulting solution was added to a pyridine/benzidine/hydrochloric acid reagent. The bromine converts the thiocyanate to bromine cyanide which splits the pyridine ring to form glutaconic dialdehyde monoenolate, which in turn gives a coloured condensation product with the diamine, benzidine (cf. "Bromine", 1.3). Aldridge estimated this after 20 min using an Ilford green filter 604.

Less than 10 μg of sample are needed but the method was criticized as requiring hazardous reagents and expert technique (Rosenblatt and Jean, 1955).

10. CHROMIUM(VI)

Chromates and dichromates are reduced to Cr(III) by various compounds including thiols. The accompanying colour change from yellow or orange to green can serve as an unspecific test for them. For example, Waksmundzki *et al.* (1963) used this for visualizing products and intermediates in the synthesis of 6-mercaptopurines on paper chromatograms.

There appear to be very few quantitative examples. Garcia-Blanco and Pascual-Leone (1955) studied the oxidation of amino acids, including those containing sulphur, with dilute potassium dichromate in concentrated sulphuric acid. Buscarons *et al.* (1960, 1961) determined thiols such as mercaptoacetic, thiomalic and -lactic acids by oxidation with excess standard potassium chromate and sulphuric acid. After reaction for 5–10 min, unused reagent was estimated colorimetrically at 470 nm after treating for 5–25 min with *o*-dianisidine in acetone/water.

11. COPPER(II)

Cupric reagents are among those most frequently used for the detection and determination of thiols. The reaction is formulated:

$$4\,RSH + 2Cu^{2+} \rightarrow RSSR + 2\,RSCu + 4H^+$$

With more complex thiols, such as reduced glutathione or mercaptopurines, further reactions may take place.

The brown colour of the cuprous mercaptide can serve as a test for the thiol group. This was used, for instance, by Joachim (1951), with a cupric oleate test paper, and by Smith and Williams (1939) with sawdust impregnated with the same compound. White and Reichardt (1949) criticized the method with cupric oleate test paper and preferred an absorbent solution of cupric butyl phthalate in acetic acid-butanol (1 + 20), through which the gas to be tested was passed. They adapted it as a semi-quantitative method by comparison with standards. Freytag (1953) also used a 5% aq solution of a copper salt, $CuCl_2 2NH_4Cl . 2H_2O$ to detect mercaptoacetic acid but found the test rather insensitive (down to 1 part in 40 000) and unsuitable for the other thiols he tried.

Metello Metto and de Figueiredo (1949) gave tests for thiouracil and its methyl and propyl derivatives in which they were dissolved in dilute alkali and then treated with cupric sulphate; the three compounds could be distinguished through the different colours and colour changes of the precipitates. One of several tests for methyl thiouracil given by Shemyakin and Berdnikov (1965) is the yellow–green colour yielded with cupric acetate.

Oguchi and Shimizu (1962) detected reduced glutathione through the red fluorescence of the products of treatment with cupric sulphate at pH 2–10. The best excitation wave length was 365 nm. Red fluorescence was the positive result of the tests for thiouracils of Bucci and Cusmano (1962) and of Berdnikov (1966) also. The former added an ethanolic extract of the sample to filter paper treated with 5% cupric sulphate; the latter dissolved the sample in a mixture of dilute sodium hydroxide and cupric acetate, and used a filter paper test also, on to which sample and drops of the alkali and copper solutions were placed.

Cupric reagents have been used in direct titration procedures and in indirect methods, mostly titrimetric, where unused reagent or a reaction product is determined. Table III contains a summary of references to direct titration of thiols (the last example depends on chelate formation rather than oxidation).

Noteworthy are the early attempts to use visual end-point indication, and the relatively recent studies of titration in dimethylformamide solution. It is perhaps surprising that titration in the presence of sulphite has not found more application. Kolthoff and Stricks state that the end-point indication is sharper than in silver titration.

Podlipskii et al. (1969) titrated thiols in motor fuels with ammoniacal copper sulphate but reported only 50% recovery with secondary and only 7% with tertiary thiols. Stekhun et al. (1967) reported low results using this titrant for thiol determination in fuels. Unused cupric reagent has generally been determined titrimetrically, either by iodometry or complexometrically with EDTA. Ellis and Barker (1951) determined thiols in hydrocarbon gases by passing into excess standard cupric acetate/acetic acid. Potassium iodide was then added and, after 20 min, the iodine liberated by the unused copper(II) was titrated with thiosulphate. In a micro method, Roth (1958) used cupric butyl phthalate as reagent. He prepared a 0·02 N solution by dissolving in glacial acetic acid and diluting to 20 times its volume with butanol, pentasol (mixture of amyl alcohols) or hydrocarbon; this reagent resembles that of Turk and Reid (see Table III). The sample (5–10 mg) was treated with 20 ml of reagent and, after 5 min, potassium iodide was added and the liberated iodine titrated with 0·01 N thiosulphate to starch.

Parushev (1963) determined 2-mercaptobenzothiazole by dissolving in ethanol and adding to 5% aq cupric sulphate. The precipitate was coagulated by heating, filtered and unused copper(II) in the filtrate was estimated by bringing to pH 8 with ammonium hydroxide and titrating with EDTA to the purple with murexide indicator. A very similar procedure was employed by

Table III. Direct titrations with copper(II)

Sample	Reagent	Conditions	End-point indication	References
Thiols in hydrocarbons	Cu oleate in kerosene		Persistent green	Bond (1933)
Mercaptobenzothiazole	Cu oleate in benzene		Persistent blue	Mogoricheva and Korsunskaya (1933)
Thiols in gasoline	Cu sulphate/NaOH/ NH_4OH	Shaken with Na_2CO_3 to remove H_2S	Persistent blue	Krause (1938)
	Cu alkyl phthalates in hydrocarbons or alcohols	In a similar solvent	Persistent blue-green	Turk and Reid (1945)
Thiols in gasoline	Cu salt in NH_4OH–water(1 + 9)		To blue	Shtram and Fateev (1945)
Cysteine, also from cystine reduction	Cu nitrate or sulphate; reaction given as: $RSH + 2Cu^{2+} + SO_3^{2-} \rightarrow RSSO_3^- + H^+ + 2Cu^+$	Ammoniacal solution + sulphite; in N_2 atm.	Amperometric with rotating Pt electrodes	Kolthoff and Stricks (1951)
Thiol groups in serum albumin	Cu acetate	In pH 4·9 acetate buffer	Amperometric	Saroff and Mark (1953)

Mercaptoacetic acid	Cu sulphate/aq. mono-ethanolamine	In presence of ammonia	To stable yellow	von Keller (1954)
Peptide and protein cysteine groups	Method of Kolthoff and Stricks (1951)			Swan (1957)
Mercaptoacetic acid	Cu sulphate, acetate, chloride, nitrate in water	At pH 4–4·5	Colour change of precipitate from deep violet to yellow	Sant and Sant (1959)
(no examples in original)	Cu acetate or chloride in DMF	In DMF (dimethyl-formamide); N_2 atm.	Potentiometric with Pt electrodes	Hladký (1965)
Thiolactic, thioglycollic acids, pentachlorothiophenol	Cu chloride in DMF	In dry DMF	Potentiometric	Hladký (1967)
As in previous publication, also thiocresols		As in previous publication		Hladký and Vřešťal (1969)
Cysteine; thiolactic acid	Cu chloride in DMF; reverse titration found best	In DMF; N_2 atm.	Potentiometric	Braun and Stock (1972)
Aromatic and aliphatic thiols up to C_{10}	Cu chloride in DMF	In DMF	Potentiometric	Haeusler et al. (1973)
Amino acids, including L-cysteine	Cu sulphate (→ insoluble complex with cysteine)	In borax buffer, pH 9	+ murexide (violet → yellowish green)	Gawargious et al. (1974)

Anokhina and Kalmykova (1970) for the same compound; they back-titrated with EDTA to Chromogen Dark Blue at pH 9.

Aronovich (1969) estimated unused copper(II) by a non-titrimetric procedure in his determination of thiol sulphur in TS-1 fuels. His reagent was aq ammoniacal cupric sulphate and after reaction the light absorbance of the lower, aqueous layer was measured and related to the residual, unused copper(II). An ingenious procedure was adopted by Mottola *et al.* (1968). The atmospheric oxidation of ascorbic acid, which they followed at 265 nm in a phosphate buffer of pH 6·4, is catalysed by copper(II). Thiols react with Cu(II), reducing the amount available for catalysis and thereby slowing the reaction to an extent related to the thiol amount in the sample used. The authors were able to determine cysteine and 2-aminoethanethiol in concentrations of the order of 10^{-5}–10^{-6} M.

Zeicu *et al.* (1972) studied photometrically the reaction between methyl thiouracil and copper(II). In a quantitative adaptation, the sample in di-methylformamide–ethanol (1 + 1) was treated with ethanol, water and cupric sulphate, and the absorbance of the 2:1 complex (compound to copper) was evaluated at 400 nm.

An alternative indirect titrimetric procedure is to determine the copper(I) formed. This has been seldom done. Schulze and Chaney (1933) oxidized thiols in benzine with cupric chloride by shaking for 30–45 sec and determined the cuprous mercaptide by titrating to blue with potassium permanganate. Frieden (1958) determined certain biological reducing agents, including cysteine and reduced glutathione in microgram amounts, by oxidation with cupric chloride/phosphate buffer of pH 7 in the presence of specific complexing agents for copper(I), such as 2,2′-biquinoline (Cuproin) and 2,9-dimethyl-1,10-phenanthroline (Neocuproin). The coloured copper(I) complexes are stable for 15 min, permitting spectrophotometric evaluation.

Cupric reagents may be used to retain thiols and prevent their participation in a reaction. Thus Kamidate *et al.* (1971), in studies of the extraction of organic sulphides with metal salts, including cupric bromide, found that thiols were converted to mercaptides at the same time. A determination of thiols by difference following such a treatment is evidently possible but no example appears to be known.

Copper(II) is also used as a component of other reagents, e.g. with ninhydrin in the detection of amino acids. This does not apply alone to sulphur-containing amino acids and scarcely qualifies for inclusion here. Nilsson (1957) detected down to 2·5 µg of thiouracil per drop through the intense blue yielded with a mixture of *o*-toluidine in 5% acetic acid with aq cupric chloride and sodium hydroxide solutions.

Greer (1929) found that aq cupric sulphate or nitrate removed thiols from naphtha. She also tested a number of amorphous sulphides (cupric. lead, stannic, cadmium and arsenious) in this capacity, on naphtha containing 0·4 to 0·5 % thiol(ethane- and s-pentane-). She found that cupric sulphide was the most effective in rendering the naphtha "sweet".

Mention can be made here of the use of copper(II) salts in the presence of reducing agents such as hydroxylamine, so that the mixture is effectively a copper(I) reagent. This is discussed in Chapter 2, Section 7.

12. COPPER(III)

Beck (1951) titrated amino acids, including cysteine, with his "percupric" reagent $K_7Cu(IO_6)_2$, prepared from cupric sulphate, potassium hydroxide, potassium iodate or periodate, and potassium persulphate. The warm alkaline solution of sample was titrated to green or violet, depending on conditions. The times of titration are characteristic of the amino acid, that for completion of reaction with cysteine being the shortest.

13. DICHLOROPHENOL-INDOPHENOL

See "Tillmans' Reagent", (Section 52).

14. DICHROMATE

See "Chromium(VI)", (Section 10).

15. o-DINITROBENZENE

See Chapter 9, "Miscellaneous Chemical Procedures" (Section 10, "Dinitrobenzenes").

16. DIPICRYLHYDRAZYL

Blois (1958) discovered that the 2,2'-diphenyl-1-picrylhydrazyl free radical interacts quantitatively with antioxidants, such as cysteine and reduced glutathione; he postulated the changes:

$$(DPPH)^. + RSH \rightarrow (DPPH):H + RS^.$$

$$RS^. + RS^. \rightarrow RSSR$$

Quantitative exploitation of the reaction is based on the strong absorption maximum at 517 nm of the radical. Thus Glavinel (1963) treated antioxidants, including reduced glutathione, with excess methanolic reagent. After 5 min, he extracted unused reagent into xylene and estimated the amount by absorbance measurements at this wavelength. Owades and Zientara (1961) likewise employed the radical to determine oxidizable substances, e.g. cysteine or glutathione, in beer. Berg (1971) recently determined some thiols of pharmaceutical interest (mercaptopurine, alkyl thiouracils, butane-thiol, BAL) by reaction with excess diphenylpicrylhydrazyl at pH 5 or 6 and carrying out absorbance measurements at 513 nm.

17. ELECTROLYTIC OXIDATION

Santhanam and Krishnan (1968) determined some sulphur-containing compounds, including cysteine and mercaptoacetic acid, by direct coulo-metry in, respectively, phosphate buffer of pH 6·2 and hydrochloric acid buffer of pH 1·3. At controlled anode potential, the current fell at the end of reaction to a negligible value. They calculated the coulomb amount to this point with the help of a hydrogen/oxygen coulometer in series.

18. ETHOXYRESAZURIN, ETHOXYRESORUFIN (7-ETHOXYPHENOXAZONE-2-10-OXIDE AND 7-ETHOXYPHENOXAZONE-2, RESPECTIVELY)

Růžička (1957) determined 2-aminothiophenol by titrating it into ethoxy-resazurin in 96% ethanol to an end-point of blue–green colour; the solution must be maintained at 50° or above. 4 moles of thiol react with 1 mole of reagent:

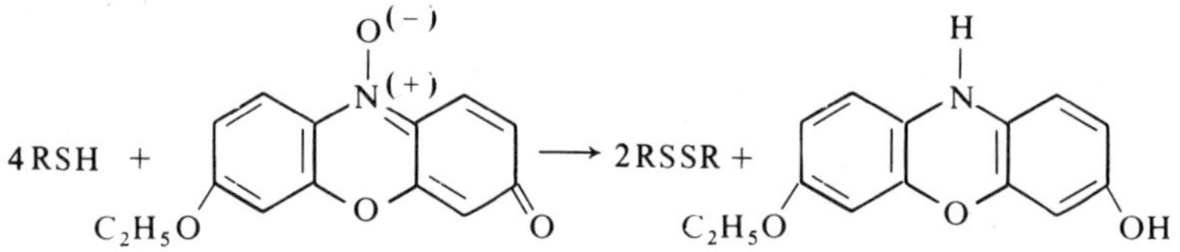

In 4 N hydrochloric acid, also at $< 60°C$, the stoichiometric ratio is 2:1, the same as that when ethoxyresorufin is used in the titration under the same condition. It is known that ethoxyresazurin is converted by hot hydrochloric acid into ethoxyresorufin, which then becomes the effective oxidizing agent:

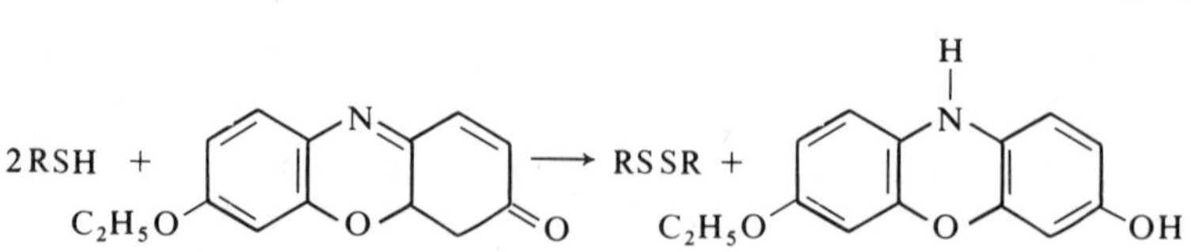

19. FERRICYANIDE

The moderately strong oxidizing agent ferricyanide has been applied analytically to thiols in three ways: direct titration; indirect quantitative methods, based on the use of excess and estimation of the unused amount; and detection and determination *via* a reaction product. The oxidation with ferricyanide normally yields disulphide:

$$2RSH + 2Fe(CN)_6^{3-} \rightarrow RSSR + 2Fe(CN)_6^{4-} + 2H^+$$

There appears to have been little systematic work on this reaction but Kolthoff *et al.* (1962) and Meehan *et al.* (1962) studied the oxidation of *n*-octanethiol in 60–70% acetone and of 2-mercaptoethanol in the pH range 1·8–4·1, discussing the kinetics in relation to the ions present.

The analytical work can be given conveniently under the three headings implied above.

19.1. Direct titration

Anson (1940, 1941) titrated biological thiol groups, e.g. in denatured egg albumin, using nitroprusside as external indicator. The end-point was that at which the indicator ceased to yield a purple colour, characteristic of its interaction with the thiol group. Lehman *et al.* (1951) titrated thiol groups in mixtures of mercaptoacetate or 2,3-dimercapto-1-propanol with mercury diuretics, such as mercuprin(mercuzanthin) with ferricyanide, taking as end-point a pale green persisting for 10 s; added cupric sulphate sharpened the end-point with the dithiol. Waddill and Gorin (1958) titrated cysteine in phosphate buffer of pH 7 and nitrogen atmosphere using a copper(II) catalyst and the "dead-stop end-point". Recently Haeusler *et al.* (1973) titrated thiols in 2 N sodium hydroxide, using potentiometric end-point indication with a platinum or graphite electrode.

19.2. Determination of unused reagent

Flotow (1928) determined glutathione by oxidizing with excess ferricyanide in alkaline solution, ultimately back-titrating the unused reagent with indigosulphonic acid. In his determination of the same compound, Gabbe (1930) estimated unused reagent by adding iodide and titrating liberated iodine with thiosulphate. Sartos Ruiz and de Franch (1943) criticized Gabbe's method as being susceptible to all oxidizable substances. Sarkar and Sivaraman (1958) determined cysteine (also derived from cystine by treatment with sulphite) with excess ferricyanide in the presence of potassium iodide/zinc sulphate/sodium chloride and sulphuric acid; iodine was titrated with thiosulphate. Khomenko and Tsinkalov'ska (1962) criticized this method and obtained their best results for glutathione and mercaptoacetic acid in neutral phosphate buffer. Recently, Chattopadhyay and Mahanti (1973) determined 2-mercapto-5-anilino-1,3,4-thiadiazole(MAT) with excess ferricyanide in 7% potassium hydroxide and in the presence of osmium tetroxide catalyst. They back-titrated unused reagent amperometrically with arsenite.

Non-titrimetric techniques for estimating unused reagent have also been used. Katyal and Gorin (1959) oxidized in phosphate buffer of pH 7 and in the presence of iodide. They used several sample aliquots to which varying amounts of reagent were added. Electrodes which were part of a DSEP-type circuit were immersed in the reaction mixture. They related the current flowing in the circuit to the amount of unused ferricyanide present, obtaining thereby a calibration curve for subsequent use. They also estimated unused ferricyanide from the diminution of light absorbance at 410 nm. In determinations of mercaptoacetic, thiomalic and thiolactic acids and of thioglycerol, Buscaróns et al. (1960, 1961) determined unused reagent by adding o-di-anisidine in aqueous acetone, with which it reacts to yield a coloured product. This was evaluated at 470 nm.

19.3. Detection or determination of a reaction product

In the presence of ferric ions, the ferrocyanide reaction product of oxidation yields Prussian blue (see also pp. 62–3 under "iron(III)"). A ferricyanide/ferric reagent has been used for visualization or location of thiol groups, especially in histochemistry. For example, Chèvremont and Frédéric (1943), Adams (1956), Freses and Winstanley (1960) and Ackerman and Sneeringer (1960) all used a mixed reagent yielding blue within 5–20 min. Quantitative methods are also based on evaluation of the blue. Mason (1930) treated cysteine for 15 min with ferricyanide in phosphate buffer, then added ferric sulphate and after 10 min compared the blue colour with those from standards. Barron et al. (1947) determined 3–20 µg dithiol amounts by oxidation in sodium hydrogen carbonate solution, subsequently adding ferric iron and evaluating the absorbance at 690 nm. Barron and Flood (1950) applied the method to determine dithiols and glutathione in a study of the oxidizing influence on them of ionizing radiation; Ullmann (1959) also employed the method in determinations of thiol groups in meat undergoing thermal denaturation and made comparisons with the silver methods. Katyal and Gorin (1959) applied the principle to determine thiol groups in ovalbumin. This was treated in a phosphate buffer of pH 7·1 with excess ferricyanide, a trace of iodide and a detergent (sodium lauryl sulphate). After 5 min, ferric reagent was added and the solution evaluated colorimetrically 20 min later.

Haas (1937) determined glutathione by oxidation with ferricyanide in neutral hydrogen carbonate solution. The hydrogen ions liberated during the oxidation liberate an equivalent amount of carbon dioxide, which he estimated manometrically. Barron *et al.* (1947) also made use of this principle in their determination of dithiols. Nanogram amounts of thiols are determined by Kidby (1969) by measuring the light absorbance of the ferrocyanide at 237 nm; up to a 50-fold excess of ferricyanide can be tolerated.

A review of ferricyanide analytical methods has been made by Sant and Sant (1959); cysteine is mentioned as one of their examples.

20. FUCHSIN/FORMALDEHYDE

Steigmann (1942) quotes a Schiff-type reagent of basic fuchsin/4·5 N sulphuric acid/formalin solution. It was used to test photographic gelatins, giving an immediate blue colour with compounds such as sulphur dioxide and thiols.

21. HALOGENS

See "Bromine" (Section 3); "Chlorine" (Section 8); "Iodine" (Section 26).

22. HETEROPOLYACIDS

Tungstophosphate is a standard reagent for reducing compounds, including thiols, with which it yields a blue product of lower tungsten valency. The main period of quantitative analytical application to thiols was that of the later twenties and thirties, although interest was at first centred on the disulphide determination. Thus Hunter and Eagles (1927) determined cystine with a reagent containing sodium hydroxide, lithium sulphate (to prevent cloudiness), sodium sulphite (to convert disulphide to thiol in a prior 1 min reaction) and the tungstophosphoric acid. The colour was evaluated after 5 min. Folin and Marenzi (1929) applied a similar procedure to cystine determination. Numerous publications followed in the next 10–12 years, mostly devoted to the biological sulphur compounds, glutathione, cystine/cysteine, and also mercaptoacetic acid. Some examples may be mentioned, such as the work of: (a) Shinohara (1935–1937); (b) Schöberl (1937, 1938), who also commented on the augmented colour intensity in the presence of sulphite, caused by the re-formation of thiol anion:

$$RS^- + reagent \rightarrow RSSR + reduced(coloured)\ reagent$$

$$RSSR + SO_3^{2-} \longrightarrow RS^- + RSSO_3^-$$

(c) Mirsky and Anson (1935) for cysteine, also for that yielded by protein thiol groups with cystine, enabling the former to be determined; (d) Langou and Marenzi (1935) for cystine and glutathione in the presence of sulphate; (e) Medes (1936) and Brand *et al.* (1940) for cysteine derived by Cu(I) reduction from cystine in its determination; (f) Pietz (1940) for cysteine obtained from protein cysteine groups by treatment with sulphite; (g) Fraenkel-Conrat (1944) used Mirsky and Anson's method for thiol determination.

Interest in the quantitative colorimetric method subsided later but the following more recent applications may be quoted; (a) Kolb and Toennies (1952) for thiols in protein hydrolysates; (b) Park and Speakman (1952) for cysteine and cystine in keratin; (c) Malikova (1954) for cystine in amino acid mixtures, after conversion to cysteine with sulphite; (d) Zahn and Traumann (1954) for cystine in wool hydrolysates, likewise treated with sulphite; (e) Tupeeva (1967) for mercaptoacetic acid in air; (f) Danchy and Elia (1972) for determinining thiols in the presence of sulphinic acids.

A major problem is the interference of other reducing compounds. Lugg (1932) carried out a parallel colorimetric determination in the presence of mercuric chloride which yields —S—Hg— bonds with the thiols and inhibits their reducing properties. A blank colorimetric value could then be obtained for the other reducing substances present. Some subsequent workers adopted this procedure, e.g. Shinohara (1936) and Kassell and Brand (1938).

A discussion of full practical details would exceed the scope of this work. Generally weakly acid (acetate buffer of pH ca. 5) or basic conditions are used, room temperature and reaction times ranging from 5 to 30 min. The blue colour has an absorption maximum at 720–730 nm. The reaction is clearly complex and has evidently been utilized empirically. Shinohara (1937) discussed the mole ratio of thiol to reagent in the colour reaction.

An interesting variant was introduced by Ionescu-Matiu and Popesco (1939, 1940). They treated glutathione with sulphite and the Folin–Wu tungstophosphate reagent and then titrated the resulting blue solution with 0·1 N ferricyanide to the disappearance of the colour.

Micheel (1939) used tungstophosphoric acid as qualitative test for free thiol groups and this became a standard method later for visualization of reducing compounds in paper and thin-layer chromatography e.g. see Stahl, 1969).

Less use appears to have been made of other heteropoly acids. Benedict and Gottschall (1933) used Benedict's reagent (1922), which contains arsenic also, to determine glutathione in deproteinized blood. Molybdophosphoric acid was evidently first proposed by Karr (1954) as a test for aliphatic and aromatic thiols. The solution had a final pH of ca. 8. A further example of its use in the visualization (of cysteine) is in the work of Hashmi *et al.* (1969). One of Tupeeva's (1967) quantitative methods for determining mercaptoacetic acid in air is based on reaction with a molybdophosphate reagent.

23. HYDROGEN PEROXIDE

The widest application of hydrogen peroxide in the analysis of sulphur-containing compounds is to oxidize to give products which can be more easily separated in subsequent chromatography. So-called "performic acid", a mixture of hydrogen peroxide and formic acid, is the standard reagent. Cysteine yields cysteic acid with it, a compound with chromatographic R_f values differing from those of cysteine itself. Cystine gives the same product. This technique is described in handbooks on chromatography and, since it concerns cysteine alone among thiols, it is not proposed to do more than quote a few references here: Wellington (1952) in paper chromatographic separation of protein hydrolysates; Schram *et al.* (1954) with protein hydrolysates and subsequent separation on Dowex 2X-10; Cole *et al.* (1956) in investigations of haemoglobin cysteine; Moore (1963) for sulphur-containing amino acids in protein, ultimately separating on Amberlite IR-120 or Dowex 1-X8; Borner (1965), using 30% hydrogen peroxide–formic acid (1 + 9) for 4 h at 0°C, for penicillamine, oxidizing it (and the corresponding disulphide) to penicillaninic acid, then separable on Dowex 1-X8 at 35° from other amino acids, including cysteic and homocysteic, likewise yielded in the oxidation. (As elution agents, he used chloroacetate buffer, pH 3·4, with plasma samples, and formate buffer, pH 4·2, for urine samples).

Szepesváry (1961) determined sulphur compounds in crude oil by oxidation and subsequent polarography of these products. The sample in benzene was treated with acetic acid and hydrogen peroxide and refluxed for two hours. An aliquot of product was polarographed in methanol–benzene (1 + 1), using tetramethylammonium hydroxide as supporting electrolyte. Thiols are presumably oxidized to reducible disulphides.

Gregg *et al.* (1961) analysed mixtures of thiols with aryl trityl sulphides by titration with a mercury(II) nitrate reagent with and without treating a sample to remove the thiols. They oxidized the thiols by heating a sample in ethanol with sodium hydroxide and 30% hydrogen peroxide to 65°C and leaving for 20 min. The authors do not say which oxidation product is formed; it may be disulphide or even contain S(VI), as in sulphonic acids. This is immaterial, since neither compound class reacts with the mercury(II) reagent.

Wojahn and Wempe (1952) determined thiouracils in tablets by oxidation with alkaline hydrogen peroxide to sulphate, then estimated gravimetrically as barium salt. This principle of drastic oxidation to sulphate, which is then detected or determined, is not specific for thiols.

24. HYPOHALITES

Hypohalites or halogens in alkaline solution have found little analytical application to thiols. In the early days of petroleum refining, sodium hypochlorite was used to separate certain sulphur compound classes, notably thiols and disulphides (for example, Birch and Norris, 1925); the amounts of these classes could be estimated through the decreased sulphur content of the sample.

Spacu and Dimitrescu (1968) determined thiols by oxidation to sulphate with excess sodium hypochlorite, followed by iodometric estimation of the unused amount.

Palma *et al.* (1971) titrated mercapto-acetic and -propionic acids (and thiocyanate ion) coulometrically at pH 8·3 to a biamperometric end-point, using bromine generated from $1F$ sodium bromide $0·016F$ in borax. The bromine must be largely present as hypobromite.

Most iodometric methods employ iodine in acidic or neutral (non-aqueous solvents) medium but some examples can be given of the determination of thiols with iodine in alkaline solution. Thus Blažek *et al.* (1957) titrated 2-mercapto-1-methylimidazole in sodium hydrogen carbonate or sodium hydroxide solution with iodine potentiometrically or using starch (then only in the hydrogen carbonate solution). Jančík *et al.* (1957) also titrated 6-mercaptopurine in sodium hydroxide with iodine potentiometrically, finding that 4 iodine atoms reacted with the thiol group.

Busev and Fang Chang (1961) determined thiomalic acid with excess iodine in concentrated alkali solution, leaving at least 5 min at ambient temperature. After acidifying the solution with sulphuric acid, unreacted iodine was back-titrated with thiosulphate. One mole of acid reacted with 6 equivalents of iodine:

$$I_2 + 2OH^- \rightarrow OI^- + I^- + H_2O$$

$$RSH + 3OI^- \rightarrow RSO_3H + 3I^-$$

Aravamundan and Rama Rao (1963) tested this method and found that the iodine consumption depended on factors such as the order of addition of reagents, temperature, time, and concentration of alkali.

Panwar *et al.* (1961) undertook an exhaustive study of the oxidation of many compounds, including the thiols cysteine, reduced glutathione and mercaptoacetic acid, with iodine under various conditions. Acidic, alkaline and nearly neutral solutions and buffers were tested. After reaction for "some time" the pH was adjusted to 7–7·5, and unused iodine was back-titrated with hydrazine biamperometrically (DSEP) or to the starch end-point.

25. IODATE

Iodate in the presence of iodide and in acidic solution yields iodine according to the equation:

$$IO_3^- + 5I^- + 6H^+ \rightarrow 3I_2 + 3H_2O$$

This has been used for direct titration in place of the rather less stable iodine solutions, although the necessary acid solution may not always be acceptable. Further, the solubility of the customary sodium and potassium iodate reagents is adequate virtually only in water; an aqueous reagent is often inconvenient with organic samples.

The first use of the titration of thiols with iodate appears to be the work of Okuda who, in the 1923–29 period, published examples of the titration of cysteine (also derived from cystine by reduction) in dil hydrochloric acid containing iodide; the end-point was a yellow colour stable for 1 min. The stoichiometry depends on conditions, notably acid concentration, and temperature, so that standardization under the same conditions was necessary.

Hess (1929) applied the same principle to reduced glutathione, keeping the temperature from exceeding 20°C and also titrating (with M/1200 potassium iodate) to a yellow colour persistent for 1 min. Woodward and Fry (1932) determined blood glutathione in the filtrate from protein precipitation with sulphosalicylic acid, hence of pH $\leqslant 2$ by titrating at ca. 20°C with 10^{-3} N potassium iodate in the presence of potassium iodide/starch, and taking the first blue as end-point. Okuda and Ogawa (1933) similarly titrated glutathione in deproteinized tissue filtrates, but with 10^{-4} iodate at 0°C. Others who used a starch end-point included Quensel and Wachholder (1935) for glutathione, to blue persisting for 30 s; Binet and Weller (1935, 1936, 1947, also Weller, 1965) for glutathione; Sabalitschka (1936) for reduced glutathione in dry yeasts after deproteinization, to red-violet; Divin et al. (1937) who used Woodward and Fry's procedure, but stated that it gave low values for blood glutathione if potassium iodide was added to the sample before titration (they thus titrated with a potassium iodate/5% potassium iodide reagent); Blyakher (1938) for blood glutathione, with a modification of Woodward and Fry's method; Fuyita and Numata (1938) for reduced glutathione in tissues; Sato et al. (1939) for cysteine from cystine/zinc/hydrochloric acid; Yamamoto (1940) for reduced glutathione, to blue persisting for 30 s; Seeligman (1962) for reduced glutathione in potatoes, using less strongly acid conditions (pH 3·2–3·8) than from sulphosalicyclic acid; Yung (1965) for blood glutathione, with 10^{-4} N reagent; Černoch (1966) for glutathione, who used Binet and Weller's method, but stated that titration values increased with the dilution of the sample, an error which could be righted by titrating at or below 5°C; Gimmerikh (1967) for blood glutathione by Woodward and Fry's method; Verma et al. (1972) for mercaptopyrimidines in N sulphuric acid.

B

End-points other than that with starch have been preferred by some. Thus Grogan *et al.* (1955) titrated polymethylene dithiols in dilute hydrochloric acid containing potassium iodide to the first yellow. Steel (1964) also used this yellow iodine end-point in his iodate titrations of a number of thiols, e.g. cysteine, glutathione, mercaptoacetic acid and mercaptrol, in dilute acetic acid containing iodide. Coulson *et al.* (1950) titrated reduced glutathione, also in the presence of ascorbic acid, at pH 3–4, using an amperometric end-point indication. Stapfer and Dworkin (1968) potentiometrically titrated organometallic (antimony, bismuth, sodium, tin) mercaptides in glacial acetic acid with aq potassium iodate. Further iodate titrations with instrumental end-point indication might be expected. Possibly oxidation methods for thiols are too sensitive to the other reducing materials which may be present, especially in biological extracts (e.g. ascorbic acid); and the development of coulometric iodine procedures has robbed iodate reagents of their main advantage of being more stable than an iodine solution.

Quantitative procedures other than direct titration are not common. Baernstein (1936) used iodate to destroy cysteine in a methionine determination. Bruno (1963) determined thiouracils in tablets by oxidation in aq acetic acid with potassium iodate. After 80 min, the light absorbance of the solution was measured at 465 nm.

26. IODINE

Iodine is one of the most extensively used reagents for thiols, especially in quantitative determination. Decoloration of an iodine reagent is a test for the thiol group, shown however by other compounds with reducing properties, e.g. hydrazines and polyphenols. It has, nevertheless, been used in chromatographic visualization, yielding pale spots on a yellow to brown background (or a blue background if starch is subsequently added). Lederer and Silberman (1952) and Giannessi (1961) visualized thiouracils in this way on paper chromatograms. Brown and Edwards (1968) exposed thin-layers to iodine vapour to visualize thiols even in the presence of other sulphur-containing compounds. White *et al.* (1974) detected BAL on silica gel G with iodine vapour after separation by TLC from propane-1,2,3-trithiol.

Most quantitative methods for thiols involve direct titration, profiting from the smooth oxidation to the disulphide stage with little danger of the reaction proceeding further; and from the easy end-point indication of the intensely coloured and electroactive reagent. Indirect titration of thiols with iodine is also well exemplified, with some procedures also in which unused reagent is determined otherwise. Reaction products of the oxidation have rarely found analytical interest, but a few examples can be given. It is convenient to classify the material under three headings:

26.1. Direct titration

The majority of examples concern the naturally occurring thiols, cysteine and reduced glutathione. Iodine reagent in aqueous and non-aqueous solution has been used, electrolytically generated reagent being as yet rare. Visual end-point indication (yellow, blue with starch, external nitroprusside) predominates in older work especially and instrumental detection has been comparatively rare. The first direct iodine titration of a thiol appears to be that of mercaptoacetic acid in 1904 by Rosenheim and Davidsohn. Much work was done in the 1925–1936 period on iodometric titration of tissue glutathione using as external indicator sodium nitroprusside (e.g. Tunnicliffe, 1925; Thompson and Voegtlin, 1926; Perlzweig and Delrue, 1927; Paradiso, 1933, Zimmet and Dubois-Ferriere, 1936) or starch (e.g. Blanchetière and Melon, 1927; Mason, 1930; Gavrilescu, 1931; and also Perlzweig and Delrue). In typical procedures with nitroprusside, a series of mixtures of tissue extract, indicator and increasing amounts of iodine reagent were made up. The iodine amount in the first mixture in the series to show no purple or red colour (characteristic of the reaction of thiol with this indicator) was taken as equivalent to the glutathione present.

Table IV gives the salient features of some direct titrations of other thiols, of determinations which were performed more recently, or of those with some special features.

26.2. Determination of unused reagent

Short reaction times have been used in most indirect determinations with some authors back-titrating immediately after having added an excess of reagent. It has also been fairly general to titrate the excess of reagent with thiosulphate, although hydrazine and electrochemical methods are known.

Most work has been performed on particular thiols, and a convenient classification is according to the compound(s) determined.

Reduced glutathione was determined by: King et al. (1930) in blood; Mason (1931) in blood, titrating to a definite blue with starch and then immediately back-titrating the excess; Moncorps and Schmid (1932) in blood and tissue, with a reaction time of 3 min (the method of King et al. was also tested); Kühnau (1931) in liver, using reaction for 2 min; Gol'dman (1937) in tissue, using Kühnau's method; Ceresa and Guala (1939) in blood, at pH 1·8–2; Kuhn et al. (1939), also for cysteine and thiomalic acid, in 70–90% acetic acid with at least 2 equivalents of iodine and reaction for 1 min; Panwar et al. (1961), also for other thiols such as cysteine and mercaptoacetic acid, in acid, near neutral and even alkaline solution (see under "Hypohalites"), titrating unused reagent at pH 7–7·5 with hydrazine biamperometrically. With this last exception, thiosulphate was used in all the above examples.

Table IV. Direct titrations with iodine

Sample	Conditions	End-point indication	References
Cysteine; glutathione	ph < 5; T < 25°C	Yellow	King and Lucas (1931)
Thiophenols, thioacids, cysteine	At pH 0–7, pH 6·5 and 0°C, and in 80% acetic acid, respectively	Yellow	Lucas and King (1932)
Glutathione	In presence of CS_2, with reverse titration	Colour fading in the CS_2 layer	DiCapua (1934)
Cysteine	+ ethanol and with <20% water to prevent further oxidation	Yellow	Bickford and Schoetzow (1973)
Mercaptoacetates	Slightly acidified with HCl	Blue with starch	Hashall (1940)
Glutathione in biological materials	Critical study of methods of Tunnicliffe, Gabbe, 1930 and Binet and Weller (see "Iodate")		Sartos Ruiz and de Franch (1943)
Mercaptoacetate, unused, from determination of mono-chloroacetic acid	In Na_2CO_3 + NaOH solution		Frankiel and Rombau (1948)
Dodecanethiol, unused from determination of $CH_2{=}CH.CN$	In ethanol or isopropanol, + acetic acid	Yellow	Beesing et al. (1949)
Thiophenol	In ether, benzene, chlorobenzene or t-butanol, + drops of pyridine	Yellow (titrated with iodine/alcohol)	Harnish and Tarbell (1949)
Mixtures of primary and tertiary thiols	In 95% ethanol, + Pb nitrate or perchlorate + $HClO_4$ (to precipitate Pb mercaptides and prevent formation of mixed disulphides; (tert-RSH → RSI; prim-RSH → RSSR; combined with determination of total SH with Ag^+)	Amperometric	Kolthoff and Harris (1949)

Compound	Conditions	Detection	Reference
Mercaptoacetic acid	In water or very dilute acetic acid	Conductometric	Freytag (1952–3)
Mercaptoacetic acid	+ H_2SO_4	Yellow	Walker (1953)
Mercaptobenzothiazole		Conductometric	Scheele and Gensch (1953)
Mercaptoacetic acid in cold-wave preparations	In dilute H_2SO_4	Blue with starch	Forster et al. (1954)
2,3-Mercapto-1-propanol (BAL)	In ethanol–$CHCl_3$(1 + 1) + pyridine	Potentiometric	Zumanová and Zuman (1954)
Mercaptoacetic acid	In 70% ethanol, + H_2SO_4 (sulphites removed by CO_2 stream)	Yellow	Strache and Mierau (1955)
Mercaptobenzothiazole	e.g. in CCl_4	Visual	Lorenz and Echte (1956)
Mercaptoacetic acid	In dilute HCl	Yellow	Pesez (1956)
2-Mercapto-1-methylimidazole	+ $NaHCO_3$ or NaOH (see also under "Hyphalites")	Potentiometric or blue to starch (only in presence of $NaHCO_3$)	Blažek et al. (1957)
Mercaptobenzothiazole	In ethanol/water, + Na acetate	Blue with starch	Blöckinger (1957)
6-Mercaptopurine	In NaOH (see also under "Hypohalites"). Reacts with 4I, probably to $-SO_2H$	Potentiometric	Jančik et al. (1957)
p-Chlorothiophenol (from disulphide by reduction)	+ ethanol + acetic acid + H_2SO_4	Yellow	Higgins and Stephenson (1957)
Long-chain mercaptoacids and esters	In alcohol; titrated with I_2/alcohol, water added near titration end to accelerate	Yellow	Koenig et al. (1958)

Table IV (*cont.*)

Sample	Conditions	End-point indication	References
Cysteine: glutathione		DSEP	Riolo and Saldi (1959)
Thiol groups in wheat gluten; tested on glutathione and 2-mercaptoethanol	Dilute acetic acid	Potentiometric	Shaefer et al. (1959)
Protein thiols	Phosphate buffer, pH 6·5, + KI, titrated at 0–5°C	Photometric, based on absorbance of triiodide ion at 355 nm	Cunningham and Nuenke (1959)
Dodecanethiol, unused from determination of epoxides	In ethanol or isopropanol, + acetic acid	Yellow for 30 s	Gudzinowicz (1960)
Benzylmercaptan, from determination of disulphide by reduction with Zn		Yellow–green	Klouček et al. (1960)
e.g. benzyl mercaptan, mercaptoacetic acid		"High frequency"	Mukherjee and Roy (1961)

Dimercaprol	Acetate buffer, pH 5 + ethanol + KI; deaerated with N_2; coulometric titration	Amperometric	Kemula and Brachaczek (1963)
(Not specified in original)	In dimethylformamide	Potentiometric	Hladký (1965)
Thiosalicyclic acid	In acetic acid-methanol (1 + 5)	Blue with starch	Brown and Fogg (1969); Agathangelou et al. (1971)
Thiols, including tertiary thiols	In presence of Pb^{2+} to suppress reactions of RSI with RSH with tertiary thiols	Potentiometric	Haeusler et al. (1972)
Cysteine/cystine mixtures	Coulometric titration, combined with bromine titration of different stoichiometry		Inkin and Kharlamov (1973)
Cysteine/ascorbic acid mixtures	Coulometric titration; likewise combined with bromine titration. Ascorbic acid reacts 1:1 with both, cysteine 2:1 with I_2, 1:2 with Br_2		Kreshkov and Oganesyan (1973)
Thiol from epithio-chlorohydrin after treatment with HCl/acetic acid		Starch	Komissarenko et al. (1974)

In addition to the last two of the references just mentioned, methods for cysteine are due to: Baernstein (1930), who reacted in 2 N acid, then added hydrazine and determined unused iodine from the volume of nitrogen evolved in a van Slyke apparatus; Virtue and Lewis (1934) (for cystine through reduction with zinc/acid) who recommended minimum excess reagent (contrast the procedure of Kuhn et al. above), fairly acidic conditions (2% hydrochloric acid) and low temperature (he froze the reaction mixture); Lavine (1935) who used a solution 1 M in hydriodic acid in which oxidation only to the disulphide stage took place during up to 3 h; Stoicescu et al. (1965) on injections in the presence of lignocaine and hexamine, reacting in dry acetic acid.

Mercaptoacetic acid has been thus determined by: Wronski (1962); Nutiu (1964), in 10% hydrochloric acid, using a reaction time of 5 min; Nutiu and Bokenyi (1969) and Nutiu (1973), who used this principle to determine the thiol in the presence of the corresponding sulphide and disulphide; the work of Panwar et al. was mentioned above under glutathione.

Thiolmalic acid was determined by Kuhn et al. (1939) (see above); Busev and Fang Chang (1961) and Aravamundan and Rama Rao (1963) employed iodine in alkaline solution and their work is given under "Hypohalites".

Alkanethiols have been the object of several investigations, mainly with a view to their determination in gases. Examples include the work of: Shaw (1940) in gas mixtures or aqueous solutions, after separation with cadmium salts and testing on C_1–C_3 thiols; Hakewill and Rueck (1946), also for mercaptans in gases after prior separation with cadmium ion; Clark (1951) for C_3–C_5 and some higher thiols in rocket propellants; Segal and Starkey (1953) for methanethiol, isolated from mixtures of sulphur compounds as the mercury(II) derivative; Hirase and Araki (1955) for ethanethiol from a sugar diethylmercaptal, and using a reaction time of 2 h with alcoholic iodine; Coope and Maingot (1955), with a reaction time of 20–30 min at room temperature; Sunner et al. (1956) for thiols up to C_7 after GLC separation and the determination of unused iodine from the potential change at a platinum redox half element immersed in the reagent of iodine/potassium iodide/70% ethanol; Torstensson (1973) who recently determined aliphatic, aromatic and heterocyclic thiols by generating iodine electrolytically from sodium iodide/sodium perchlorate in water or aqueous methanol at controlled potential, then adding the sample and reducing excess iodine electrolytically; the differential coulomb consumption gave the iodine reacting with the thiol. Apart from these last two procedures, thiosulphate was the reagent for back-titration.

Other determinations, all with thiosulphate as back-titrant, have been of: thiols from thioketonic esters, by Mitra (1938), reacting at −7°C for 25 s; thiols in a tobacco mosaic virus (Fraenkel-Conrat, 1955); 2-amino-5-

mercapto-1,3,4-thiadiazole in dilute acetic acid, using 1 min reaction time (Simionovici *et al.*, 1960); and organometallic mercaptides (antimony, bismuth, sodium, tin) in benzene or hexane, also with 1 min reaction time.

26.3. Identification or determination *via* a reaction product

Disulphides are not customary derivatives for identification of thiols but Jones (1944) oxidized mercaptoacetic acid with iodine to the disulphide, ultimately extracting with ether and crystallizing from benzene–ethyl acetate(9 + 1); he identified by means of the melting point or equivalent weight from titration with alkali to phenolphthalein. Sporek and Danyi (1963) identified and determined thiols by oxidizing with iodine to the disulphides, extracting with ether and carrying out gas chromatography on silicone oil or silicone gum rubber on 60–80 mesh Chromosorb at 60–150°C, depending on the sample.

The hydriodic acid formed in the oxidation

$$2RSH + I_2 \rightarrow RSSR + 2HI$$

has been determined only very rarely as the final quantitative stage. Sampey and Reid (1932) oxidized thiols in benzene with aqueous iodine reagent. Virtually quantitative oxidation was observed within 3 h with primary thiols from C_2 to C_9 but other (secondary) types needed longer. The acid in the aqueous layer was then titrated with alkali to litmus or bromocresol purple. Similar in principle is a patent of Morriss and Meiners (1962) who determined concentration changes of low molecular weight thiols in inert gases by passing through aqueous iodine and measuring the pH differences between the original and the reagent solutions so treated.

Willemart (1956) determined cysteine by oxidizing in hydrochloric acid with excess iodine. After removing unused reagent with thiosulphate he determined the cystine formed through its optical rotation.

27. IODINE MONOCHLORIDE

This reagent oxidizes according to:

$$I^+ + 2e \rightarrow I^-$$

Thiols are oxidized to disulphides:

$$2RSH + ICl \rightarrow RSSR + I^- + Cl^- + 2H^+$$

Číhalík and Růžička (1957) titrated 2,3-dimercapto-1-propanol(BAL) and 5-mercapto-3-phenyl-2-thio-1,3,4-thiadiazolone-2 (bismuthone) potentiometrically. The solution of BAL was freed of oxygen with a current of nitrogen, and saturated with sodium acetate. The best conditions for bismuthone were an initial pH of 4–5 (not more acidic than 0·1 N acid). Číhalík and Terebová (1957) titrated mercaptobenzothiazole (the excess after precipitation in determination of metals) either potentiometrically or to the blue of added starch indicator. Zhdanov et al. (1961) also titrated mercaptobenzothiazole with iodine monochloride in a solution containing sodium acetate by means of amperometric end-point indication.

Rapaport and Raznatovska (1960) determined some sulphur compounds by adding an excess of reagent to the boiling solution of sample. After 15–20 min, the solution was cooled, iodide was added, and the liberated iodine was titrated with thiosulphate. Their examples included 1,2-dimercaptopropane-3-sulphonic acid and methyl thiouracil.

A new test for methyl thiouracil was given by Tsareva and Kuleshova (1969), based on the violet colour, presumably iodine, yielded within 1–2 min on treatment with iodine monochloride.

See Section 3 for reference to titration with iodine cyanide.

28. IODOSO COMPOUNDS

Hellerman et al. (1941) introduced iodosobenzoate as a reagent for thiols. It evidently oxidizes according to:

$$2RSH + ArIO \rightarrow RSSR + ArI + H_2O$$

They added the sample (cysteine, reduced glutathione) to an excess of reagent at pH 7. After 30 s at room temperature, iodide was added and the iodine liberated from the unused reagent was titrated with thiosulphate. This has remained the standard principle. Further oxidation does not take place at this pH. Hellerman and co-workers (1943) later used the method to determine thiol groups in denatured albumin in the presence of guanidine hydrochloride as denaturing agent, allowing a reaction time of 2 min. Barron and Singer (1945) used iodosobenzoate in analytical work on enzyme thiols, and Larson and Jenness (1950) to determine protein thiols, with biamperometric back-titration. In their investigation of amperometric titration of cysteine with silver ions, Sarkar and Sivaraman (1956) obtained data for comparison by means of the iodosobenzoate method. Merville et al. (1959) applied the procedure to reduced glutathione and thiols in denatured urease, using 30 s reaction as in the original method of Hellerman et al.; and in a later publication the same authors (1960) determined thiol groups in biological, pharmaceutical and toxicological procedures, at pH 7·1, but with reaction for 1 min.

Nordin and Spencer (1951) determined thiol groups in flour, adding sodium iodide containing radioactive iodine after reaction of the excess iodosobenzoate. The freed iodine was taken up by the starch and the excess of sodium iodide was estimated in the supernatant with the help of a Geiger counter.

Recently, Verma and Bose (1973c) titrated cysteine in acetic acid–water directly with phenyl iodoacetate, $C_6H_5I(OCOCH_3)_2$ in anhydrous acetic acid in the presence of iodide ion, taking the blue with starch as the end-point. The same authors later (1974) discussed the indirect method with an excess of 2-iodosobenzoic acid, pointing out the risk of further oxidation through the iodine liberated on adding iodide in the back-titration stage. They circumvented this by using ascorbic acid as an intermediate reagent. From 0·1 to 2 mmol of cysteine are treated in pH 7 phosphate buffer with 0·2 M reagent (neutralized) in ca. 50% excess. After swirling for 30 s, a measured excess of standard (0·05 M) ascorbic acid is added, then 5 ml of M hydrochloric acid and the unconsumed ascorbic acid is finally titrated with 0·05 M iodine.

29. LEAD(IV)

This oxidizing agent, functioning according to the change

$$Pb(IV) + 2e \rightarrow Pb(II)$$

has been used in some titrimetric procedures, although these are studies rather than practical applications.

Tomíček (1952) investigated titrations in glacial acetic acid with lead tetraacetate of numerous compounds, including benzyl mercaptan. He took as end-point the colour change from red to blue of the indicator quinalizarin. Zýka and Berka (1962) determined cysteine and mercaptoacetic acid also in acetic acid with lead tetraacetate, this time using measured excess which they back titrated with hydroquinone. The thiols were oxidized in 4-electron changes to sulphinic acids:

$$RSH + 2H_2O \rightarrow RSO_2H + 4H^+ + 4e$$

Suchomelová and Zýka (1963) tried titration of thiols in acetic acid solutions with lead tetraacetate and with a potentiometric or indicator (quinalizarin) end-point. In pure acid, mercaptoacetic acid was oxidized to the disulphide. In the presence of water, oxidation went further: reaction for 15 min in 30% acetic acid yielded the sulphinic acid product in a 1:2 stoichiometry; in 50% acetic acid plus some hydrochloric acid, sulphochloride was formed in 1:3 stoichiometry involving a 6-electron oxidation:

$$RSH + HCl + 2H_2O \rightarrow RSO_2Cl + 6H^+ + 6e$$

Cysteine reacted too slowly for direct titration.

Hladký and Ruza (1969) used lead tetrachloride in dimethylformamide for carrying out potentiometric titrations of glutathione, pentane- and dodecanethiols, thiophenols, and mercaptoacetic acid in dimethyl formamide as solvent. Yurow and Sass (1971) also used lead tetraacetate, but in trifluoroacetic acid solvent, to titrate non-aromatic thiols in trichloroacetic acid–methylene dichloride(1 + 1) to a bipotentiometric end-point. They found that the thiol group reacted with 1·6 equivalents of Pb(IV). The initial reaction was probably the formation of $RSPb(OCOCH_3)_3$, ultimately yielding RSSR *via* RS^+. The disulphide is also oxidized. Agasyan and Sira-kanyan (1971) developed a coulometric procedure, generating Pb(IV) in anhydrous acetic acid. Their thiol examples were 2-mercapto-N-2-naphthyl-acetamide (thionalide) and α-dimethylcysteine. In recent work, Verma and Bose used lead tetraacetate reagent to determine thiols. In one publication (1973a) they titrated directly in acetic acid or mixtures of it with benzene, toluene or petroleum ether, detecting the end-point potentiometrically or with quinalizarin. The water content must not exceed 6%. In another publication (1973b) they determined 3-mercaptopropionic, mercapto-succinic and o-mercaptobenzoic acids with an excess of reagent at 27°C, then estimating unused reagent iodometrically.

The attempt to determine thiols through their "active hydrogen" may be mentioned here. Gilman and Nelson (1937) treated numerous compounds such as aliphatic and aromatic thiols, mercaptoacetic acid, mercaptobenzo-thiazole, in dibutyl ether or dioxan with lead tetraethyl (also bismuth triethyl) in dibutyl ether at 25°C. It was hoped to obtain ethane quantitatively accord-ing to the equation,

$$4RSH + Pb(C_2H_5)_4 \rightarrow 4C_2H_6 + Pb(SR)_4$$

The ethane was determined gas-volumetrically. Unfortunately all values were low, ranging from 25 to 75% of theoretical.

30. MOLYBDATE

Freytag (1953) tried a number of tests for thiols, including that of Richter (1949). Richter found that molybdate yielded a yellow product with mercapto-acetic acid which could be used for molybdate determination. It is possible that a molybdenum(VI) compound is formed, $Mo(-S-CH_2-COO-)_3$ although Meyer (1950) considers it more likely that oxidation occurs to thio- or dithiodiacetic acid, which then reacts with the molybdenum(III) resulting from this oxidation. This is the justification for inclusion here under "Oxidation". Freytag found that the test was only of moderate value for mercaptoacetic acid and cysteine.

31. MOLYBDOPHOSPHATE

See "Heteropolyacids" (Section 22).

32. NITRATE

A solitary reference can be made here, to work of Tsareva and Kuleshova (1969). They published some new tests for 6-methyl-2-thiouracil, including treatment with sodium nitrate and then concentrated sulphuric acid, which yielded a yellow colour that quickly turned to red.

33. NITRITE

A moderately intense red colour with nitrous acid is characteristic of thiols. Possibly the initial reaction is:

$$RSH + HNO_2 \rightarrow RSNO + H_2O$$

Oxidation to disulphide may occur under certain conditions. Thunberg (1911) mentioned this "nitrosyl" reaction, addition of sodium nitrite and then concentrated hydrochloric acid to give a violet–pink colour. Lecher and Siefken (1920) used alkyl (ethyl, amyl) nitrites as the source of nitrous acid but commented on the greater sensitivity of a turbidity test with mercury(II).

The test is sometimes connected with the name of Rheinbold (1927). He covered solid sodium nitrite in a test tube with a solution of the sample and then added sulphuric acid cautiously. The sample can also be added to a mixture of sodium nitrite solution and ethanol, then acidifying. Primary and secondary thiols yield a red coloration; tertiary and aromatic give green turning to red. In earlier work (1926) he had tested the action of nitrosyl chloride which furnished wine-red solutions with thiols; these were stable at low temperature, but evolved nitric oxide and formed disulphides at room temperature.

Authors who used this colour reaction as a test later were: Ruemele and Walker (1953) for identifying thiols in hair-waving lotions; Karr (1954), in an extensive investigation of methods for alkane and arene thiols in petroleum samples, used 2 drops of acetic acid and 1 drop of 4 N aq sodium nitrite, and this gave a green or red upper (petroleum) layer; Mapstone (1954) as a test for thiols in gasoline; Nakamura (1959) for thiols in wool.

Quantitative procedures are based directly on evaluation of the comparatively stable colour which has a strong ultraviolet band at 330–340 nm. Thus Woodward (1949) determined 2-thioethanol in thiodiglycol by treatment in absolute ethanol with 4 N aqueous sodium nitrite and acetic acid; he evaluated the colour after 3 min. Hirsch (1951) applied a closely similar method with sodium nitrite/acetic acid to determine mercaptoacetic acid in cold-wave solutions, comparing the red colour with standards. Walker (1953) tried Woodward's method but preferred the iodometric determination. The test of Ruemele and Walker (1953) mentioned above (addition of 5 ml of wave lotion to 0·2–0·3 g sodium nitrite and subsequently of some drops of sulphuric acid) permitted a rough quantitative assessment. Later publications by Walker (1955a) and Walker and Freeman (1955) referred more favourably to the nitrite method for determining mercaptoacetic acid in hair waving solutions; absorbance was measured at 330 nm or, in more concentrated solutions, at 545 nm. Walker (1955b) subsequently modified the procedure by using amyl nitrite as reagent. A mixture of it with ethanol was added to the sample at pH 9·2–9·5 (ammonia).

Kizer and Howell (1963) also made evaluations at 334 nm in an assay of glutathione reductase activity of rat liver, although they admitted that it was not a specific method for this thiol alone. Tertiary thiols were determined by Ashworth and Keller (1967). They treated the sample in carbon tetra-chloride with 20% sodium nitrite and 6 N hydrochloric acid; after shaking twice for 30 sec, the organic layer was washed with water and ammonium carbonate, filtered, and the absorbance was read at 344 nm. Recently Singh and Mukherjee (1972) tested the influence of some compounds in enhancing the absorbance at 550 nm of 2-mercaptoethanol/nitrite mixtures in acid solution. They found that magnesium chloride, for example, almost doubled it. The colour could be extracted with ether. Their work was aimed more at nitrite determination. Bhuchar and Amar (1972) also published work recently on the reaction between nitrite and mercaptoacetic acid. The red product in acid solution (up to pH 4) was completely extractable into tributyl phosphate, and the solution had absorption maxima at 232, 332, 547 nm, with a molar absorptivity of 775 at 332 nm. Beer's law held at this wavelength from 0·2 to 40 µg nitrite/ml. As in the work mentioned previously, interest was centred on nitrite and not thiol.

Von Wacek and Smitt-Amundsen (1958) tried to apply the Fischer and Schmidt method for lower alcohols to thiol determination. This method depends on conversion to the volatile, easily separable nitrite esters, subse-quently hydrolysing them to nitrous acid and estimating this iodometrically. The thionitrites proved too resistant to hydrolysis. The catalytic effect of mercury(II) might help here, since this is exploited in a procedure of Saville (1958) for thiols; after converting to the thionitrite, he added an acidic sulphanilamide/mercury(II) reagent. The nitrous acid liberated by hydrolysis diazotizes the sulphanilamide and the diazonium salt is coupled with N-1-naphthylethylenediamine to yield an azo compound. This is evaluated colorimetrically. Saville formulates the catalytic action of mercury as:

$$R\text{---}S\text{---}N{=}O + Hg^{2+} \rightarrow R\text{---}\overset{+}{S}\text{---}N{=}O$$
$$\underset{Hg^+}{|}$$

$$R-\overset{+}{\underset{\underset{Hg^+}{|}}{S}}-N=O + H_2O \rightarrow R-\overset{+}{\underset{\underset{Hg}{|}}{S}} \quad + \quad \overset{H}{\underset{H}{>}}\overset{+}{O}-N=O$$

$$H^+ + H-O-N=O$$

Experimental details are given: To 5 ml of a solution of 0·01 M aq sodium nitrite + 0·2–1·0 N sulphuric acid (1 + 9) in a 25 ml graduated flask is added 1 ml of thiol solution (2–50 × 10^{-5} M). After 0·5 to 5 min, depending on the thiol, 1 ml of 0·5% aq ammonium sulphamate is added to destroy the excess of nitrous acid. After 1–2 min, 10 ml of a 1% aq mercuric chloride–3·4% sulphanilamide in 0·4 N hydrochloric acid (1 + 4 vol) are added and the volume is made up to the 25 ml mark with a 0·1% solution of N-1-naphthyl-ethylenediamine dihydrochloride in 0·4 N hydrochloric acid (made up freshly each day). Colour development takes 3–5 min and the absorbance is evaluated after 10 min (yellow–green Ilford No. 605 filter). Saville mentioned that silver also catalyses the ester hydrolysis.

Later workers utilized this method, e.g. Liddell and Saville (1959) for cysteine; Ogawa (1960) for glutathione extracted from tissues; Bethge et al. (1968) for methanethiol in industrial effluent; Todd and Gronow (1969) and also Gronow and Todd (1969) for protein thiol groups. Recently, Pesez and Bartos (1972) evidently coupled with 4-aminoantipyrine instead.

Danchy and Elia (1972) adopted a wholly different technique and titrated thiols directly in dilute sulphuric acid with sodium nitrite; potassium iodide was used as external indicator. Sulphinic acid was titrated simultaneously. It should be possible to use instrumental end-point indication.

34. p-NITROSODIMETHYLANILINE

Feigl and Goldstein (1957) give a spot test for thiols with this reagent, which oxidizes them and is itself reduced to the p-amino-dimethylaniline. A test paper bathed in a solution of the reagent and p-N-dimethylamino-benzaldehyde is employed. This second component reacts with the amino-dimethylaniline to yield a red Schiff's base:

$$(CH_3)_2N-\!\!\left\langle\bigcirc\right\rangle\!\!-CHO \quad + \quad H_2N-\!\!\left\langle\bigcirc\right\rangle\!\!-NCH_3)_2 \quad \longrightarrow$$

$$(CH_3)_2N-\!\!\left\langle\bigcirc\right\rangle\!\!-CH=N-\!\!\left\langle\bigcirc\right\rangle\!\!-N(CH_3)_2$$

Down to a few micrograms of compounds such as cysteine, mercaptoacetic acid, thiosalicyclic acid, mercaptobenzothiazole, thionalide and toluene-2,3-dithiol could be detected.

35. NORADRENOCHROME

Roston (1963) determined cysteine through reaction with noradreno-chrome with which it yields a colour change from pink to yellow. The reagent was prepared from noradrenaline and ferricyanide in phosphate buffer of pH 7·5. After reaction for 2 min, excess yellow ferricyanide was removed with ascorbic acid. The reaction time with the thiol was 10 min. Sodium hydrogen sulphite was then added to remove excess reagent and the light absorbance was measured at 414 nm. The molar absorptivity is rather low but other thiols, such as reduced glutathione, yield no colour.

Schneider et al. (1968) adapted the method to determine cystine, which was reduced in a previous stage.

36. OSMIUM TETROXIDE

Wawrzyczek (1962) detected cysteine by adding sodium acetate to the sample, then 2% osmium tetroxide solution. This gave an immediate pale pink colour (of osmium dioxide), darkening to reddish brown with a violet tinge. Cysteine is oxidized to cysteic acid. Neither cystine, methionine nor taurine interfere, although other reducing agents evidently would.

37. OXIDATION, ELECTROLYTIC

See "Electrolytic Oxidation" (Section 17).

38. OXIDIZED THIOFLUORESCEIN

See "Thiofluorescein, Oxidized" (Section 51).

39. OXYGEN UPTAKE

Okuda and Katai (1929) measured the oxygen uptake of proteins (muscle) which they attributed to reaction of the thiol groups. Cysteine reacts at pH 7–8. This does not appear to have been exploited analytically, probably because of technical difficulties.

Martin and Grant (1965) determined thiols, separated by GLC, by oxidizing the effluent in oxygen to sulphur dioxide which was then estimated coulometrically with electrolytically generated iodine.

40. PER-ACIDS (PERFORMIC ACID)

See "Hydrogen Peroxide" (Section 23).

41. PERIODATE

Periodate is a powerful oxidizing agent which has found some limited use in analytical work on sulphur compounds. The sodium or potassium salt in aq solution or dilute acid or alkaline solution is used as a spray reagent for reducing compounds on paper and thin-layer chromatograms. Thus Mitchell and Metzenberg (1954) sprayed paper chromatograms with 0·01 M aqueous potassium periodate, allowed them to dry for 8–10 min at room temperature, and then sprayed with a borate/boric acid/potassium iodide/starch reagent. Periodate yields a blue colour by oxidizing the iodide to iodine, so that the paper becomes blue. Zones of periodate–oxidizable substances show white, since the iodate reduction product of periodate does not yield iodine under the too weakly acid conditions. Their examples are of amino acids, including methionine and cystine. Cifonelli and Smith (1955) suggested several periodate procedures for detecting amino acids on paper chromatograms. In one, they followed periodate treatment by spraying with an aqueous–ethanolic, very weakly acidic benzidine reagent, which gave white zones on a blue blackground. The blue product results from oxidation of the benzidine by periodate which is absent in the areas of amino acid. Cysteine was among their examples. They also sprayed with a more strongly acid benzidine reagent which gave yellow spots with cysteine and blue spots with other amino acids. In yet another method, they heated the paper chromatogram for 20–30 min at 50–60° after periodate spraying, and then sprayed with 5% aqueous sodium iodide. This yielded brown spots, turning blue with starch. Stephan and Erdman (1964) likewise detected sulphur(II) compounds on chromatograms by spraying with 0·1% sodium periodate in water or 8 N acetic acid, then 4 min later with benzidine, giving white zones against a blue background. Wolf et al. (1971) employed an almost identical procedure for visualizing some alkane- and arenethiols on porous glass powder layers.

Berka and Zýka (1956, 1958) titrated cysteine and other compounds in ca. 9 N hydrochloric acid with periodate, using potentiometric end-point indication. The products of reaction were disulphide and iodine. Recent direct titrations have been performed by Ahmed and Bose (1973) on milligram amounts of thiol in 30% acetic acid or water with potassium periodate/0·05 N sulphuric acid starch indicator): and by Bhatti *et al.* (1973) on 2-mercapto-4,4,6-trimethylpyrimidines in 0·5 to 1 M sulphuric acid with 0·01 M sodium or potassium periodate, also to starch indicator.

Others have used an excess of reagent. Thus Sandri (1957–58) oxidized some sulphur compounds, including mercaptobenzothiazole and dimercaptothiadiazole, in hydrochloric acid with 0·1 N periodic acid. After 20–30 min, unused reagent was estimated by adding potassium iodide and titrating the iodine liberated. Matsuoka (1961) estimated unused potassium periodate in a quantitative procedure for amino acids (including cysteine) through the diminution in size of its polarographic reduction wave. Aravamudan and Rama Rao (1963) observed that the stoichiometry of oxidation of thiomalic acid with numerous oxidizing agents, among them periodate, depended on factors such as pH, temperature and time; an accurate quantitative procedure for it thus seems difficultly attainable.

42. PERMANGANATE

Like periodate, the strong oxidizing agent, permanganate, has not been used greatly in analytical work with thiols, especially in quantitative procedures. This is probably because of frequent uncertainty about the reaction stoichiometry.

Decoloration of permanganate is a standard test, albeit unspecific, for reducing compounds. Alkaline (sodium carbonate) permanganate as a spray reagent enables such compounds to be visualized on chromatograms as yellow to brown spots on a red or pink background. Thus Dalgleish (1950) visualized amino acids, including cysteine, on paper chromatograms in this way. There are certainly many examples of this sort where a thiol happened by chance to be among the compounds. Beer *et al.* (1967) located L-cysteine and L-glutathione on glass fibre paper chromatograms by treatment with 1% potassium permanganate–1% sulphuric acid (1 + 1). The chromatogram was then exposed to hydrogen chloride containing $H^{36}Cl$ which yielded $Mn^{36}Cl_2$ from the manganese dioxide zones; these were detected with a Geiger–Müller counter.

Quantitative application probably dates from the titration of *p*-nitrothiophenol in sulphuric acid medium to the first pink, done by Willgerodt in 1885. Kalinowski *et al.* (1957) determined methyl thiouracil with an excess of potassium permanganate in a reagent containing bromide and acid. After reaction for 1 h in darkness, a measured excess of arsenite was added and the unused part was titrated with permanganate to methyl red. Finkel (1967) oxidized 6-mercaptopurine in serum to the sulphonate with permanganate for 10 min in a borate buffer of pH 9·2. The reaction product was sulphonate. They then added alkaline hydrogen peroxide; after 15 min they centrifuged and evaluated the fluorescence of the supernatant at 400 nm (excitation wavelength = 305 nm).

43. PERSULPHATE

Nakamura and Binkley (1948) determined 0·5 to 5 milliequivalent amounts of cysteine by treating in ca. 50% sulphuric acid with a brucine/glycine/ potassium persulphate reagent. After 30 min at 30°C, the light absorbance was measured at 600 nm. Many other sulphur compounds, e.g. homocysteine, glutathione and mercaptoacetic acid did not react in this way. Freytag (1954) included the brucine/persulphate reaction among the thiol tests which he studied critically. He considered it too insensitive (cysteine could be detected in concentrations down to ca. 1 in 16 000 and mercaptoacetic acid did not yield a colour). Il'ina *et al.* (1965) determined micro amounts of cysteine and glutathione according to a different principle. They mixed the sample with hydrochloric acid, starch, potassium iodide and persulphate and measured the times from the moment of addition of persulphate to that of appearance of blue. This time was found to be proportional to the thiol concentration.

44. PLATINUM(IV)

Platinum(IV) reagents are among the standard reagents for chromatographic visualization of organic sulphur compounds with oxidizable sulphur. They are generally made up from iodide and hexachloroplatinate and contain the PtI_6^{2-} ion. A typical composition is 1·1% aqueous potassium iodide and 0·135% hexachloroplatinic acid, mixed 1:1 before use (Merck). The oxidized substances appears as pale zones on a pink background.

Winegard *et al.* (1948) appear to be the first to have used the reagent, namely for detecting sulphur-containing amino acids (eg cysteine) on paper chromatograms. They sprayed with an iodoplatinate reagent and then held the damp strip in hydrogen chloride vapours, so that the zones were visible as mentioned above. Toennies and Kolb (1951) tested the reagent of Winegard *et al.* among others for thiols and slightly modified it for use as a dipping reagent. Wong (1971) prepared a similar spray reagent for use with thin-layer and paper chromatograms. It contained 14·5 g chloroplatinic acid, 311 mg of potassium iodide and 2·8 ml of 2 N hydrochloric acid, made up to 500 ml with anhydrous ethanol. The layers were dried after spraying and then sprayed with water which yielded white spots on purple with thiols and other sulphur compounds. Millingen (1974) recently visualized vulcanization accelerators in TLC with iodoplatinate; mercaptobenzothiazole gave a yellow spot.

Fowler and Robins (1972) determined sulphur compounds, including cysteine and homocysteine, in physiological fluids in a continuous flow system by oxidation with iodoplatinate; the decrease in absorbance at 500 nm was measured.

45. PORPHYREXIDE, PORPHYRINDINE

These diimino-iminazolidines are air-stable, water-soluble free radicals with strong oxidizing properties, whereby they are converted to leuco-porphyridines and porphyrexine, respectively. Porphyridine has an E_0 of

$$
\begin{array}{c}
(CH_3)_2 C - \overset{(+)}{N}{}^{O^{(-)}} \\
\mid \quad\quad\; > C = NH \\
HN = C - NH
\end{array}
\xrightarrow[O]{H}
\begin{array}{c}
(CH_3)_2 C - N{}^{OH} \\
\mid \quad\quad\; > C = NH \\
HN = C - NH
\end{array}
$$

$$\text{Porphyrexide} \qquad\qquad\qquad \text{Porphyrexide}$$

$$
\left(
\begin{array}{c}
(CH_3)_2 C - \overset{(+)}{N}{}^{O^{(-)}} \\
\mid \quad\quad\; > C = N- \\
HN = C - NH
\end{array}
\right)_2
\xrightarrow[O]{H}
\left(
\begin{array}{c}
(CH_3)_2 C - N{}^{OH} \\
\mid \quad\quad\; > C = N- \\
HN = C - NH
\end{array}
\right)_2
$$

$$\text{Porphyrindine} \qquad\qquad\qquad \text{Leucoporphyrindine}$$

+0·57 V at pH 7, compared with only +0·22 for Tillmans' reagent, 2,6-dichlorophenol–indophenol. Reduction is accompanied by a colour change from deep blue to faint yellow.

Use of the reagent evidently dates from the work of Kuhn and Desnuelle (1938). Published applications of the redox reaction come from the years up to ca. 1945, and concern studies of thiol groups in proteins. Examples are Greenstein (1938, 1939); Hellerman (1939); Greenstein and Edsall (1940); Perez and Sandor (1940) as a test; Anson (1941); Greenstein and Jenrette (1942); Barron and Singer (1945). One titration technique evidently depends on using the colour change of the reagent as end-point. Another technique makes use of nitroprusside as a type of external indicator. Sample and a certain amount of reagent were shaken together for 1–2 min, a drop of ammonium hydroxide was then added, followed by 0·5 ml of a sodium nitroprusside solution. If it yielded the coloration for thiol, the procedure was repeated with more reagent and so on until the nitroprusside no longer responded.

46. SELENIUM DIOXIDE

Selenium dioxide oxidizes according to the equation:

$$SeO_2 + 4H^+ + 4e \rightarrow Se + 2H_2O$$

The formation of purplish selenium in acidified solutions of the dioxide is a sign that reducing compounds are present. Most organic reductants, including thiols, do not accomplish this at ambient temperatures (Feigl, 1966), although Werner (1941) quotes the reaction as a basis for a test for thiols (which are oxidized to disulphides) and also for thioureas, noting that many other sulphur compounds do not react. Dewey and Gelman (1942) studied the colour reactions of organic nitrogen compounds with a selenium dioxide reagent. Their examples included sulphur compounds, such as thioureas, and also cysteine hydrochloride. They treated 1 mg of sample on a spot plate with 1 drop of 0·5% selenium dioxide in concentrated sulphuric acid. Cysteine hydrochloride yielded a yellow to pink colour. Levine and Nachman (1963) observed that many organic compounds reacted with selenious acid/ sulphuric acid ("Mecke reagent") to yield green products. Their examples included 6-mercaptopurine, β-mercaptovaline (penicillamine) and thiouracil.

Franchi's method (1952) for quantitative determination of the $-\overset{|}{N}=C-SH$ group does not strictly qualify for inclusion here. He applied it to heterocyclic compounds containing this grouping Oxidation by reflux for 4 h in acetic acid solution proceeds according to:

$$2 -\overset{|}{N}=C-SH + 3 SeO_2 \rightarrow 3Se + 2SO_2 + 2 -\overset{|}{N}=C-OH$$

After filtering from selenium, unused selenium dioxide was determined iodometrically by adding iodide/sulphuric acid and titrating with thiosulphate to a starch end-point. This procedure is presumably applicable to alkenethiols, and to others which are dealt with in the present book.

47. SULPHOXIDE, DIMETHYL

Sulphoxides possess oxidizing properties according to the equation:

$$R_2SO + 2H^+ + 2e \rightarrow R_2S + H_2O$$

Spencer and Wold (1969) recently published an example of the oxidation of cysteine (and cystine) in proteins by dimethyl sulphoxide, to yield cysteic acid as is accomplished by "performic acid" (see "Hydrogen Peroxide" above, p. 19). They hydrolysed the proteins with 6 N hydrochloric acid, 0·2 to 0·3 M in the sulphoxide; the cysteic acid was identified by high voltage paper electrophoresis and paper chromatography.

48. SULPHUR

Sulphur oxidizes thiols under basic conditions in accordance with:

$$2RSH + S \rightarrow RSSR + H_2S$$

Holmberg (1908) observed this change in 20% sodium hydroxide. McMillan and King (1948) found that certain amines, such as morpholine or n-butylamine catalysed the reaction, for example with n-octanethiol, benzyl mercaptan and β-phenylethanethiol. Good yields of the disulphides were thereby obtained.

In his investigation of tests for thiols in petroleum samples, Karr (1954) used a reaction mixture containing methanol, 28% ammonia as base, and sulphur. The positive outcome of the test for thiols was the formation of bright yellow in the lower (aqueous) layer after shaking for 1 min; the product is presumably ammonium polysulphide.

Two tests in which heavy metal mercaptides undergo the reaction instead of thiols can be classified here. One is the so-called "Doctor" test for detecting thiols in motor fuels, kerosine and similar petroleum samples (cf. Boyd, 1933; Henderson et al., 1940; Le Rosen et al., 1952). Shaking for ca. 15 s with an alkaline sodium plumbite solution indicates thiols present as yellow lead(II) mercaptide, $Pb(SR)_2$. This is confirmed by adding sulphur to the suspension and shaking again for ca. 15 s, whereby brown to black lead sulphide is formed:

$$Pb(SR)_2 + S \rightarrow PbS + RSSR$$

The other example is one of several spot tests for microgram amounts of thiols, given by Feigl et al (1969). The sulphur is yielded in situ from thiosulphate. They heat the sample under alkaline conditions with mercuric cyanide and sodium thiosulphate for a few min at 100°. The presence of a thiol is shown by the formation of black mercuric sulphide. Evidently the thiosulphate supplies sulphur which oxidizes the mercuric mercaptide to disulphide, liberating hydrogen sulphide, which then reacts with the mercury:

$$Hg(SR)_2 + S_2O_3^{2-} \rightarrow HgS + SO_3^{2-} + RSSR$$

Arnold et al. (1952) made use of the reaction to isolate naphthalenethiols from virgin naphtha. After extraction with methanolic potassium hydroxide, hydrolysing and washing out the methanol with water, they treated with sodium sulphide/ethanol/excess sulphur for 4 h to yield the corresponding disulphides of higher boiling point than the thiols. Phenol and aromatic hydrocarbons could then be separated by distillation. The residual disulphides were purified by distillation and reduced to the original thiols with lithium aluminium hydride.

Guthrie and co-workers (1932, 1938, 1941) adapted the reaction to the quantitative determination of thiols, such as glutathione, based on the method of Heffler (1908) for thiol determination in egg albumin. The sample was digested in alcohol with colloidal sulphur at pH 6·7. Hydrogen sulphide was aerated out into zinc acetate solution, yielding zinc sulphide. The sulphide was estimated iodometrically or colorimetrically by conversion to methylene blue with p-dimethylaminoaniline and ferric iron:

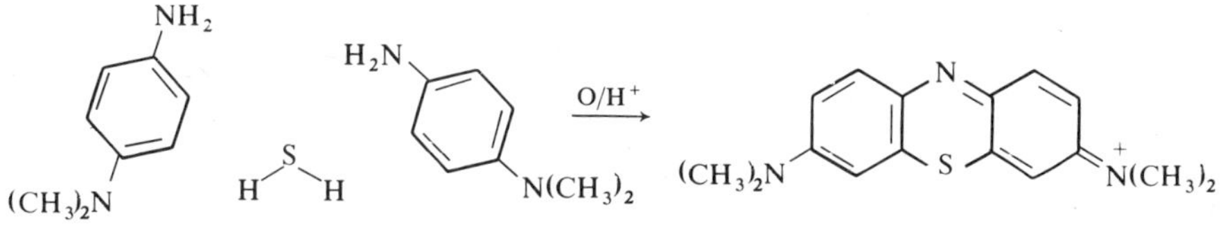

Brenner *et al.* (1954) similarly determined volatile thiols in beer. These were distilled out under reducing conditions and treated with colloidal sulphur for 1 h at 30°C in nitrogen. They collected the evolved gas (hydrogen sulphide) in zinc acetate, then added p-dimethylaminoaniline and ferric sulphate and measured the absorbance at 745 nm.

49. TETRATHIONATE

Tetrathionate is a mild oxidizing agent:

$$S_4O_6^{2-} + 2e \rightarrow 2S_2O_3^{2-}$$

It has been seldom used for thiol determination. Baernstein (1936) quoted its oxidation of a thiol in his procedure for the determination of methionine by demethylation with hydrogen iodide. This yields methyl iodide and the stable thiolactone of homocysteine,

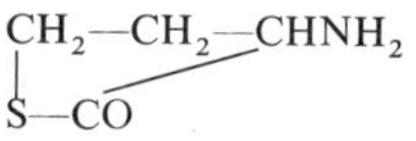

The thiolactone is converted to the thiol by treatment for 15 min in a solution made alkaline by the addition of concentrated ammonium hydroxide. He oxidized it with excess tetrathionate, back-titrating the excess with iodate. Schormüller and Ballschmieter (1951) adapted this procedure to the micro scale for determining methionine in protein foodstuffs. The thiolactone was left with excess tetrathionate for 15 min in concentrated ammoniacal solution at 40–45°C. They acidified subsequently and titrated with biiodate.

Anson (1940, 1941) determined protein thiols *via* the amount of tetrathionate necessary to give the point where no colour reaction with nitroprusside was evident.

Catalytically active thiol groups of enzymes have been shown to react quantitatively with tetrathionate to give inactive sulfenylthiosulphate derivatives:

$$E\text{—}SH + S_4O_6^{2-} \rightarrow E\text{—}S\text{—}S_2O_3^- + S_2O_3^{2-} + H^+$$

(e.g. Pihl and Lange, 1962; Parker and Allison, 1969). Such work has not generally been directly analytical.

Inglis and Liu (1970) (see also Liu and Inglis, 1972) determined cysteine in acidic protein hydrolysates (and cystine after reduction to it with dithiothreitol) by treatment with tetrathionate for 16 h at 25°, followed by bubbling oxygen through the solution for 7 min. The S-sulphocysteine formed could be separated by ion exchange in the Aminco-Bowman 120C analyser by elution with sodium citrate of pH 3·25, then 4·25, at 55°C. It moved more rapidly than other amino acids.

50. TETRAZOLIUM SALTS

The tetrazolium cation is a reagent that has been introduced comparatively recently for demonstrating the presence of reducing compounds, with which it yields coloured formazans:

Tetrazolium cation   Formazan

It has become a standard reagent for some compound classes, notably reducing sugars. The best-known reagent is probably TTC, 2,3,5-triphenyltetrazolium chloride.

Applications to thiols have been rare, however. Fairbridge et al. (1951) determined cysteine with TTC or the corresponding bromide salt. The reaction is slower than with sugars and they heated a 50–600 μg sample in alkaline solution for 25 min at 100°C. After cooling, acidifying and adding isopropanol, the absorbance was measured at 485 nm. Freytag (1955) considered that the method was too insensitive for thiol determination in female human hair. It is better known as a test, especially in histochemical work. Rogers (1953) used 2-(p-iodophenyl)-3-(p-nitrophenyl)-5-phenyltetrazolium chloride to demonstrate thiol groups in wool and hair follicles. Barrnett and Seligman (1954), Roberts and Lucchese (1955) and Gomori (1956) employed tetrazolium salts for histochemical demonstration of thiols. Verne (1956) reviewed histological applications of tetrazolium salt reduction, including that to thiols. Freytag (1954) undertook a critical study of several tests for thiols, taking cysteine and mercaptoacetic acid as examples. He found that TTC was an insensitive reagent and needed a minimum of 3 mg ml^{-1} of cysteine hydrochloride and 1 mg ml^{-1} of mercaptoacetic acid. Reaction for 4 h at room temperature was necessary and the reagent yielded a coloured product in light in the absence of any reducing compound. Jámbor *et al.* (1957) were also critical of tetrazolium salts in histochemistry, finding inverse proportionality between thiol amount (estimated by amperometric titration with silver ion) and formazan values.

Martinez-Rodriguez *et al.* (1972) demonstrated the presence of thiols with a menadione–nitro blue tetrazolium (NBT) reagent {NBT = 3,3'-(3,3'-dimethoxy-4,4'-biphenylylene)-bis[2-(*p*-nitrophenyl)-5-phenyl-2H-tetrazolium chloride]}. The menadione acts as a proton-transporter.

51. THIOFLUORESCEIN, OXIDIZED

Dubouloz *et al.* (1962) decolorized an alkaline thiofluorescein reagent by exposure for ca. 15 min to a 100 W lamp at 15 cm distance. The colour is restored by thiols, e.g. cysteine. They used reaction conditions of 15 h/ 4°, 30 min/50° and 4 min/70° in ethyl acetate + phosphate buffer, pH 8 and evaluated the colour either at 585 nm or with an appropriate filter to give a measure of the thiol amount.

This is an unusual principle, but the reagent may be difficult to prepare reproducibly.

52. TILLMANS' REAGENT

This oxidizing agent, 2,6-dichlorophenol-indophenol, usually reacts in a 2-electron reaction according to the equation:

Blue (red in acid solution)   Colourless

It is best known in the detection and determination of ascorbic acid (and other ene-diols). Colourless zones on a blue background are seen on chromatograms; direct titration, usually to the persistent blue given by an excess of reagent, is a standard procedure for ascorbic acid.

Thiols are oxidized by the reagent to disulphides, although this reagent/

thiol ratio of 1:2 is not always observed. Thus Todrick and Walker (1937) found a 1:1 ratio which they considered to be caused by the formation of RS̄ sulphenium ion. Basford and Huennekens (1955) followed the oxidation of thiols spectrophotometrically in pH 7 phosphate buffer through the fall in absorbance at 600 nm after the addition of thiol. They found a 1:1 ratio if the reaction was fast, but 1:2 if it was slower, e.g. on addition of metal ions. Hadler *et al.* (1963) explained the 1:1 ratio with cysteine, for example, as due to the formation of a 1:1 conjugate between the reaction product and cysteine.

Basford and Huennekens compared thiol titrations with Tillmans' reagent and nitroprusside. As advantages of the former, they gave slightly higher sensitivity, absence of rapid reoxidation and the possibility of determining thiols in the presence of acyl thiol esters which hydrolyse in the basic conditions of the nitroprusside method. On the other hand, the indophenol method is sensitive to other reagents.

Terada and Nomiyama (1957) determined thioglycol by reaction with 2,6-dichlorophenol-indophenol in acetate buffer + xylene. After shaking for 20 sec, the mixture was centrifuged and the xylene layer was evaluated colorimetrically at 500 nm.

53. TUNGSTOPHOSPHATE

See "Heteropoly Acids" (Section 22).

54. VANADATE

This moderately strong oxidizing agent has found relatively little use and only a few examples of its application to thiols have been found. These are evidently based on the formation of coloured lower vanadium valencies. Thus Levine and Hubbell (1930) detected cysteine by treating with ammonium metavanadate in aqueous sodium carbonate, then acidifying with hydrochloric acid to yield a green colour. Garcia-Blanco and Pascual-Leone (1953) also detected cysteine (and cystine and methionine) through the blue colour given with a solution of ammonium vanadate in sulphuric acid. Tsareva and Kuleshova (1969) described new tests for 6-methyl-2-thiouracil, including the light green or blue yielded on heating with an ammonium vanadate/sulphuric acid reagent for 1–2 min.

Solsona (1956) analysed a mixture of cysteine, cystine and methionine through differential oxidation with three vanadate reagents containing different concentrations of sulphuric acid. The blue colour produced after 30 min was evaluated. One reagent reacted only with cysteine; a second only with cysteine and methionine; and the third with all three sulphur compounds.

REFERENCES

Ackerman, G. A. and Sneeringer, S. C. (1960). *Lab. Invest.* **9**, 356.

Adams, C. W. M. (1956). *J. Histochem. Cytochem.* **4**, 23.

Adams, D. F. (1969). *Tappi,* **52**, 53.

Agasyan, P. K. and Sirakanyan, M. A. (1971). *Zh. Anal. Khim.* **26**, 1404.

Agathangelou, S. P., Brown, D. R., Fogg, A. G. and Burns, D. T. (1971). *Lab. Pract.* **20**, 115.

Ahmed, J. and Bose, S. (1973). *J. Indian Chem. Soc.* **50**, 506.

Aldridge, W. N. (1948). *Biochem. J.* **42**, 52.

Alexander, W. A., Mash, C. J. and McAuley, A. (1969). *Talanta,* **16**, 535.

Anokhina, D. I. and Kalmykova, R. S. (1970). *Fiz.-Khim. Metody Anal.* No. 1, 117; *Chem. Abs.* **77**, 13744.

Anson, M. L. (1940). *J. Biol. Chem.* **135**, 797; also (1941). *J. Gen. Physiol.* **24**, 399.

Aravamudan, G. and Rama Rao, C. (1963). *Talanta,* **10**, 231.

Arnold, R. C., Launer, P. J. and Lien, A. P. (1952). *Anal. Chem.* **24**, 1741.

Aronovich, Kh. A. (1969). *Neftepererab. Neftekhim.* (Moscow), 6–8; *Chem. Abs.* **72**, 91902.

Ashworth, G. W. and Keller, R. E. (1967). *Anal. Chem.* **39**, 373.

Austin, R. R., Percy, L. E. and Escher, E. E. (1950). *Gas,* **26**, No. 5, 47; No. 6, 33.

Avakyants, S. G. and Murtazaev, A. M. (1969). *Dokl. Akad. Nauk Uzbek. SSR,* **26**, 35.

Bachhawat, J. M., Ramegowda, N. S., Koul, A. K., Narang, C. K. and Mathur, N. K. (1973). *Indian J. Chem.* **11**, 614.

Baernstein, H. G. (1930). *J. Biol. Chem.* **89**, 125.

Baernstein, H. G. (1936). *J. Biol. Chem.* **115**, 25, 33.

Bakes, J. M. and Jeffery, P. G. (1961). *Talanta,* **8**, 641.

Barrnett, R. J. and Seligman, A. M. (1954). *J. Natl. Cancer Inst.* **14**, 769.

Barron, E. S. G. and Flood, V. (1950). *J. Gen. Physiol.* **33**, 229.

Barron, E. S. G. and Singer, T. P. (1945). *J. Biol. Chem.* **157**, 221.

Barron, E. S. G., Miller, Z. B. and Kalnitzky, G. (1947). *Biochem. J.* **41**, 62.

Basford, R. E. and Huennekens, F. M. (1955). *J. Amer. Chem. Soc.* **77**, 3873.

Bayfield, R. F. and Cole, E. R. (1969). *J. Chromatogr.* **40**, 470.

Bayfield, R. F., Clarke, V. and Cole, E. R. (1965). *J. Chromatogr.* **19**, 370.

Beck, G. (1951). *Mikrochim. Acta,* **38**, 1.

Beer, J. Z., Budzynski, A. Z. and Malwinska, K. (1967). *Chem. Anal.* (*Warsaw*), **12**, 1055.

Beesing, D. W., Tyler, W. P., Kurtz, D. M. and Harrison, S. A. (1949). *Anal. Chem.* **21**, 1073.

Benedict, S. R. (1922). *J. Biol. Chem.* **51**, 187.

Benedict, S. R. and Gottschall, G. (1933). *J. Biol. Chem.* **99**, 729

Berdnikov, A. I. (1966). *Lab. Delo,* 85.

Berg, B. H. (1971). *Acta Pharm. Suecica,* **8**, 453.

Berka, A. and Zýka, J. (1956). *Chem. Listy* **50**, 314; also (1958). *Pharmazie* **13**, 81.

Beskova, G. S., Kontorovich, L. M., Kutilina, R. A. and Bolrova, V. P. (1971). *Tr. Nauchn.-Issled. Proekt. Inst. Azotn. Prom. Prod. Org. Sin.* No. 6, 367; *Chem Abs.* **77**, 37268.

Bethge, P. O., Carlson, M. and Radestrom, R. (1968). *Svensk. Papperstidn.* **71**, 864.

Bhatti, A. M., Sondhi, S. M. and Ralhan, N. K. (1973). *J. Indian Chem. Soc.* **50**, 406.

Bhuchar, V. M. and Amar, V. K. (1972). *Indian J. Technol.* **10**, 433.

Bickford, C. F. and Schoetzow, R. E. (1937). *J. Amer. Pharm. Assoc.* **26**, 409.
Binet, L. and Weller, G. (1935). *Compt Rend. Soc. Biol.* **119**, 939; also (1936). *ibid.*, **18**, 358 and (1947). *Compt Rend.* **224**, 870.
Birch, S. F. and Norris, W. F. (1925). *J. Chem. Soc.* 898, 1934.
Blanchetière, A. and Mélon, L. (1927). *Compt. Rend. Soc. Biol.* **97**, 242.
Blažek, J., Kraćmář, J. and Stejskal, Z. (1957). *Česk. Farm.* **6**, 441.
Blöckinger, G. (1957). *Chem. Zvesti*, **11**, 340.
Blois, M. S. (1958). *Nature (London)*, **181**, 1199.
Blyakher, Ya. I. (1938). *Lab. Prakt. (USSR)*, **13**, No. 4, 15.
Bond, G. R. (1933). *Ind. Eng. Chem., Anal. Ed.* **5**, 257.
Borner, K. (1965). *Z. Physiol. Chem.* **341**, 264.
Boyd, G. A. (1933). *Oil and Gas J.* **32** No. 8, 16, 31; *Chem. Abs.* **27**, 5525.
Brand, E., Cahill, G. F. and Kassell, B. (1940). *J Biol. Chem.* **133**, 431.
Braun, R. D. and Stock, J. T. (1972). *Anal. Chim. Acta*, **60**, 167.
Brenner, M. W., Owades, J. L., Gutcho, M. and Golyzniak, R. (1954). *Amer. Soc. Brew. Chem., Proc. Ann. Meeting*, 88.
Brown, P. R. and Edwards, J. O. (1968). *J. Chromatog.* **38**, 543.
Brown, D. R. and Fogg, A. G. (1969). *Loughboro' Univ. Technol. Dept. Chem., Sum. Final Year Stud. Proj. Theses*, **10**, 10; *Chem. Abs.* **73**, 52118.
Bruno, S. (1963). *Boll. Chim. Farm.* **102**, 468.
Bucci, F. and Cusmano, A. M. (1962). *Rend. Ist. Super Sanitá*, **25**, 518.
Bucher, K. (1951). *Pharm. Helv. Acta*, **26**, 145.
Buděšinský, B. and Vaníčková, E. (1957). *Česk. Farm.* **6**, 308.
Buscaróns, F, Artigas, J. and Rodriguez-Roda, C. (1960). *Anal. Chim. Acta*, **23**, 217; also (1961). *Chim. Anal. (Paris)*, **43**, 228.
Busev, A. I. and Fang Chang (1961). *Talanta*, **8**, 470.
Ceresa, F. and Guala, P. (1939). *Giorn. Accad. Med. Torino*, **102**, 193; *Chem. Abs.* **35**, 6995.
Černoch, M. (1966). *Coll. Czech. Chem. Commun.* **31**, 782.
Chattopadhyay, S. S. and Mahanti, H. S. (1973). *J. Inst. Chem. Calcutta*, **45**, 104.
Chèvremont, M. and Frédéric, J. (1943). *Arch. Biol.* **54**, 589.
Cifonelli, J. A. and Smith, F. (1955). *Anal. Chem.* **27**, 1501.
Číhalík, J. and Růžička, J. (1957). *Chem. Listy*, **51**, 264; also (1957). *Coll. Czech. Chem. Commun.* **22**, 764.
Číhalík, J. and Terebová, K. (1957). *Chem. Listy*, **51**, 272.
Clark, J. D. (1951). *U.S. Naval Air Rocket Test Sta., Rept. No.* 11, 1; *Chem. Abs.* **48**, 4840.
Cole, R. D., Stein, W. H. and Moore, S. (1958). *J. Biol. Chem.* **233**, 1359.
Coope, J. A. R. and Maingot, G. J. (1955). *Anal. Chem.* **27**, 1478.
Coulson, D. M., Crowell, W. R. and Friess, S. L. (1950). *Anal. Chem.* **22**, 525.
Cunningham, L. W. and Nuenke, B. J. (1960). *J. Biol. Chem.* **235**, 1711.
Dalgleish, C. E. (1950). *Nature (London)*, **166**, 1076.
Danchy, J. P. and Elia, V. J. (1972). *Anal. Chem.* **44**, 1281.
Dewey, B. T. and Gelman, A. H. (1942). *Ind. Eng. Chem., Anal. Ed.* **14**, 361.
DiCapua, C. B. (1934). *Boll. Soc. Ital. Biol. Sper.* **9**, 86, 88.
Divin, I. A., Grechanovskii, V. P. and Malusis, Ya. I. (1937). *Lab. Prakt. (USSR)*, No. 1, 25.
Dubouloz, P., Fondarai, J. and Pavone-Marville, R. (1962). *Anal. Chim. Acta*, **26**, 249.
Ellis, E. W. and Barker, T. (1951). *Anal. Chem.* **23**, 1777.

Fairbridge, R. A., Willis, K. I. and Booth, R. G. (1951). *Biochem. J.* **49**, 423.
Fecko, J. and Zaborniak, F. (1968). *Chem. Anal.* (*Warsaw*), **13**, 659.
Feigl, F. (1966). *Spot Tests in Organic Analysis, 7th Ed.* p. 136, Elsevier, London.
Feigl, F. and Goldstein, D. (1957). *Michrochem. J.* **1**, 177.
Feigl, F., Goldstein, D. and Libergott, E. K. (1969). *Anal. Chim. Acta,* **47**, 553.
Finkel, J. M. (1967). *Anal. Biochem.* **21**, 362.
Flotow, L. (1928). *Biochem. Z.* **194**, 132.
Folin, O. and Marenzi, A. D. (1929). *J. Biol. Chem.* **83**, 103.
Forster, H., Meyer, A. and Volkart, H. (1954). *Mitt. Lebensm. u. Hyg.* **45**, 490.
Fowler, B. and Robins, A. J. (1972). *J. Chromatog.* **72**, 105.
Fraenkel-Conrat, H. (1944). *J. Biol. Chem.* **152**, 385.
Fraenkel-Conrat, H. (1955). *J. Biol. Chem.* **217**, 373.
Franchi, G. (1952). *Ann. Chim.* (*Rome*), **42**, 701.
Frankiel, J. and Rombau, P. (1948). *Chim. Anal.* (*Paris*), **30**, 60.
Freses, A. T. and Winstanley, D. M. (1960). *Rev. Cienc.* (*Lima*), **62**, No. 11, 24.
Freytag, H. (1952–3). *Z. Anal. Chem.* **137**, 331.
Freytag, H. (1953). *Z. Anal. Chem.* **138**, 259.
Freytag, H. (1954). *Z. Anal. Chem.* **143**, 401.
Freytag, H (1955). *Parfums Cosmét. Savons,* No. 112, 19; *Chem. Abs.* **50**, 9485.
Frieden, E. (1958). *Biochim. Biophys. Acta* **27**, 414.
Fuyita, A. and Numata, I. (1938). *Biochem. Z.* **299**, 249.
Gabbe, E. (1930). *Klin. Wochschr.* **9**, 169.
Garcia-Blanco, J. and Pascual-Leone, A. M. (1955). *Rev. Españ. Fisiol.* **11**, 143.
Gavrilescu, N. (1931). *Biochem. J.* **25**, 1190.
Gawargious, Y. A., Besada, A. and Hassouna, M. E. M. (1974). *Mikrochim. Acta,*
    1003.
Giannessi, P. (1961). *Ann. Staz. Chim.-Agrar. Sper. Roma Ser. III Publl. No.* 190;
    Chem. Abs. **58**, 3830.
Gilman, H. and Nelson, J. F. (1937). *J. Amer. Chem. Soc.* **59**, 935.
Gimmerikh, F. I. (1967). *Lab. Delo,* 564.
Glavinel, J. (1963). *Acta Chem. Scand.* **17**, 1635
Gol'dman, S. (1937). *Ukr. Bioch. Zh.* **10**, 595; *Chem. Abs.* **32**, 5021.
Gomori, G. (1956). *Quart. J. Microscop. Sci.* **97**, 1.
Greenstein, J. P. (1938). *J. Biol. Chem.* **125**, 501; also (1939). *ibid.,* **128**, 233.
Greenstein, J. P. and Edsall, J. T. (1940). *J. Biol. Chem.* **133**, 397.
Greenstein, J. P. and Jenrette, W. W. (1942). *J. Biol. Chem.* **142**, 175.
Greer, E. J. (1929). *Ind. Eng. Chem.* **21**, 1033.
Gregg, D. C. and Blood, jr., C. A. (1951). *J. Org. Chem.* **16**, 1255.
Gregg, D. C., Bouffard, P. E. and Barton, R. (1961). *Anal. Chem.* **33**, 269.
Grogan, C. H., Rice, L. M. and Reed, E. E. (1955). *J. Org. Chem.* **20**, 50.
Gronow, M. and Todd, P. (1969). *Anal. Biochem.* **29**, 540.
Grossert, J. S. and Langler, R. F. (1974). *J. Chromatog.* **97**, 83.
Gudzinowicz, B. J. (1960). *Anal. Chem.* **32**, 1520.
Guthrie, J. D. (1938). *Contrib. Boyce Thompson Inst.* **9**, 223.
Guthrie, J. D. and Allerton, J. (1941). *Contrib. Boyce Thompson Inst.* **12**, 103.
Guthrie, J. D. and Wilcoxon, F. (1932). *Contrib. Boyce Thompson Inst.* **4**, 99.
Haas, E. (1937). *Biochem. Z.* **291**, 79.
Hadler, H. I., Erwin, M. J. and Lardy, H. A. (1963). *J. Amer. Chem. Soc.* **85**, 458.
Haeusler, K. G., Geyer, R. and Rennhak, S. (1972). *Z. Chem.* (*Leipzig*), **12**, 339.
Haeusler, K. G., Geyer, R. and Rennhak, S. (1973). *Z. Chem.* **13**, 196.

Hakewill, H. and Rueck, E. M. (1946). *Proc. Amer. Gas Assoc.* **28**, 529.

Harnish, D. P. and Tarbell, D. S. (1949). *Anal. Chem.* **21**, 968.

Hartner, F. and Schleiss, E. (1936). *Mikrochemie,* **20**, 63.

Hashall, E. M. (1940). *J. Assoc. Offic. Agr. Chemists,* **23**, 727.

Hashmi, M. H., Chughtai, N. A., Ahmad Shahid, M., Iftikhar Ajmal, A. and Chughtai, M. I. D. (1969). *Mikrochim. Acta,* 36.

Heffler, A. (1908). *Med. Naturw. Arch.* **1**, 81.

Hellerman, L. (1939). *Cold Spring Harbor Symposia Quant. Biol.* **7**, 165; *Chem. Abs.* **35**, 6985.

Hellerman, L., Chinard, F. P. and Ramsdell, P. A. (1941). *J. Amer. Chem. Soc.* **63**, 2551.

Hellerman, L., Chinard, F. P. and Deitz, V. R. (1943). *J. Biol. Chem.* **147**, 443.

Henderson, L. M., Agruss, M. S. and Ayers, jr., G. W. (1940). *Ind. Eng. Chem., Anal. Ed.* **12**, 1.

Hess, W. C. (1929). *J. Washington Acad. Sci.* **19**, 419.

Higgins, D. J. and Stephenson, W. H. (1957). *Analyst* (*London*), **82**, 435.

Hirase, S. and Araki, C. (1955). *Bull. Chem. Soc. Japan,* **28**, 481.

Hirsch, F. (1951). *Seifen-Öle-Fette-Wachse,* **77**, 457.

Hladký, Z. (1965). *Z. Chem.* (*Leipzig*), **5**, 424.

Hladký, Z. (1967). *Wiss. Z. Tech. Hochsch. Chem. "Carl Schlorlemner", Leuna-Merseburg,* **9**, 5.

Hladký, Z. and Ruza, J. (1969). *Chem. Zvesti,* **23**, 336.

Hladký, Z. and Vřešťal, J. (1969). *Coll. Czech. Chem. Commun.* **34**, 984.

Holmberg, B. (1908). *Ann.* **359**, 81.

Hunter, G. and Eagles, B. A. (1927). *J. Biol. Chem.* **72**, 177.

Il'ina, Yu. N., Talmud, B. A. and Afanas'ev, P. V. (1965). *Prikl. Biochim. i Mikrobiol.* **1**, 352; *Chem. Abs.* **63**, 11985.

Inglis, A. S. and Liu, T.-Y. (1970). *J. Biol. Chem.* **245**, 112.

Inkin, A. A. and Kharlamov, V. T. (1973). *Zh. Anal. Khim.* **28**, 2037.

Ionescu-Matiu, A. I. and Popesco, A. (1939). *Bull. Acad. Méd. Roumanie,* **4**, 385; also (1940). *Bull. Soc. Chim. Biol.* **22**, 474; *Chem. Abs.* **33**, 8143 and **35**, 3559.

Jámbor, B., Devay, M. and Roberts, L. W. (1957). *Nature* (*London*), **180**, 997.

Jančík, F., Buděšinský, B. and Činková, O. (1957). *Česk. Farm.* **6**, 108.

Jaselskis, B. (1959). *Anal. Chem.* **31**, 928.

Joachim, J. L. (1951). *Riv. Combustibili,* **5**, 404; *Chem. Abs.* **46**, 4198.

Jones, J. H. (1944). *J. Assoc. Offic. Agr. Chemists,* **27**, 574.

Kalinowski, K. and Piotrowska, A. (1959). *Acta Polon. Pharm.* **16**, 107.

Kalinowski, K., Bersztel, J., Fecko, J. and Zwierzchowski, Z. (1957). *Acta Polon. Pharm.* **14**, 77.

Kamidate, T., Yatsunayagi, T. and Aomuri, K. (1971). *Kogyo Kagaku Zasshi,* **74**, 1275.

Karr, jr., C. (1954). *Anal. Chem.* **26**, 528.

Kassell, B. and Brand, E. (1938). *J. Biol. Chem.* **125**, 115.

Katyal, J. M. and Gorin, G. (1959). *Arch. Biochem. Biophys.* **82**, 319.

von Keller, D. (1954). *Seifen-Öle-Fette-Wachse,* **80**, 560.

Kemula, W. and Brachaczek, W. (1963). *Chem. Anal.* (*Warsaw*), **8**, 579.

Khomenko, O. K. and Tsinkalovs'ka, S. M. (1962). *Ukr. Biokhim. Zh.* **34**, 888.

Kidby, D. K. (1969). *Anal. Biochem* **28**, 230.

King, E. J. and Lucas, C. C. (1931). *Biochem. Z.* **235**, 66.

King, E., Baumgärtner, L. and Page, I. H. (1930). *Biochem. Z.* **217**, 389.

Kizer, D. E. and Howell, B. A. (1963). *Proc. Soc. Exp. Biol. Med.* **112**, 967.

Klouček, B., Jehlička, V. and Gasparić, J. (1960). *Chem. Průmysl.* **10**, 624.

Koenig, N. H., Sasin, G. S. and Swern, D. (1958). *J. Org. Chem.* **23**, 1525.

Kolb, J. J. and Toennies, G. (1952). *Anal. Chem.* **24**, 1164.

Kolthoff, I. M. and Harris, W. E. (1949). *Anal. Chem.* **21**, 963.

Kolthoff, I. M. and Stricks, W. (1951). *Anal. Chem.* **23**, 763.

Kolthoff, I. M., Meehan, E. J., Tsao, M. S. and Choi, Q. W. (1962). *J. Phys. Chem.* **66**, 1233.

Komissarenko, V. P., Molat, L. A. and Petrunina, A. V. (1974). *Zh. Anal. Khim.* **29**, 408.

Kosower, N. S. with Song, K.-R., Ernst, M. J., Tsien, D. M. C., Correa, W. and Kosower, E. M. (1969). *Biochim. Biophys. Acta,* **192**, 1, 8, 15, 23.

Krause, W. (1938). *Chemist-Analyst,* **27**, 14.

Kreshkov, A. P. and Oganesyan, L. B. (1971). *Zh. Anal. Khim.* **26**, 614.

Kreshkov, A. P. and Oganesyan, L. B. (1973). *Zh. Anal. Khim.* **28**, 2260.

Kühnau, J. (1931). *Biochem. Z.* **230**, 353.

Kuhn, R. and Desnuelle, P. (1938). *Z. Physiol. Chem.* **251**, 14.

Kuhn, R., Birkofer, L. and Quackenbush, F. W. (1939). *Chem. Ber.* **72**, 407.

Kul'berg, L. M. and Presman, Sh. E. (1940). *Lab. Prakt. (USSR),* **15**, No. 4, 15.

Langou, A. and Marenzi, A. D. (1935). *Anales Farm. Bioquím. (B. Aires),* **6**, 70; *Chem. Abs.* **30**, 3007.

Larson, B. L. and Jenness, R. (1950). *J. Dairy Sci.* **33**, 890, 896.

Lavine, T. (1935). *J. Biol. Chem.* **109**, 141.

Lecher, H. and Siefken, W. (1926). *Chem. Ber.* **59**, 2597, 2600.

Lederer, M. and Silberman, H. (1952). *Anal. Chim. Acta,* **6**, 133.

Lehman, J. F., Barrack, L. P. and Lehman, R. A. (1951). *Science,* **113**, 410.

Le Rosen, A. L., Moravek, R. T. and Carlton, J. K. (1952). *Anal. Chem.* **24**, 1335.

Levine, V. E. and Hubbell, A. M. (1930). *Proc. Soc. Exp. Biol. Med.* **28**, 199.

Levine, V. E. and Nachman, M. (1963). *J. Forensic Med.* **10**, 65.

Liberti, A. (1951). *Ann. Chim. (Rome),* **41**, 363.

Liddell, H. F. and Saville, B. (1959). *Analyst (London),* **84**, 188.

Liu, T.-Y. and Inglis, A. S. (1972). *Meth. Enzym.* **25**, 55.

Lorenz, O. and Echte, E. (1956). *Kautschuk u. Gummi,* **9**, WT 300.

Lucas, C. C. and King, E. J. (1932). *Biochem. J.* **26**, 2076.

Lugg, J. W. H. (1932). *Biochem. J.* **26**, 2144, 2160.

McMillan, F. H. and King, J. A. (1948). *J. Amer. Chem. Soc.* **70**, 4143.

Mahadevappa, D. S. (1965). *Current Sci.* **34**, 530.

Malikova, E. M. (1954). *Ref. Zh. Khim. Abstr. No.* 34241; *Chem. Abs.* **49**, 9447; *Anal. Abs.* **2**, 1623.

Mapstone, G. E. (1954). *Chem. & Ind. (London),* 1113.

Martin, R. L. and Grant, J. A. (1965). *Anal. Chem.* **37**, 1644.

Martinez-Rodriguez, R., Toledano, A. and Gonzalez, M. (1972). *Ann. Histochem.* **17**, 215.

Mason, H. L. (1930). *J. Biol. Chem.* **86**, 623.

Mason, H. L. (1931). *Proc. Staff Meetings, Mayo Clinic,* **6**, 168; *Chem. Abs.* **25**, 4572.

Matsuoka, M. (1961). *Nippon Kagaku Zasshi,* **82**, 1193.

Matsuoka, M. (1961). *Nippon Kagaku Zasshi,* **82**, 1193.

Medes, G. (1936). *Biochem. J.* **30**, 1293.

Meehan, E. J., Kolthoff, I. M. and Kakiuchi, H. (1962). *J. Phys. Chem.* **66**, 1238.

Merck, E. (Darmstadt, W. Germany). *Anfärbereagenzien für Dünnschicht- und Papier-Chromatographie,* p. 29.

Merville, R., Dequidt, J. and Corteel, M. L. (1959). *Bull. Soc. Pharm. Lille,* **57**; also (1960). *Ann. Pharm. Franc.* **18**, 625.

Metello Metto, J. M. and de Figueiredo, A. P. (1949). *Rev. Brasil. Farm.* **31**, 17; *Chem. Abs.* **43**, 8311.

Meyer, F. O. W. (1950). *Pharm. Zentralh.* **89**, 3.

Micheel, F. (1939). *Chem. Ber.* **72**, 68.

Millingen, M. B. (1974). *Anal. Chem.* **46**, 746.

Mirsky, A. E. and Anson, M. L. (1935). *J. Gen. Physiol.* **18**, 307.

Mitchell, H. K. and Metzenberg, R. L. (1954). *J. Amer. Chem. Soc.* **76**, 4187.

Mitra, S. K. (1938). *J. Indian Chem. Soc.* **15**, 205.

Mogoricheva, I. and Korsunskaya, E. (1933). *J. Rubber Inst. (USSR),* **10**, 341.

Moncorps, C. and Schmid, R. (1932). *Z. Physiol. Chem.* **205**, 141.

Moore, S. (1963). *J. Biol. Chem.* **238**, 235.

Morriss, F. V. and Meiners, A. F. (1962). *U.S. Patent* 3,061,414, Oct. 30.

Mottola, H., Haro, M. and Freiser, H. (1968). *Anal. Chem.* **40**, 1263.

Mukherjee, S. N. and Roy, B. (1961). *Sci. & Culture (Calcutta),* **27**, 402.

Nakamura, Y. (1959). *Sen-i-Gakkaishi,* **15**, 635; *Chem. Abs.* **53**, 22695.

Nakamura, K. and Binkley, F. (1948). *J. Biol. Chem.* **173**, 407.

Nilsson, G. (1957). *Sci. Rev. (Holland),* **89**, 86; *Chem. Abs.* **51**, 8577.

Nordin, P. and Spencer, E. Y. (1951). *Cereal Chem.* **29**, 29.

Nutiu, R. (1964). *Rev. Chim. (Bucharest),* **15**, 114.

Nutiu, R. (1973). *Rev. Chim. (Bucharest),* **24**, 916.

Nutiu, R. and Bokenyi, A. (1969). *Rev. Chim. (Bucharest),* **20**, 637.

Ogawa, M. (1960). *Nippon Nogai Kagaku Zasshi,* **34**, 729.

Oguchi, M. and Shimizu, Y. (1962). *Sogo Igaku,* **19**, 923; *Chem. Abstr.* **65**, 12972.

Okuda, Y. (1923). *J. Sci. Agr. Soc. Japan,* No. 253, 1; *Chem. Abs.* **18**, 2014, 3613.

Okuda, Y. (1924). *J. Chem. Soc. Japan,* **45**, 1, 18.

Okuda, Y. (1924b). *J. Biochem. (Japan),* **5**, 217; *Chem. Abs.* **20**, 1252.

Okuda, Y. (1927). *Proc. Imp. Acad. (Tokyo),* **3**, 287; *Chem. Abs.* **21**, 3212.

Okuda, Y. and Katai, K. (1929). *J. Dep. Agr., Kyusu Imp. Univ.* **2**, No. 5, 133; *Chem. Abs.* **21**, 3212.

Okuda, Y. and Ogawa, M. (1933). *J. Agr. Chem. Soc. Japan,* **9**, 655.

Owades, J. L. and Zientara, F. (1961). *Amer. Brewer* **94**, 33, 36, 38, 41; *Chem. Abs.* **55**, 12762.

Palma, R. J., Gupta, H. K. L. and Boltz, D. F. (1971). *Anal. Letters,* **4**, 277.

Panwar, K. S., Rao, S. D. and Gaur, J. N. (1961). *Anal. Chim. Acta,* **25**, 218.

Paradiso, G. (1933). *Arch. Farmacol. Sper.* **56**, 487; *Chem. Abs.* **28**, 1729.

Park, G. S. and Speakman, J. B. (1952). *Bull. Inst. Textile France,* No. 30, 255.

Parker, D. J. and Allison, W. S. (1969). *J. Biol. Chem.* **244**, 180.

Parushev, M. (1963). *Khim. i Ind. (Sofia),* **35**, 49.

Paul, R. C., Aggarwal, R. C., Kumar, N. and Parkash, R. (1974). *Indfan J. Chem.* **12**, 986.

Perez, J. J. and Sandor, G. (1940). *Bull. Soc. Chim. Biol.* **33**, 149.

Perlzweig, W. A. and Delrue, G. (1927). *Biochem. J.* **21**, 1416.

Pesez, M. (1956). *Ann. Fals. Fraudes,* **49**, 403.

Pesez, M. and Bartos, J. (1972). *Analusis,* **1**, 257.

Pietz, J. (1940). *Jahresber. Inst. Bäckerei Reichsanst. Getreideverarb.* **6**, 18; *Chem. Abs.* **37**, 6353.

Pihl, A. and Lange, R. (1962). *J. Biol. Chem.* **237**, 1356.

Piotrowska, A. (1970). *Acta Polon. Pharm.* **27**, 131.

Podlipskii, L. A., Ropyanaya, M. A. and Kruglova, L. I. (1969). *Khim. Tekhnol. Topl. Masel,* **14**, 61; *Chem. Abs.* **70**, 69842.

Podurovskaya, O. M., Bogdanova, N. I., Beskova, G. S. and Chizkov, L. V. (1966). *Zav. Lab.* **32**, 1455.

Prochukhan, A. S., Baranovskaya, E. M., Bikbulatova, G. M. and Zainullina, R. V. (1969). *Zh. Anal. Khim.* **24**, 1261.

Quensel, W. and Wachholder, K. (1935). *Z. Physiol. Chem.* **231**, 65.

Rapaport, L. I. and Raznatovska, V. F. (1960). *Farm. Zh. (Kiev)*, **15**, 22.

Rheinboldt, H. (1926). *Chem. Ber.* **59**, 1311.

Rheinboldt, H. (1927). *Chem. Ber.* **60**, 184.

Richter, F. (1949). *Chem. Technik*, **1**, 31.

Riolo, C. B. and Saldi, T. (1959). *Ann. Chim. (Rome)*, **49**, 382.

Roberts, L. W. and Lucchese, G. (1955). *Stain Tech.* **30**, 291.

Rogers, G. E. (1953). *Quart. J. Microscop. Sci.* **94**, 253.

Rosenblatt, A. H. and Jean, G. N. (1955). *Anal. Chem.* **27**, 951.

Rosenheim, A. and Davidsohn, I. (1904). *Z. Anorg. Chem.* **41**, 231.

Rosenthaler, L. (1938). *Z. Vitaminforsch.* **7**, 126; also *Mikrochim. Acta*, **3**, 190.

Roston, S. (1963). *Anal. Biochem.* **6**, 486.

Roth, H. (1958). *Mikrochim. Acta*, 767.

Ruemele, T. and Walker, G. T. (1953). *Soap, Perfumery & Cosmetics*, **26**, 461; *Chem. Abs.* **47**, 10178.

Růžička, E. (1957). *Chem. Listy*, **50**, 969.

Sabalitschka, T. (1936). *Microch. Festschrift von H. Molisch*, 387.

Sampey, J. R. and Reid, E. E. (1932). *J. Amer. Chem. Soc.* **54**, 3404.

Sandri, G. (1957–8). *Atti Accad. Sci. Ferrara*, **35**, 105.

Sant, S. B. and Sant, B. R. (1959). *Anal. Chem.* **31**, 1879.

Santhanam, K. S. V. and Krishnan, V. R. (1968). *Z. Anal. Chem.* **234**, 256.

Sarkar, B. C. R. and Sivaraman, R. (1956). *Analyst (London)*, **81**, 668.

Saroff, H. A. and Mark, H. J. (1953). *J. Amer. Chem. Soc.* **75**, 1420.

Sartos Ruiz, A. and de Franch, M. R. (1943). *Trabajos Inst. Cajal. Invest. Biol.* **1**, 49; *Chem. Abs.* **43**, 7531.

Sato, M., Hirano, T. and Kan, T. (1939). *J. Agric. Chem. Soc. Japan*, **15**, 783.

Saville, B. (1956). *Chem. and Ind. (London)*, 660.

Saville, B. (1958). *Analyst (London)*, **83**, 670.

Scheele, W. and Gensch, C. (1953). *Kautschuk u. Gummi*, **6**, WT 147.

Schneider, J. A., Bradley, K. H. and Seegmiller, J. E. (1968). *Anal. Biochem.* **23**, 129.

Schneider, P. O., Thibert, R. J. and Walton, R. V. (1972). *Mikrochim. Acta*, 925.

Schöberl, A. (1937). *Chem. Ber.* **70**, 1186.

Schöberl, A. and Ludwig, E. (1937). *Chem. Ber.* **70**, 1422.

Schöberl, A. and Rambacher, A. (1938). *Biochem. Z.* **295**, 377.

Schormüller, J. and Ballschmieter, H. (1951). *Z. Anal. Chem.* **132**, 1.

Schram, E., Moore, S. and Bigwood, E. J. (1954). *Biochem. J.* **57**, 33.

Schulze, W. A. and Chaney, L. V. (1933). *Natl. Petroleum News*, **25**, No. 34, 37.

Seeligman, P. (1962). *Arch. Bioquim. Quim. Farm. Tucuman*, **10**, 105; *Chem. Abs.* **60**, 2256.

Segal, W. and Starkey, R. L. (1953). *Anal. Chem.* **25**, 1645.

Shaefer, W. C., Wilham, C. A., Dimler, R. J. and Senti, F. R. (1959). *Cereal Chem.* **36**, 431.

Shaw, J. A. (1940). *Ind. Eng. Chem., Anal. Ed.* **12**, 668.

Shemyakin, F. M. and Berdnikov, A. I. (1965). *Aptech. Delo*, **14**, 63.

Shinohara, K. (1935). *J. Biol. Chem.* **109**, 665.

Shinohara, K. (1936). *J. Biol. Chem.* **112**, 671, 683, 709.

Shinohara, K. (1937). *J. Biol. Chem.* **120**, 743.
Shinohara, K. and Padis, K. E. (1936). *J. Biol. Chem.* **112**, 697.
Shtram, D. A. and Fateev, A. V. (1945). *Zavod. Lab.* **11**, 861.
Simionovici, R., Stoianovici, M. and Ioan, C. (1960). *Rev. Chim. (Bucharest)*, **11**, 591.
Singh, D. V. and Mukherjee, P. P. (1972). *Indian J. Technol.* **10**, 469.
Smith, O. K. and Williams, G. T. (1939). *Gas Age,* **84**, No. 3, 14.
Solsona, M. (1956). *Rev. Españ. Fisiol.* **12**, 29.
Spacu, P. and Dimitrescu, H. (1968). *An. Univ. Bucuresti, Ser. Stiint. Natur., Chim.* **17**, 112; *Chem. Abs.* **72**, 28799.
Spencer, R. I. and Wold, F. (1969). *Anal. Biochem.* **32**, 185.
Sporek, K. F. and Danyi, M. D. (1963). *Anal. Chem.* **35**, 956.
Spray, G. H. (1947). *Biochem. J.* **41**, 360.
Srivastava, A. and Bose, S. (1974). *J. Indian Chem. Soc.* **51**, 736.
Stahl, E. (1969). "Thin-Layer Chromatography", p. 903. Springer.
Stapfer, C. H. and Dworkin, R. D. (1968). *Anal. Chem.* **40**, 1891.
Steel, K. J. (1964). *J. Pharm. Pharmacol.* **10**, 574.
Steigmann, A. (1942). *J. Soc. Chem. Ind.* **61**, 18.
Stekhun, A. I., Starikova, A. I., Bondarenko, L. P. and Nosal, T. P. (1967). *Khim. Tekhnol. Topl. Masel,* **12**, 56; *Chem. Abs.* **67**, 75093.
Stephan, R. and Erdman, J. G. (1964). *Nature (London),* **203**, 749.
Stoicescu, V. and Beral, H. (1969). *Pharm. Zentralh.* **108**, 466.
Stoicescu, V., Beral, H. and Ivan, C. (1965). *Pharm. Zentralh.* **104**, 776.
Strache, F. and Mierau, H. J. (1955). *Deut. Apotheker-Z.* **95**, 55.
Suchomelová, J. and Zýka, J. (1963). *J. Electroanal. Chem.* **5**, 57.
Sunner, S., Karrman, K. J. and Sundén, V. (1956). *Mikrochim. Acta,* 1144.
Swan, J. M. (1957). *Nature (London),* **180**, 643.
Szepesvary, P. (1961). *Magy. Asvanyolaj Foldgaz Kiserl. Int. Kozlem.* **2**, 50; *Chem. Abs.* **57**, 2491; *Anal. Abs.* **10**, 1068.
Terada, S. and Nomiyama, H. (1957). *Eisei Kagaku,* **4**, 31.
Thibert, R. J. and Sarwar, M. (1969). *Mikrochim. Acta,* 259.
Thibert, R. J., Sarwar, M. and Carroll, J. E. (1969). *Mikrochim. Acta,* 615.
Thompson, J. W. and Voegtlin, C. (1926). *J. Biol. Chem.* **70**, 793.
Thunberg, T. (1911). *Skand. Arch. Physiol.* **25**, 343.
Todd, P. and Gronow, M. (1969). *Anal. Biochem.* **28**, 369.
Todrick, A. and Walker, E. (1957). *Biochem. J.* **31**, 292.
Toennies, G. and Kolb, J. J. (1951). *Anal. Chem.* **23**, 823.
Tomíček, O. (1952). *Sborník Celostatní Pracovní Konf. Anal. Chemiků,* **1**, 246; *Chem. Abs.* **49**, 15602.
Tomíček, O. and Valcha, J. (1950). *Chem. Listy,* **44**, 283.
Torstensson, L.-G. (1973). *Talanta,* **20**, 1319.
Trop, M. Sprecher, M. and Pinsky, A. (1968). *J. Chromatog.* **32**, 426.
Tsareva, V. A. and Kuleshova, M. I. (1969). *Farmatsiya (Moscow),* **18**, 76.
Tunnicliffe, H. E. (1925). *Biochem. J.* **19**, 194.
Tupeeva, R. B. (1967). *Gig. i Sanit.* **32**, 62; *Chem. Abs.* **67**, 5517.
Turk, E. and Reid, E. E. (1945). *Ind. Eng. Chem., Anal. Ed.* **17**, 713.
Ullmann, J. (1959). *Spisy Přirodověd. Fak. Univ. Brně,* No. 400, 45; *Chem. Abs.* **54**, 15747.
Varga, E. and Zöllner, E. (1955). *Acta Pharm. Hung.* **25**, 150.
Verma, K. K. and Bose, S. (1973a). *Anal. Chim. Acta,* **65**, 236.

Verma, K. K. and Bose, S. (1973b). *J. Indian Chem. Soc.* **50**, 367.
Verma, K. K. and Bose, S. (1973c). *J. Indian Chem. Soc.* **50**, 542.
Verma, K. K. and Bose, S. (1974). *Anal. Chim. Acta,* **70**, 227.
Verma, B. C., Ralhan, S. M. and Ralhan, N. K. (1972). *Z. Anal. Chem.* **259**, 367.
Verne, J. (1956). *Ann. Histochem.* **1**, 199.
Virtue, R. W. and Lewis, H. B. (1934). *J. Biol. Chem.* **104**, 415.
von Wacek, A. and Smitt-Amundsen, J. (1958). *Monatsh.* **80**, 110.
Waddill, H. G. and Gorin, G. (1958). *Anal. Chem.* **30**, 1069.
Waksmundzki, A., Wawrzynowicz, T. and Wolski, T. (1963). *Acta Polon. Pharm.* **20**, 259.
Walker, G. T. (1953). *Mfg. Chemist,* **24**, 376.
Walker, G. T. (1955a). *Seifen-Öle-Fette-Wachse,* **81**, 117.
Walker, G. T. (1955b). *Industr. Parfum.* **10**, 236; *Chem. Abs.* **54**, 12482.
Walker, G. T. and Freeman, F. M. (1955). *Mfg. Chemist,* **26**, 11.
Wawrzyczek, W. (1962). *Z. Anal. Chem.* **185**, 446.
Weller, G. (1965). *Rev. Franç. Etudes Clin. Biol.* **10**, 547.
Wellington, E. F. (1952). *Can. J. Chem.* **30**, 581.
Werner, A. E. A. (1941). *Sci. Proc. Roy. Dublin Soc.* **22**, 387.
White, D. L. and Reichardt, F. E. (1949). *Gas,* **25**, No. 6, 38.
White, L. B., Torosian, G. and Becker, C. H. (1974). *Anal. Chem.* **46**, 143.
Willemart, R. (1956). *Ann. Pharm. Franç.* **14**, 718.
Willemart, R. and Fabre, P. (1958). *Ann. Pharm. Franç.* **16**, 676.
Willgerodt, C. (1885). *Chem. Ber.* **18**, 331.
Winegard, H. M., Toennies, G. and Block, R. J. (1948). *Science,* **108**, 506.
Wojahn, H. and Wempe, E. (1952). *Arch. Pharm.* **285**, 280.
Wojahn, H. and Wempe, E. (1953). *Pharm. Zentralh.* **92**, 124; also *Arch. Pharm.* **286**, 344; also Wojahn, H. (1953). *Pharm. Acta Helv.* **28**, 336.
Wolf, F., Kotte, G. and Hannemann, J. (1971). *Chem. Tech. (Berlin),* **23**, 550.
Wong, F. F. (1971). *J. Chromatog.* **59**, 448.
Woodward, F. N. (1949). *Analyst (London),* **74**, 179.
Woodward, G. E. and Fry, E. G. (1932). *J. Biol. Chem.* **97**, 465.
Wronski, M. (1962). *Chem. Anal. (Warsaw),* **7**, 851.
Yamamoto, K. (1940). *Mitt. Med. Akad. Kyoto,* **29**, 379; *Chem. Abs.* **35**, 3663.
Yamazaki, K. (1930). *J. Biochem. (Tokyo),* **12**, 207; *Chem. Abs.* **25**, 127.
Yung, Y.-H. (1965). *Acta Biochim. Biophys. Sin.* **5**, 531; *Chem. Abs.* **64**, 14585.
Yurow, H. W. and Sass, S. (1971). *Anal. Chim. Acta,* **56**, 297.
Zahn, H. and Traumann, K. (1954). *Melliand. Textilber.* **35**, 1069.
Zeicu, V., Rus-Sirbat, M. and Craciuneanu, R. (1972). *Farmacia (Bucharest),* **20**, 615.
Zhdanov, A. K., Khadeev, V. A., Kubrakova, A. I. and Bondarenko, N.V. (1961). *Uzbek. Khim. Zh.* No. 2, 44; *Anal. Abs.* **9**, 3031.
Zimmet, D. and Dubois-Ferrière, H. (1936). *Compt. Rend. Soc. Phys. Hist. Nat. Genève,* **53**, 132; *Chem. Abs.* **31**, 5826.
Zöllner, E. and Varga, E. (1957). *Acta Chim. Acad. Sci. Hung.* **12**, 1.
Zumanová, R. and Zuman, P. (1954). *Pharmazie,* **9**, 554.
Zýka, J. and Berka, A. (1962). *Microchem. J. Symp. Ser.* **2**, 789.

2. MERCAPTIDE FORMATION

Many metal-containing compounds react with thiols, at least initially, to yield mercaptides:

$$RSH + Met^+ \rightarrow RSMet + H^+$$

$$RS^- + Met^+ \rightarrow RSMet$$

Twenty-two such reagent types are classified alphabetically in the present chapter; for five of these, cross references are given to Chapter 1, where they are discussed on account of their mainly oxidizing behaviour with thiols. Cross references to this chapter are given also for four oxidizing reagents with metal-containing anions.

The remaining thirteen include the two most extensively used reagents in analytical work on thiols, namely mercury(II) and silver(I). Some reagents, e.g. iron(III), appear to display varied behaviour, including oxidation, but not predominantly enough to merit inclusion in Chapter 1.

1. ARSENIC(III)

Trivalent arsenic forms stable rings with 1,2-dithiols but little analytical profit appears to have been derived from this. Barron and Singer (1945) eliminated such thiols with the help of an arsenic(III) reagent. Arsenite has also been used to complex excess dithiol where used to reduce disulphides to thiols; e.g. dithioerythritol or dithiothreitol by Zahler and Cleland (1968).

2. BISMUTH(III)

A small number of references to the use of reagents containing trivalent bismuth can be given. Harris (1922) noted the characteristic yellow yielded on adding cysteine to a suspension of bismuth hydroxide in ammoniacal solution. McAllister (1952) prepared an iodobismuthite reagent (BiI_4^-) from acid bismuth sulphate, potassium iodide and sulphuric acid; it yielded a scarlet or orange colour with 1- and 4-methyl-2-mercaptoimidazoles, respectively. Many other sulphur compounds, e.g. thiouracils, gave no colour. Zijp (1956) visualized similar compounds, such as mercapto-thiazoline, -benzothiazole and -iminazole on paper chromatograms with 5% bismuth(III) nitrate in 0·5 N nitric acid; small amounts could still be seen in UV-light. A bismuth nitrate reagent was used also by Stepień and Gaczyński (1961) to visualize mercaptobenzothiazole on paper chromatograms as an orange spot; the spots were cut out and weighed in a semi-quantitative procedure.

Gilman and Nelson (1937) tried to determine thiols through reaction with bismuth triethyl to yield an equivalent amount of ethane:

$$3RSH + Bi(C_2H_5)_3 \rightarrow 3C_2H_6 + Bi(SR)_3$$

This "active hydrogen" method, carried out in the inert solvent dibutyl ether, proved unsuitable; yields lay between 25% and 75% of theory.

3. CADMIUM(II)

Yoshida and Kurihara (1952) titrated mercaptobenzothiazole in ammonium hydroxide using cadmium chloride and an amperometric end-point indication. Calzolari and Donda (1954) titrated 2,3-dimercapto-1-propanol (BAL) directly in 10% isopropanol + sodium acetate, also amperometrically with cadmium(II). Otherwise the cadmium(II) reagent appears to have been used only as a precipitating agent for separation of thiols from biological material (mostly glutathione) or from petroleum fractions, industrial effluents and gases. Hydrogen sulphide is usually present in these samples and is coprecipitated with the thiols. Differential precipitation is possible, although Nametkin et al. (1943) were sceptical about the results of thiol determination in petroleum through initial precipitation with cadmium(II) (also silver(I) and lead(II)) if other sulphur-containing compound classes were present. The cadmium mercaptides evidently remain in solution under slightly more acidic precipitating conditions. Thus Shaw (1940) precipitated hydrogen sulphide and lower alkanethiols (tested on C_1–C_3) with 10% cadmium chloride, ca. 0·3 N in sodium carbonate, ultimately acidifying the precipitate, adding excess iodine reagent and finally titrating the excess with thiosulphate. This gave a value for the sum of sulphide and thiol. He repeated the determination with a reagent just acid to methyl orange, which precipitated only sulphide. Thiols were thus obtained by difference. Lur'e et al. (1963), working on cellulose plant waste, precipitated hydrogen sulphide with slightly acid cadmium chloride and then thiols with cadmium carbonate suspension, using a final iodometric determination.

Hakewill and Rueck (1946) also precipitated thiols from gases with a cadmium chloride/sodium carbonate reagent, then acidifying the precipitate, adding excess iodine and finally back-titrating with thiosulphate. Harding *et al.* (1964) used cadmium chloride to separate thiols (and hydrogen sulphide) from process gas streams and likewise concluded with an iodometric procedure. Riesz and Wohlberg (1943) determined thiols (ethanethiol) in gases by difference, measuring the sulphur content of the gas (by combustion) before and after removal of thiols with cadmium chloride.

Most methods for glutathione determination have been concluded iodometrically, e.g. Binet and Weller (1934–36, 1938) on tissue, precipitating with cadmium acetate at pH 6–7 or with the lactate at pH 6·8–7; Ceresa and Guala (1939), using cadmium sulphate to determine oxidized glutathione in fact, first reducing with zinc dust; Contopoulos and Anderson (1950); also Ghiglione and Bozzi-Tichadou (1954) for cysteine in protein hydrolysates, who precipitated with alkaline cadmium sulphate and dissolved the precipitate in sulphuric acid.

Hartner and Schleiss (1936) determined glutathione in biological material using cadmium sulphate/sodium hydroxide, oxidizing the precipitate with bromide/bromate/hydrochloric acid for 8 min, then adding disodium hydrogen phosphate and potassium iodide and finally titrating with thiosulphate. Kul'berg and Presman (1940) used this procedure for blood glutathione. Černoch (1966) precipitated with cadmium lactate like Binet and Weller, and his final stage was direct titration with potassium iodate to starch after acidifying with phosphoric acid and adding iodide.

Feldstein *et al.* (1965) determined thiols by GLC after separating them as cadmium (or mercury) derivatives. Recently, Gershkovich (1971) determined ethanethiol in air by precipitation with cadmium hydroxide and then utilizing catalysis of the iodine/azide reagent (see p. 195). After contact with this reagent for 1 h he titrated with arsenite.

Only a single example of qualitative application of cadmium ions could be found. Joyet-Lavergne (1938) detected glutathione in tissue sections by treating with 1% cadmium lactate; the cadmium derivative could be observed under the microscope.

4. CERIUM(IV)

See Chapter 1, "Oxidation" (Section 6).

5. CHROMIUM(VI)

See Chapter 1, "Oxidation" (Section 10).

6. COBALT

Both cobalt(II) and cobalt(III) salts have been used in analytical work on thiols and it is sometimes not clear from the literature which salt was used as reagent. For convenience, all methods with cobalt salts are treated together here.

The procedures fall into two groups: those based on colour formation (presumably complex formation perhaps following initial reaction to yield a mercaptide); and titrations.

Shinohara and Kilpatrick (1934) used a colorimetric method to determine cysteine in the presence of cystine. They treated the sample with cobalt(II) chloride, sodium hydroxide and then 3% hydrogen peroxide. The precipitated cobalt(III) hydroxide was removed by centrifuging and the residual brownish yellow coloured solution evaluated colorimetrically. The product, stable for over 12 h, is evidently a Co(III)–cysteine complex.

Spray (1947) gives a sensitive method for 2,3-dimercapto-1-propanol, which is probably applicable to other 1,2-dithiols. Details are quoted here: 1 ml of 0·5% cobalt nitrate (presumably divalent cobalt, although the author does not expressly say so) is mixed with 0·5 ml of 2% gum arabic, and 0·1 N borate solution of pH 9 is added to bring the final total volume to 10 ml after addition of the sample. This reagent mixture is warmed to 45° on a water-bath, the sample added and the whole maintained for 10 min at this temperature. The absorbance is then read at 470 nm. Beer's law holds between 10 and 200 μg of dithiol.

Holt and Mattson (1949) developed a colorimetric method for uracil, thiouracil and related compounds containing the —CO—NH—CS— group. This does not fall into the categories of compounds treated in this book but the method is briefly described. The reaction medium is 25% isopropylamine in chloroform and the reagent is methanolic cobalt(II) acetate. Absorbance is measured at 530 nm immediately after mixing. The authors made no suggestion about the chemistry.

Marras and Dall'olio (1953) estimated 2,3-dimercapto-1-propanol by absorbance measurements at 650 nm after treating with cobalt(II) nitrate. Recently, Stebletsova and Evstifeev (1971) determined mercaptobenzothiazole in wastewater (after extracting with benzene) by adding a cobalt oleate reagent to the extract and evaluating colour at 590 nm after 15 min. Another recent procedure is that of Namikoshi and Yamakawa (1973), for n-butanethiol in hydrocarbons. The sample in heptane is treated with 0·01 M cobalt(II) naphthenate in the same solvent, ether is added and the absorbance is measured at 500 nm after 2 h. Neither the sec- nor tert-butane compounds react.

Shemyakin and Berdnikov (1965) give colour reactions with several cations for identifying methyl thiouracil. A 2% solution of thiol in 0·15 N sodium hydroxide gave green with cobalt(II) nitrate. This test was more sensitive than similar tests with other cations, such as nickel (II), copper(II), mercury(II) and vanadyl, VO^{2+}. It enabled thiouracil concentrations down to $4·8 \times 10^{-5}$ g l^{-1} to be detected.

Direct titration procedures have been published recently. Both are evidently based on oxidation with cobalt(III). Pszonicka and Skwara (1970) titrated 0·5–50 mg samples of cysteine (and some other compound classes, such as hydrazines) in 25 ml of 4–5 N hydrochloric acid with 0·1 N cobalt(III) acetate in acetic acid. The end-point was potentiometric. Yurow and Sass (1971) tried titrations of sulphur compounds using a bipotentiometric end-point indication with two platinum electrodes and a current of 7 μA. Their results were better with lead(IV) (see pp. 31–2) in non-aqueous solution. Cobalt-(III) reagent was less successful with the non-aromatic thiols and some sulphides and disulphides.

Mention must be made of the Brdička polarographic procedure for cysteine and cystine in the presence of cobalt(III) salts. This is dealt with under "Polarography", p. 219.

7. COPPER(I)

Thiols yield difficultly soluble mercaptides with Cu(I) reagents:

$$RSH + Cu^+ \rightarrow RSCu + H^+$$

Feigl et al.(1969) used a copper(I) reagent in a spot test for thiols. It is a mixture of cupric chloride/ammonium hydroxide/ammonium chloride and hydroxyl-ammonium chloride and yields a yellow or brown colour or precipitate with microgram amounts of thiols. Tupeeva (1967) had used a similar reagent for determining mercaptoacetic acid in air, estimating the colour intensity after 10 min. Bhatia et al. (1972) visualized thiols in TLC by spraying first with an ammoniacal cupric reagent and then with 20% hydroxyl ammonium chloride; yellow, green or yellow-brown colours were yielded. Swan's (1965) titration or thiols in presence of ascorbic acid is mentioned under "Copper-(II)", p. 11.

Misra and Tandon (1970) titrated many aliphatic and aromatic thiols in deaerated (nitrogen) solutions, using a cuprous chloride/N potassium chloride reagent and potentiometric end-point indication. Another direct quantitative method is described in a patent of Berg et al. (1970). They prepared a sensitive element from bronze screen grid electrodes, separated by a copper(I)-containing mixture, e.g. polyester previously treated with a mixture of glycerol, water, cuprous chloride and acetamide. The electrical conductivity of this layer decreases if it comes into contact with a thiol, through formation of insoluble cuprous mercaptide. This principle enables propanethiol to be determined in gas passed through the element.

Thiols can be separated from other compound classes *via* their cuprous derivatives. Hopkins (1929) appears to have been the first to apply this principle. He isolated glutathione (also Coenzyme A and ergothioneine) by adding a slight excess of cuprous oxide at 50°C in a type of titration. The thiols can be regenerated from the cuprous derivatives by treatment with acid, e.g. hydrochloric acid or hydrogen sulphide. Pirie (1931) prepared cuprous derivatives from many thiols and discussed their analytical application. Copper(I) reagents were subsequently used for separating cysteine *via* precipitation, usually as the final stage of a cystine estimation. Some investigators reduced the cystine with metals in acid solution, e.g. Vickery and White (1932–3) who employed tin, and Graff *et al.* (1937), Beach and Teague (1942) and Schultz and Vars (1947) who all used zinc. They added a cuprous oxide suspension at pH 4–5, after which the cysteine was determined through the sulphur content (gravimetrically as barium sulphate). Graff *et al.* also concluded with a Kjeldahl nitrogen determination but Schultz and Vars criticized this as susceptible to interference by other possibly precipitated nitrogen-containing compounds such as purines.

The cuprous reagent does in fact also fulfil the role of reducing agent for cystine. Thus Rossouw and Wilken-Jorden (1935), Medes (1936), Brand *et al.* (1940) and Rossouw (1940) used a cuprous chloride reagent in hydrochloric acid/potassium chloride solution. After 40–60 min reaction, they separated the copper–cysteine product, dissolved it in acid, precipitated copper as thiocyanate and then determined the freed cysteine colorimetrically using molybdophosphate (see Chapter 1, "Heteropoly Acids", Section 22) or 1,2-naphthoquinone-4-sulphonate (see Chapter 4, "Quinones", Section 1.5.2). Zittle and O'Dell (1941) used the cuprous oxide suspension at pH 4 for reduction and precipitation, concluding with the quinone colorimetric method or sulphur determination.

Cuprous mercaptides are of course formed and precipitated in the reaction of thiols with copper(II). This has been dealt with above in Chapter 1, "Oxidation", Section 11.

8. COPPER(II)

See Chapter 1, "Oxidation" (Section 11).

9. COPPER(III)

See Chapter 1, "Oxidation" (Section 12).

c*

10. IRON(III)

Coloured products are yielded by many thiols with ferric reagents. Andreasch (1879) observed that mercaptoacetic acid gave a transient blue with ferric salts; in alkaline (ammonia) solution, it yielded a dark red. As mentioned by Claësson (1881), other thiols also gave dark red but other intense colours may arise, e.g. deep green with α-mercaptocinnamic acid (Andreasch, 1889). The reaction of thiols with iron(III) has been investigated by numerous authors, e.g. Cannan and Richardson (1929), Michaelis and Barron (1929) and Leussing et al. (1960). Oxidation and also complex formation, evidently without oxidation, appear to take place, so that the reagent is disqualified from inclusion under "Oxidation" and is better classified here.

The formation of colour has been applied as a test in both directions for thiol or ferric iron. Dubský and Šindelář (1938) claimed that down to 60 μg of mercaptoacetic acid can be detected with ferric salts. Semco (1950) patented a test for mercaptoacetic acid with a test paper saturated with 10% ferric chloride and 0·5% sodium hydroxide. Freytag (1953) studied several tests for thiols, including that with ferric chloride (0·01 M), and found it satisfactory for mercaptoacetic acid and cysteine, although less sensitive than with nitroprusside.

Direct titration of biological thiol groups with iron(III) to external nitro-prusside indicator was suggested by Anson (1940, 1941). Hladký (1965) carried out potentiometric titrations of thiols and some other compounds in dimethylformamide in oxygen-free atmosphere and using ferric chloride in the same solvent. Hladký and Vřešťál (1969) performed similar titrations of mercaptoacetic and thiolactic acids, and thiophenols with ferric chloride in pyridine.

Recently Jaselskis and Schlough (1974) determined cysteine through oxidation with iron(III) and colorimetric estimation of the iron(II) formed with the disodium salt of 3-(2-pyridyl)–5,6-bis(4-phenylsulphonic acid)–1,2,4-triazine ("Ferrozine"). A mixture is made of 2 ml buffer of pH 3·2–4·3 (monochloroacetate–acetate), 2 ml of ca. 0·003 M ferric perchlorate and 1 ml of ca. 0·015 M, freshly prepared Ferrozine. The solution is then diluted to 25 ml and evaluated at 562 nm after 20 min reaction. They were able to determine cysteine amounts between 10 and 150 μg/25 ml.

Ferric iron is used also in oxidizing combinations with other reagents, such as p-N,N-dimethylaminoaniline (see Chapter 9, "Miscellaneous Chemical Procedures", Section 8) and ferricyanide. Both components of this latter combination probably oxidize the thiol but the reduction product is in both cases blue, whether ferric ferrocyanide (Prussian blue) or ferrous ferricyanide (Turnbull's blue). The arbitrary decision has been taken of quoting examples of use of this dual reagent under "Ferricyanide, Chapter 1, Section 19.3" on "Oxidation".

Ferricyanide itself contains iron(III) but, as just indicated, is accorded the separate heading of Section 19 under "Oxidation" in Chapter 1.

Another iron(II) coordination compound, nitrilotriacetatoferrate(III), was used by Bydalek and Poldoski (1968) to determine cysteine in the 10^{-4}–10^{-5} M range. The method depends on colorimetric determination of the iron(II) reduction product using 1,10-phenanthroline and is given here in detail: To a 100 ml graduated flask are added, in succession: 10 ml of 1 M tris buffer, 7 ml of 1 M hydrochloric acid (giving a final pH of 7·2–7·7) and 7 ml of 10^{-2} M nitrilotriacetate reagent (yielding 1–3% excess over the iron amount, and prepared by dissolving 1·95 g of nitrilotriacetic acid in 200 ml of distilled water containing 0·129 g of sodium hydroxide, diluting to 1 litre and standardizing by titrating with standard copper(II) to murexide). This mixture is then treated with 7 ml of 10^{-2} M ferric chloride, made by adding 20 ml of concentrated hydrochloric acid to 2·703 g of ferric chloride hexahydrate and diluting to 1 litre. After washing down the sides of the flask, the mixture is left for 2–3 min to enable the nitrilotriacetatoferrate(III) to form. Seven ml of 5×10^{-3} M 1,10-phenanthroline (0·248 g of monohydrate dissolved in 250 ml of dissolved water and not more than 3 d old) are added, followed immediately by the cysteine sample to give a final concentration of 10^{-4}–10^{-5} M. After 10 min the solution is made up to the 100 ml mark and the absorbance read at 510 nm 30 min after the addition of cysteine.

11. LEAD(II)

Thiols form yellow to brown, poorly soluble lead(II) mercaptides:

$$2RSH + Pb(II) \rightarrow Pb(SR)_2 + 2H^+$$

This is the basis of the first half of the "doctor" test for thiols in petroleum products. An alkaline plumbite reagent, prepared from a lead(II) salt such as the acetate and usually sodium hydroxide, is used. The sample is shaken with the reagent for ca. 15 s and a yellow colour is strong evidence of the presence of a thiol. This is confirmed by subsequent shaking with sulphur which, as mentioned in "Oxidation", under "Sulphur", Chapter 1, Section 53, reacts to form black or brown lead sulphide. Hydrogen sulphide also reacts with the lead reagent and is generally first removed with cadmium(II) if its presence in the sample is suspected. Boyd (1933) tried the test on natural gasolines and was able to detect C_1, C_4 and C_7 thiols in 0·002%, 0·0002% and 0·00009% amounts, respectively. Gabrielyantz and Artem'eva (1934) detected thiols in distillates, using alcoholic plumbite or basic lead acetate suspended in benzene. Henderson *et al.* (1940) also examined the application to 7 thiols. LeRosen *et al.* (1952) prepared a chromatographic streak reagent from lead(II) oxide and sodium hydroxide, which yielded yellow with thiols and hydrogen sulphide. As in the original "doctor" test, overstreaking with a solution of sulphur in a hydrocarbon solvent gave a colour change to grey, then black. This test was compared with others by Mapstone (1946b). Ohara *et al.* (1947) used a reagent made up of lead acetate, hydantoin, calcium hydroxide and water for histochemical demonstration of labile-bound sulphur, e.g. cysteine. A brown to black stain appeared in 12–20 h at 60°C.

Wertheim (1929) proposed metal derivatives of thiols for their identification. To prepare lead(II) derivatives he mixed ca. 2 ml of alcohol containing 6 drops of sample with an excess of 20% lead acetate. After shaking for a few minutes, the solution was cooled, the mercaptide filtered and dried on a porous plate. Melting points of only four derivatives are quoted. Challenger and Greenwood (1949) identified propanethiol, isolated from onions, *via* metal derivatives. They first precipitated the mercury(II) compound and from it, as confirmation, the lead mercaptide by decomposing it with hydrochloric acid and passing the vapours into 20% lead acetate.

Direct titration of thiols appears to have been rarely performed. Oehme (1960) determined thiols (and hydrogen sulphide) in technical gases by absorption in 30% sodium hydroxide, removing excess alkali with the Merck I ion exchange in the H^+-form and then titrating with lead acetate using the "high frequency" end-point indication. As mentioned below, he then extracted the lead mercaptide with collidine and estimated it colorimetrically.

Sen and Bahadur (1974) recently titrated cysteine in the presence of other amino acids, in a borate buffer of pH 9·2, taking as end-point the colour change from pink to blue with catechol violet as indicator.

Some indirect procedures, based on titration of unused reagent, can be given. Thus Faragher *et al.* (1927) precipitated thiols from naphtha solutions and petroleum distillates by adding basic lead acetate to a benzene solution of sample which had been freed from sulphur and hydrogen sulphide. The lead mercaptide was separated and mixed with an excess of standard sulphuric acid and the unused acid was back-titrated with alkali. Lennartz and Middeldorf (1949) added excess sodium plumbite reagent to the thiol sample (e.g. cholesteryl mercaptan) in alcohol and left for 10 min. After filtering off the product, unused lead in the filtrate was precipitated as sulphate, this dissolved in ammonium acetate/acetic acid and the lead content was titrated with standard ammonium molybdate to a tannin external indicator. Shakh and Kagan (1962) precipitated unithiol, 1,2-mercapto-propane-3-sulphonic acid, with excess lead nitrate in the presence of 5% pyramidone solution. Unused lead was back-titrated with EDTA. Oelsner and Huebner (1964) also concluded with a stage involving EDTA in their determination of thiols in gasoline. They shook the sample for 3 min with aqueous sodium plumbite and treated the aqueous phase with 10 % nitric acid, concentrated ammonium hydroxide, pH 10 buffer and excess EDTA. The unused EDTA was titrated with 0·01 M zinc reagent to the eriochrome black T end-point.

Others have preferred to estimate a reaction product. Moncorps and Schmid (1932) warmed glutathione with 30 % sodium hydroxide and saturated lead acetate to yield lead sulphide. They decomposed this with hydrochloric acid and oxidized the hydrogen sulphide to sulphate with ammoniacal hydrogen peroxide. Their final stage was precipitation and gravimetric determination of barium sulphate. In Oehme's method (1960) for thiols in technical gases, mentioned above, the lead mercaptide precipitate was dissolved in picoline or collidine at 50°C to separate from lead sulphide which is insoluble. After centrifuging, the lead mercaptide solutions were estimated colorimetrically using a blue filter and comparing with standards. Mizuguchi *et al.* (1962) converted cysteine into a lead complex which could be estimated by light absorbance measurements at 267 nm in acetate buffer of pH 5–5·5. Khusmitdinova (1965) determined ^{35}S-labelled cysteine also *via* absorbance of the lead complex in acetate buffer at 267 nm.

12. LEAD(IV)

See Chapter 1, "Oxidation" (Section 30).

13. MANGANESE(VII)

See Chapter 1, "Oxidation" (Section 42).

14. MERCURY(II)

Together with Ag(I) this reagent has been more extensively used in analytical work on thiols, especially for quantitative determination, than any other. The customary three headings have been adopted, of: direct titration; use of excess reagent and determination of unused; detection, identification, separation and determination *via* a reaction product.

As will be seen from the tables and other information below, inorganic and organo-mercury reagents have been used. Doubts were expressed about the uniform stoichiometry of the reactions with inorganic reagents; one or two thiol groups could react. This ambiguity is removed by using organo-mercury reagents containing only one available mercury valency:

$$-\overset{|}{\underset{|}{C}}-Hg-X + RS^- \rightarrow -\overset{|}{\underset{|}{C}}-Hg-SR + X^-$$

Special mercury reagents have been prepared which yield favourably coloured or fluorescent products.

Most of the analytical work has been devoted to biological thiol groups. Good results have generally been obtained with purer samples of thiols such as cysteine, homocysteine, glutathione or penicillamine; but determinations of thiols in intact or hydrolysed proteins have been in many cases less satisfactory.

14.1. Direct titration

Table V contains ca. 100 references mostly from the last twenty-five years. A summary of end-point indications may be given: Most indicators respond to the first excess of mercury(II) by forming characteristically coloured complexes, e.g. diphenylcarbazone ($\rightarrow$ blue); diphenylthiocarbazone(dithizone) ($\rightarrow$ red); 4,4′-bis(dimethylamino)-thiobenzophenone, thio-Michler's ketone, ($\rightarrow$ blue, the absorption maximum of which, 580 nm, severed for photometric titration); N,N-dimethyl-4-(2-pyridylazo)aniline giving a product of absorption maximum 550 nm, for visual or photometric titration); p-N-dimethylaminobenzylidinerhodanine ($\rightarrow$ red or purple).

3,6-Dimercaptofluoran (thio- or dithiofluorescein), used in much work by Wronski, is a blue indicator which yields a colourless product with mercury(II) and hence an end-point involving decolorization. Similar in function is EDTA, which forms weak complexes of absorption maxima near 400 nm with certain organomercury reagents; these are decomposed by reverse titration with thiol, with accompanying change of absorbance.

Electrometric end-point indication, especially amperometric, also depends on electrode response to mercury(II).

Response to the thiol is rarer. The classical nitroprusside indicator, usually employed externally, is the most prominent example. Another is cobalt which yields an intense blue colour with dithiols.

Tetrakis(acetoxymercuri)fluorescein(TMF) is a reagent with indicator function. Thiols may quench its fluorescence or give highly fluorescent products. In the former case, fluorescence of the reaction mixture increases after the equivalence point and extrapolation of fluorescence values can furnish an end-point by intersection of two lines; an alternative is to use a TMF/prominal titrant which fluoresces powerfully, giving a clearer intersection (prominal = 3-ethyl-1-methyl-5-phenylbarbituric acid). In the latter case, a TMF/oxine reagent is better; fluorescence first increases, then decreases after the equivalence point because of light absorption by the red, non-fluorescent TMF/oxine complex. The end-point is the curve intersection or, if the straight lines are poorly defined, the point of maximum fluorescence.

Photometric titration has also been based on absorbance measurements (at ca, 250 nm) of the reaction product of thiols with *p*-chloromercuribenzoate titrant.

14.2. Determination of unused reagent

Indirect procedures falling under this heading are far rarer than direct titrations but a number of interesting principles can be mentioned. These serve as sub-headings:

14.2.1. *Reaction with a Standard Thiol*

In their determination of ovalbumin thiols, MacDonnell *et al.* (1951) appear to have been the first to back-titrate unused reagent (*p*-chloromercuribenzoate at pH 5·3 in acetate buffer, reacting for 5–15 min) with a thiol. They used cysteine, with nitroprusside as external indicator. Pihar (1953) also employed cysteine, back-titrating potentiometrically the unused *p*-chloromercuribenzoate in determining protein thiols by reaction in a phosphate buffer of pH 7·3 to 8·5. Lontie and Beckers (1956) adapted the method of MacDonnell *et al.* to the micro scale in a study of protein thiols. Calcutt and Doxey (1959) used back-titration of the same reagent with cysteine in their determination of tissue thiols.

To determine cystine, Wronski (1962) converted it to cysteine in the usual way with sulphite in monosodium phosphate buffer and then added an excess of standard *o*-hydroxymercuribenzoate in the presence of EDTA to minimize air oxidation. After the addition of ammonium hydroxide, unused reagent was back-titrated with 0·01 N mercaptoacetate to the blue of thiofluorescein indicator. Fernandez Diez *et al.* (1964) determined thiol groups in ovalbumins, β-lactoglobulins and bovine serum albumin by treating for 2 h with excess *p*-chloro- or *p*-hydroxymercuribenzoate, then back-titrating with 2-mercaptoethanol until the absorbance at 225 nm(sic) ceased to increase. Through comparison with the titration of a control reagent amount, they obtained the amount of mercury reagent reacting with the sample.

Frater and Hird (1965) determined protein thiol (and disulphide after treatment with sulphite) by reaction with excess methylmercuric iodide in dimethylformamide/phosphate buffer, pH 9. The polarographic diffusion current of the reaction was recorded, again after each of two extra additions of reagent, then after adding excess standard reduced glutathione and finally after a further addition of reagent. The reagent used in the original reaction could be estimated from the plot of the current values. Hofmann (1971a) determined 0·2 to 0·8 μmol amounts of thiols by reacting for 2 h with excess *p*-chloromercuribenzoate. 1·0 μmol of glutathione was then added and the remainder of this after reaction with unused mercury reagent he titrated amperometrically with 10^{-4} M silver nitrate.

Holzapfel and Stottmeister (1967) used *S*-methylisothiouronium sulphate reagent for amperometric back-titration of unused mercuric chloride in a

Table V. Direct titrations with mercury(II).

Sample	Titrant	Conditions	End-point indication	References
Denatured egg albumin	p-Chloromercuribenzoate	+ guanidine hydrochloride	Cessation of response to nitroprusside	Anson (1940, 1941)
Cysteine . HCl; 2,2-dithioisobutyric acid	p-Chloromercuribenzoate	pH 5·3	External nitroprusside →no colour	Jansen (1948)
Cysteine, glutathione, thiol groups in egg albumin	p-Chloromercuribenzoate	pH ca. 7	Amperometric	Hata (1951a)
Cysteine, glutathione	p-Chloromercuribenzoate	Denatured with guanidine hydrochloride	External nitroprusside	Jansen and Jang (1952)
β-Lactoglobulin	p-Chloromercuribenzoate	+ guanidine hydrochloride pH 5 (acetate)	External nitroprusside	Fraenkel-Conrat et al. (1952)
Free protein thiol groups	p-Chloromercuribenzoate	Acetate buffer, pH 4–5	Amperometric	Matoušek and Laučíková (1953)
Albumin	Methylmercuric nitrate	In guanidine bromide, ca. 0°C	Internal nitroprusside	Edelhoch et al. (1953)
Serum albumin	Mercuric chloride	Acetate buffer, pH 4·9	Amperometric	Saroff and Mark (1953)
Thiouracils	Mercuric nitrate	Acetic acid/sodium acetate	Diphenylcarbazone →rose-violet for 2 min	Abbott (1953)
Methyl thiouracil	Mercuric chloride	Acetic acid/acetate	Diphenyl carbazone	Cynajek and Szlanga (1954)
Biological thiols (glutathione, cysteine)	Mercuric chloride or acetate	+ borax + KCl (+ phosphate for proteins); in nitrogen	Amperometric (rotating Pt wire)	Kolthoff et al. (1954)

	Reagent	Conditions	Method	Reference
Protein thiols (tested on cysteine)	Organic reagents, especially p-chloromercuribenzoate	pH 4·6 or 7 (study of rates of reaction, etc.)	Absorbance of reaction product at ca. 250 nm (proteins may also absorb strongly here)	Boyer (1954)
Cysteine	Mercuric chloride	Nitrate, acetate or phosphate buffer, pH 1·2 to 6·7	Potentiometric	Cecil (1955)
Gluten thiol groups	Mercuric chloride (10^{-4} M)	pH 4·4 to 8·7, best between 4 and 5; +KCl + urea to prevent coagulation	Amperometric	Matsumoto and Shimada (1955)
Thiol groups in tobacco mosaic virus	Method of Boyer (1954) at pH 5 and 7			Fraenkel-Conrat (1955)
Haemoglobin thiol groups (number)	Mercuric chloride	Ammonium hydroxide/ chloride buffer	Amperometric	Ingram (1955)
Serum thiol groups		Phosphate buffer pH 7·4	Amperometric	Robert and Robert (1956)
Protein thiols	Mercuric nitrate	Carbonate buffer, pH 10·6, + guanidine bromide; at 0°C	sodium nitroferricyanide, to colour (pink) discharge	Pepe and Singer (1956)
A dehydrogenase	Method of Boyer (1954)			Koeppe et al. (1956)
Thiol from disulphide in determination with sulphite	Mercuric chloride	+ nitric acid to give pH 1·8 to 2	Potentiometric	Cecil and McPhee (1957)
Thiols; also from disulphides with sulphite	Mercuric chloride	In nitrogen atm.	Amperometric	Csagoly (1957)

Table V (cont.)

Sample	Titrant	Conditions	End-point indication	References
Thiols of bovine serum albumin	Mercuric chloride	+ guanidine. HCl + NH_4OH/NH_4NO_3 or phosphate buffers pH 9 and 7, respectively	Amperometric	Kolthoff et al. (1957)
Wheat flour thiols	Mercuric chloride	Method of Kolthoff et al. (1954)		Kong et al. (1957)
Toluene-3,4-dithiol derivatives of various metals	Mercuric chloride	In pyridine	+ trace of Co salt, titrated to discharge of intense blue	Clark (1957)
Aldolase thiol groups	p-Chloromercuribenzoate	pH 7	Photometric (method of Boyer, 1954)	Swenson and Boyer (1957)
Thiol groups of sickle cell haemoglobin	Mercuric titrate	NH_4OH/NH_4NO_3	Amperometric	Ingram (1957)
Thiol groups of normal and sickle cell haemoglobin		Method of Kolthoff et al. (1954)		Murayama (1957)
		Study of effect of pH	Amperometric	Burton (1958)
Thiols in wool hydrolysates; also from disulphides + sulphite	Mercuric chloride	KCl/borate buffer	Amperometric	Human (1958)
Non-protein thiols in muscle	Mercuric chloride	Borate buffer, pH 9	Amperometric	Oganesyan and Dzhbanibekova (1958)
Thiols and other sulphur compounds	o-Hydroxymercuriphenol or -benzoate and other organomercury compounds	+ alkali	Nitroprusside, diphenylcarbazone, dithizone, thiofluorescein	Wronski (1958a); Wronski and Burkart (1958)

Thiophenols; mercaptobenzothiazoles	Mercuric nitrates; bis(hydroxymercuri)thymol	+ alkali	Diphenylcarbazone	Wronski (1958b)
Cysteine; mercaptoacetic acid	Electrolytically generated Hg^{2+}	Borate buffer, pH 9·2	Potentiometric	Przybylowicz and Rogers (1958)
Thiol groups of various animal haemoglobins	Method of Kolthoff et al. (1954)			Murayama (1958)
Serum albumin thiols	Methylmercuric nitrate; mercuric chloride		Method of Edelhoch et al. (1953) / Method of Saroff and Mark (1953)	Simpson and Saroff (1958)
Thiol groups of normal adult human haemoglobin	Mercuric chloride, phenylmercuric chloride	In borax/KCl/sulphite, nitrogen atm.	Amperometric (dropping Hg)	Allison and Cecil (1958)
Bovine plasma albumin thiols	Mercuric chloride	Method of Kolthoff et al. (1954) in phosphate buffer, pH 7·6		Werner and Levy (1958)
Thiols in flour, gluten	Mercuric nitrate	Many buffers	Amperometric	Bloksma (1959)
Zinc complex of toluene-3,4-dithiol	Mercuric chloride	In pyridine	Decoloration of blue with Co salt	Clark and Neville (1959)
E.g. cysteine; also thiols from disulphides + SO_3^{2-}	Phenylmercuric chloride; mercuric chloride	Acetate buffer, pH 5, also in borax, pH 9·2	Amperometric / Amperometric	Kapoor (1959)
Thiols in skin homogenates	Mercuric chloride	Phosphate buffer, pH 7·3, + KCl	Amperometric	Montagnani et al. (1959)
Myosin thiol groups	Mercuric chloride		Amperometric	Torchinskii (1959)
Thiols in intact proteins	Mercuric chloride	pH 9	Amperometric	Leach (1959)
Thiols + S^{2-}; (e.g. cysteine, mercaptoacetic acid, ethanethiol)	o-Hydroxymercuribenzoate	In ammoniacal solution + SO_3^{2-}	Discharge of blue of thiofluorescein	Wronski (1960a)

Table V (*cont.*)

Sample	Titrant	Conditions	End-point indication	References
Mercaptoacetic acid	o-Hydroxymercuri-benzoate	Alkaline (NaOH) solution; reverse titration	To blue of thiofluor-escein	Wronski (1960b)
Penicillamine (from alkalin hydrolysis of penicillin)	Mercuric chloride	Borate buffer, pH 9·2	Amperometric	Grafnetterová (1960)
Cysteine	o-Hydroxymercuri-benzoate; mercuric nitrate	Alkaline solution	Dithizone → red	Wronski (1960c)
Cysteine	p-Chloromercuribenzoate	$T°$ not above 20° to prevent reaction of the —NH₂ and —COOH groups	Potentiometric	Calcutt (1960)
E.g. cysteine, mercaptoacetic acid, BAL, glutathione	Mercuric nitrate	Acetate buffer, pH 5·6; fuchsin as maximum suppressor	Amperometric	Oganesyan and Zaminyan (1960)
Cysteine, mercaptoacetic acid in presence of CN⁻	Mercuric nitrate	+ NaOH + FeSO₄ + EDTA	Discharge of blue of thiofluorescein	Wronski (1960d)
Study of reaction with thiols, e.g. cysteine, mercaptoethanol, glutathione	p-Chloromercuribenzoate	pH 7·5; (influence of temperature found on stoichiometry)	Absorbance at 255 nm (Boyer, 1954)	Hoch and Vallee (1960)
Cysteine.HCl, glutathione, BAL, mercaptoethylamine HCl, etc.; also thiol groups in albumins	Methylmercuric iodide/DMF; mercuric chloride		Amperometric	Leach (1960)

Substance determined	Reagent	Conditions	End-point	Reference
Thiol groups; standardized with cysteine	3,6-bis(acetoxymercuri-methyl)dioxan; also methylmercuric bromide, mercuric nitrate	Carbonate, pH 10·6; + guanidine bromide	Discharge of pink colour with nitroferricyanide (Pepe and Singer, 1956)	Singer et al. (1960).
Thiols in oxidized wheat flour	Mercuric chloride	Borax buffer + KCl	Amperometric	Sullivan et al. (1961)
E.g. methanethiol, thiophenol (also S^{2-}; cysteine less good, scarcely reducing fluorescence of titrant	Tetrakis(acetoxymercuri) fluorescein in dil. NaOH	+ HCHO to yield $RSCH_2OH$ to prevent oxidation, then + alkali which reforms thiol	Fluorescence values → intersection end-point	Wronski (1961a)
Cysteine, from reduction of cystine	o-Hydroxymercuri-benzoate	+ excess alkali	Discharge of blue of thiofluorescein	Wronski (1961b)
Higher alkane thiols, thiophenols, thioacohols, mercapto acids	Mercuric perchlorate in dilute $HClO_4$	In water or acetone, + 1% pyridine	4,4'-bis(dimethylamino) thiobenzophenone → blue; also photometric at 580 nm; also potentiometric	Fritz and Palmer (1961)
Thiophenols, butanethiol	Mercuric nitrate in dilute HNO_3; acetate in dilute acetic acid	In ethanol, some slightly warmed to prevent precipitation	Diphenylcarbazone → blue-violet	Gregg et al. (1961)
E.g. glutathione, mercaptobenzothiazole, cysteine	Mercuric nitrate	Various conditions, e.g. pH 2 and 9	Amperometric (rotating Pt wire or Hg pool)	Kolthoff and Eisenstädler (1961a)
Cysteine, glutathione; thiol groups in amino acids, peptides, proteins	Ethyl- or phenylmercuric chloride in dilute NaOH	Various buffers—acetate, borate, phosphate, ammonium	Amperometric (rotating dropping Hg)	Stricks and Chakravarti (1961)

Table V (cont.)

Sample	Titrant	Conditions	End-point indication	References
E.g. cysteine, glutathione, mercaptosuccinic acid	Anhydride of 2 [3-(hydroxymercuri)-2-methoxypropyl]-carbamoyl phenoxyacetic acid ("salyrganic or mersalyl acid")	pH 6·5; nitrogen atmosphere	N,N-dimethyl-4-(2-pyridylazo)aniline; visual or photometric at 550 nm (abs. max. of complex with excess titrant)	Klotz and Carver (1961)
Methane-, ethanethiols; cysteine	o-Hydroxymercuri-benzoate	+ NH_4OH + Na_2SO_3	Discharge of blue of thiofluorescein	Wronski (1961c)
Mixtures of thiols with, e.g. SCN^-; also two thiols	o-Hydroxymercuri-benzoate	NH_4OH/NH_4Cl	Discharge of blue of thiofluorescein	Wronski (1961d)
	Mercuric perchlorate	Neutral aqueous or acetone solution	Potentiometric	Palmer (1962)
Cysteine from reaction of cystine with SO_3^{2-}	o-Hydroxymercuri-benzoate	NH_4OH/NH_4HPO_4 from cystine reaction	Discharge of blue of thiofluorescein	Wronski (1962)
Thiols in wool hydrolysates with $6N\ H_2SO_4$	Phenylmercuric hydroxide	+ guanidine·HCl + Na_2CO_3	1% nitroprusside (violet → colourless)	Zahn et al. (1962)
Thiols in flour	Ethylmercuric chloride	Dispersed in 6–8 M urea, + EDTA to minimize effect of metal ions	Amperometric (dropping Hg)	Tsen and Anderson (1963)
Cysteine + S^{2-}; also cystine + SO_3^{2-}	o-Hydroxymercuri-benzoate	In NaOH	Discharge of blue of thiofluorescein	Wronski (1963a)

Sample	Reagent	Conditions	Method	Reference
Cysteine + mercaptoacetic acid	o-Hydroxymercuri-benzoate	+ NaOH	Discharge of blue of thiofluorescein	Wronski (1963b)
Thiomalic acid	o-Hydroxymercuri-benzoate	In water, + NH_4OH + SO_3^{2-}	Discharge of blue of thiofluorescein	Chromý and Svoboda (1963)
Thiols in organic solvents	Mercuric chloride	Neutral or slightly acid	Amperometric (rotating Hg-pool)	Singh and Varma (1963–4)
	p-Tolylmercuric chloride; o-hydroxymercuri-benzoate	In toluene + ethanol + aq KOH in toluene, shaken with aq NH_4OH	Discharge of blue of thiofluorescein	Wronski (1964a)
Cysteine, mercaptoacetic acid; also redn. prods. of diformylcystine and bis(tosyl)cystine	o-Hydroxymercuri-benzoate	Also after adding HCHO to bind cysteine	Discharge of blue of thiofluorescein	Wronski (1964b)
Microthiol in plant extracts	Compared several methods and criticized Wronski's (1958a) as too insensitive			Spanyar et al. (1964)
Thiol groups in bovine serum albumin (critical study of method)	Methyl-, ethylmercuric chlorides, mercuric chloride	Tris, ammonium buffers; 8 M urea, 1% Na dodecylsulphate	Amperometric (rotating Hg-pool)	Kolthoff et al. (1965)
E.g. cysteine, glutathione, thiophenols, mercaptoheterocycles	Mercuric perchlorate in $HClO_4$	Aq. ethanol + pH 6 buffer (pyridine–acetic acid); also in acetone	4,4'-bis(dimethyl-amino)-thiobenzo-phenone → blue; also diphenylcarbazone	Belcher et al. (1965)
Glutathione/cysteine mixtures	o-Hydroxymercuri-benzoate	In NH_4OH; also after adding HCHO to remove cysteine	Discharge of blue of thiofluorescein; glutathione alone to dithizone → purple	Wronski (1965a)

Table V (cont.)

Sample	Titrant	Conditions	End-point indication	References
Alkane-, arenethiols, mercaptoacids	o-Hydroxymercuri-benzoate	Aq ethanol	p-N-Dimethyl-aminobenzylidene-rhodanine → red or purple	Wronski (1965b)
Thiols from disulphides by reduction with BuLi	o-Hydroxymercuri-benzoate	+ ethanol at 0°C	Discharge of blue of thiofluorescein	Veibel and Wronski (1966)
Equivalent weight determination of thiols	p-Chloromercuri-benzoate/NaOH	In ethanol or water, + NaOH or NH_4OH	Discharge of blue of thiofluorescein	Wronski (1966a)
Thiols from determination of thiol esters by hydrolysis with NaOH/dimethylamine; also thiols in presence of thiol esters	o-Hydroxymercuri-benzoate	in presence of ethanol/triethylamine/$HClO_4$	Discharge of blue of thiofluorescein	Wronski (1966b)
Thiols; thiol/S^{2-} mixtures	Mercuric chloride in methanol	In ethanol	"High frequency"	Serrano Bergés and Fernández (1966)
5-Thiouracil: 5-thiodeoxyuridine		Based on method of Klotz and Carver (1961)		Bardos and Kalman (1966)
D-penicillamine hydrochloride	Mercuric nitrate	+ NaOH + ethanol	Dithizone	Polaczek and Kuszczak (1966)
Thiol groups, e.g. in actinomycin		Based on method of Klotz and Carver (1961)		Ehrlich (1967)
L-Cysteine, penicillamines	Phenylmercuric acetate	Phosphate buffer, pH 8, + $CHCl_3$; N_2 atmosphere	Diphenylcarbazone → violet in $CHCl_3$	Doornbos (1967)

Sample	Reagent	Conditions	Method	Reference
Thiol groups in yeast cells	p-Chloromercuri-benzoate	Method of Boyer (1954), compared with Ag$^+$ titration		Wei (1967)
Haptoglobin thiols	p-Chloromercuribenzoate	In 7·2 M guanidine hydrochloride	Photometric (Boyer, 1954)	Tattrie and Connell (1967)
Wheat gluten protein thiols	Ethylmercuric chloride	Tris buffer, pH 7·5, + KNO$_3$, EDTA and 8 M urea	Amperometric (Hg electrode)	Okada and Yonezawa (1967)
Cysteine from wool cystine by reduction	o-Hydroxymercuri-benzoate	Alkali	Discharge of blue of thiofluorescein	Wronski and Goworek (1968)
Protein thiols; tested on glutathione, cysteine	Methylmercuric iodide		Amperometric (found very slow)	Forbes and Hamlin (1968)
Thiols in gastric contents	Mercuric chloride	Phosphate buffer pH 7·1	Amperometric	Korobeinik (1968)
Cysteine in wheat flour; also cystine after reduction	Ethylmercuric chloride	+ 8 M urea, KCl, NH$_4$OH/NH$_4$Cl, EDTA	Amperometric (dropping Hg)	Hack and Allan (1968)
6-Methyl-2-thiouracil	Mercuric chloride	pH 4·6	Amperometric (Pt electrode)	Fecko and Zaborniak (1968)
Cysteine in wool samples		Method of Zahn et al. (1962)		Asquith et al. (1968)
Decanethiol, cyclopentonethiol, mercaptoacetic acid and its anilide	p-Dimethylamino-phenylmercuric acetate; also p-diethyl analogue	Weakly acid, neutral or alkaline medium (pH 5 to 11)	Amperometric; potentiometric; diphenylcarbazone	Busev and Teternikov (1969)
Protein thiols	Phenylmercuric hydroxide	pH 2	Amperometric	Ambrosino et al. (1969)

Table V (cont.)

Sample	Titrant	Conditions	End-point indication	References
Protein thiols; tested with mercaptoacetic acid	Chloromercuri-nitrophenols	pH 7·9; + EDTA yielding weak complex with reagent; reverse titration	Photometric (sharp fall in absorbance at ca. 410 nm due to release of EDTA)	McMurray and Trentham (1969)
Penicillamine	Mercuric perchlorate	Aqueous pyridine	Potentiometric to 1st inflection	Körbl and Vaníček (1969)
β-Lactoglobulin thiol groups	4-(p-Sulphophenylazo)-2-mercuriphenol	pH 8·3	Photometric at 440 nm	Joniau et al. (1970)
Cysteine, albumin	4-Chloromercuri-benzenesulphonate; mercuric chloride	pH 7 to 9·5; pH 4·5 to 9·5; inert atm. at pH above 7	Potentiometric	Toribara and Koval (1970)
E.g. methanethiol, thiophenols, mercaptoacetic acid	Mercuric chloride	In NaOH	Potentiometric with sulphide-selective electrode	Papp and Havas (1970)
E.g. butanethiol, thiomalic acid in organic solvents	Tetrakis(acetoxy-mercuri)fluorescein, + prominal (highly fluorescent product) or + oxine (non-fluorescent product)	Ethanol–isopropanol–water, + triethylamine;	Fluorimetric, with fluorescence as end-point (extrapolated)	Wronski (1970)
	p-Chloromercuribenzoate (critized low solubility and stability and susceptibility to anions in the medium)	Method of Boyer (1954)		Muftic (1970)

Application	Reagent	Conditions	Detection	Reference
Cysteine · HCl in bread improver formulations	o-Hydroxymercuri-benzoate	Alkaline medium (+ NH₄OH)	Discharge of blue of thiofluorescein	Barrett et al. (1971)
Thiols in alkaline pulping liquors	Mercuric chloride		Potentiometric with a sulphide–selective electrode	Papp (1971)
Thiols in air after after removal of H_2S and CS_2	Tetrakis (acetoxymercuri)fluorescein	Hexanol/triethyl-amine/Et_3PbCl (to prevent air oxidation)/oxine/96% ethanol	Fluorimetric, to maximum fluorescence	Wronski (1971a)
Thiols in motor oils	p-Dimethylamino-phenylmercuric acetate and diethyl analogue	Ethanol/sodium acetate	Diphenylcarbazone → lilac	Busev et al. (1971)
β-Mercaptopyruvic acid	o-Hydroxymercuri-benzoate	pH 9; + HCHO to inactivate cysteine and other compounds	Discharge of blue thiofluorescein	Thibert and Ke (1971)
Study of extr. coeffs. of thiols e.g. cysteine, homocysteine, penicill-amine	Boyer's (1954) method compared with 4 others and considered reliable			Wenck et al. (1972)
	o-Hydroxymercuri-benzoate	+ NaOH	Dithizone → purple	Wronski (1972)
Methanethiol	o-Hydroxymercuri-benzoate	In ethanol/NH₄OH	Discharge of blue of thiofluorescein	Bald (1972)
Ultramicro thiols in beer	Tetrakis (acetoxymercuri) fluorescein + oxine	Method of Wronski (1970)		Brenner and Laufer (1972) Brenner and Khan (1974).
Cysteine in dyed wool, hydrolysed with 30% H_2SO_4	Ethylmercuric chloride	Solution containing 10% DMF, pH 7·7	Amperometric	Alfredo et al. (1973)

Table V (cont.)

Sample	Titrant	Conditions	End-point indication	References
	Mercuric perchlorate	Best in acetone/ethanol; some in dioxan	Potentiometric with bromide-selective electrodes based on Ag_2S matrixes	Selig (1973)
Thiols in gases	o-Hydroxymercuri-benzoate	Alkali + EDTA	Discharge of blue of thiofluorescein	Staszewski and Zygmunt (1973)
Silanethiols, $(RO)_3SiSH$, $(RO)_2Si(SH)_2$	o-Hydroxymercuri-benzoate	Isopropanol + NH_4OH	Discharge of blue of thiofluorescein	Wojnowska and Wojnowski (1973)
E.g. cysteine, glutathione, Na thiopental	Electrolytically generated Hg^{2+}	pH 9·3, containing borax and KNO_3	Potentiometric with Hg-indicator electrode	Mairesse-Ducarmois et al. (1973)
Mercaptobenzothiazole	Mercuric nitrate	In $CHCl_3$	Diphenylcarbazone	Rublev et al (1974)
Cysteine in presence of other amino acids	Mercuric chloride	pH 6·2 (citrate) pH 9·2 (carbonate/ bicarbonate)	4-(2-Pyridylazo)-resorcinol, yellow → pink; diphenylcarbazone, yellow → violet	Sen and Bahadur (1974)

determination of methane- and ethanethiol. It yields thiol immediately through hydrolysis in the 0·3 M sodium hydroxide solution used and has the advantage of being a more stable reagent, less susceptible to oxidation, than a thiol:

$$\left[CH_3-S-C\begin{array}{c} \nearrow NH_2 \\ \searrow NH_2 \end{array} \right]^+ \xrightarrow{OH^-} CH_3SH + O=C\begin{array}{c} \nearrow NH_2 \\ \searrow NH_2 \end{array}$$

Recently Mildner *et al.* (1972) determined thiols in proteins and enzymes by treating with excess phenylmercuric acetate in tris buffer of pH 7·4 for 15 min. They then added excess standard glutathione and, after 5 min, back-titrated with standard silver nitrate.

14.2.2. *Polarography*

Unused mercury can be estimated through the diminution of the polarographic wave height of its reduction. Benesch and Benesch (1951, 1952) found that organomercury compounds are also polarographically reducible in a 2-electron step; they are thus as potentially useful as inorganic mercury(II) compounds. Hata (1951b) observed that the diffusion current of its well-defined polarographic reduction wave at pH 7 was directly proportional to the concentration of *p*-chloromercuribenzoate and that the wave-height was lowered by successive additions of thiols. This reagent was used to determine cysteine by Proctor (1956) and by Nakamura and Nemoto (1961) (in wool). Human (1958) determined thiols in intact wool by 10–12 days reaction in a borate/acetate/phosphate mixture with a mercury(II) reagent, followed by polarographic evaluation of the unused reagent. He used *p*-chloromercuribenzoate or "Neohydrin", $Cl-Hg-CH_2-CH(OCH_3)-CH_2-NH-CO-NH_2$.

Leach (1960) appears to have been the first to use methylmercuric iodide reagent. He determined the thiol groups in fibrous keratin by allowing to stand for 4–24 h with an excess of reagent in a medium containing tris, potassium chloride and dimethylformamide and measuring diffusion currents; he also used mercuric chloride. Leach's method was employed by Wolfram and Lennhoff (1967) for hair cysteine after treatment with sulphite; and by Sakamoto *et al.* (1969) to determine thiols in modified cotton. Others using methylmercuric iodide in this way have been: Maclaren *et al.* (1960) for protein disulphide, treating with sulphite; Forbes and Hamlin (1968) for protein thiols (tested on cysteine and glutathione) and disulphide; and Shah and Gandhi (1970) for thiols and disulphide in untreated and modified wools.

Mrowetz and Klostermeyer (1972) recently determined thiols in milk through reaction with excess methylmercuric chloride in a tris buffer of pH 7 containing potassium chloride and gelatine.

14.2.3. *Reaction with Diphenylcarbazone and Diphenylthiocarbazone*

The sulphur-containing reagent (dithizone) has most been applied to determine unused mercury(II) colorimetrically after extraction into a solvent. Hughes (1949) was probably the first to apply this method, namely to a study of mercaptalbumin. He used excess methylmercuric iodide and extracted the unused reagent into an organic solvent and treated it with dithizone in the presence of a base such as amylamine. Fridovich and Handler (1957) determined cysteine in 5–50 nmol amounts with excess *p*-chloromercuribenzoate in neutral or weakly acid solution. They then added a dithizone reagent in carbon tetrachloride, shook for 1 min and measured the light absorbance in the organic layer at 625 nm, the absorption maximum of the reaction product of dithizone with *p*-chloromercuribenzoate. A similar procedure was used for thiols in protein solutions. Mussini (1958) applied an analogous method (with chloroform as solvent instead of carbon tetrachloride) to determine thiols in biological preparations (e.g. cysteine, glutathione); and Sasago *et al.* (1963) used Fridovich and Handler's procedure for determining milk thiols.

An interesting variant is due to Simpson and Saroff (1958). They investigated the changes of thiol content of serum albumin with time. The sample, containing ca. 1 μmol of thiol, in a phosphate buffer of pH 7–7·5, was mixed with an excess of methylmercuric iodide in toluene for ca. 1 h. They then added EDTA (to complex interfering metals) and pyridine–acetic acid (1 + 1) to an aliquot of the toluene layer and titrated unused mercurial with 0·2 M dithizone in chloroform or carbon tetrachloride to the first green.

Busev and Teternikov (1971) describe a recent example of the use of diphenylcarbazone. Their reagent was excess N-dimethyl- (and diethyl-) aminophenylmercuric acetate for determining various thiols, such as cysteine, decanethiol or benzyl mercaptan. The reaction mixture had pH 6·5–10 (for the dimethyl–reagent) or 5·5 to 8·5 (for the diethyl reagent) in aqueous ethanol. Diphenylcarbazone in benzene was then added and the absorbance of the benzene layer was evaluated at 540 nm. Amounts of 5–200 μg could be determined with 4–8% accuracy.

14.2.4. *Reaction with 3,6-Dimercaptofluoran (thiofluorescein)*

Many examples are given above under 14.1, "Direct Titration" of the use of this compound as a colour indicator, based on the disappearance of its blue colour through reaction with the mercury(II) titrant to yield a colourless product. Wronski (1967) determined thiols and thiol esters after hydrolysis by reacting with excess *o*-hydroxymercuribenzoate reagent in a basic medium containing triethylamine and phosphoric acid at 0°C. After 5 min, he added a measured amount of the thiofluorescein and evaluated the absorbance of the solution at the absorption maximum of this compound,

588 nm. The diminution in absorbance compared to that of a control without original sample is a measure of the mercury reagent consumed by the thiol. Wronski later applied the method (1968a) to determine thiols in wool hydrolysates, reacting with the o-hydroxymercuribenzoate for 10 min at pH 9·6.

14.2.5. Photometric Determination of Unused Reagent

Isles and Jocelyn (1963) determined protein thiols in aqueous solution by reacting for 10 min with excess p-chloromercuribenzoate in phosphate buffer of pH 7·4. Proteins were then precipitated with phosphoric acid, the precipitate was centrifuged and the residual reagent was estimated in the supernatant through absorbance at 247 nm. The authors stated that the precipitate carried some mercurial reagent with it. Hamm and Hofmann (1967) used a similar procedure for determining thiol groups in myofibrillae. They treated with excess of the same reagent at pH 7 (phosphate), shaking for 7 h. After centrifuging, they evaluated absorbance of the supernatant at 232 nm.

These photometric utilizations of the p-chloromercuribenzoate reagent are less well-known than the direct titration of Boyer (1954), where the absorbance of the reaction product is estimated at ca. 250 nm.

A mercurial containing a chromophore lends itself better to photometric methods. Thus Flesch and Kun (1950) used 1-(4-chloromercuriphenylazo)-2-naphthol to estimate thiols in tissue homogenates. An aqueous solution or suspension of the sample was shaken with a solution of the reagent in amyl acetate. The mercury derivative of the thiol precipitated. They centrifuged this and measured the decrease in absorbance of the amyl acetate layer with reference to the original value for the reagent solution. Burley (1956) used the same dye to estimate thiols in wool samples. He treated these with the dye solution in formamide at pH 7·8–8·2, at 30°C, examining the mixtures from time to time until the colour intensity in the filtrate became constant. From its value he estimated residual, and hence consumed, dye (standardization with glutathione). Horowitz and Klotz (1956) used 4-(4-dimethylaminobenzene-azo) phenylmercuric acetate,

$$(CH_3)_2N - \langle \rangle - N=N - \langle \rangle - HgOCOCH_3$$

also with the azo chromophore, for work on protein thiols. The sample in aqueous, glycine-containing buffer of pH 9·3 was mixed with the reagent in heptanol in a ratio of 98:2 (to counter the unfavourable partition coefficient of the reagent). They shook the mixture for 4 h, then pipetted off an aliquot of the aqueous phase, clarified it by centrifuging and measured the absorbance at 460 nm, the reagent maximum. Evaluation of the organic phase at 414 nm was also possible but the results were less good. The reagent has a molar absorptivity of ca. 26 000. Glycine improves its solubility in water.

14.2.6. *Fluorometric Evaluation of Unused Reagent.*

The principal reagent used in these procedures is tetrakis (acetoxy-mercuri) fluorescein (TMF). Wronski (1960e) applied it (and also mercuriated phenolphthalein) to determine thiols, such as methanethiol, cysteine, thiophenol, mercaptobenzothiazole and mercaptoacetic acid. He added the sample to excess reagent in dilute sodium hydroxide and evaluated the residual fluorescence after 5 min. The difference in intensity from a standard without sample was a measure of the mercury reagent reacting with the thiol and hence of the thiol. Recently, Wronski (1971b) fashioned detector tubes for thiols (also hydrogen sulphide and hydrocyanic acid) containing cellulose powder impregnated with TMF/borax/boric acid/biuret mixture. Fluorescence decay indicated the presence of thiols (or the other compounds) in a gas passed through, and semi-quantitative estimation was possible from the length of the quenched zone.

Vakaleris and Pofahl (1968) likewise observed the reduction of fluorescence of TMF at 520 nm (excitation wave-length of 499 nm) in proportion to the amount of added thiol and tested this on glutathione in tris buffer of pH 7·5. They applied it to determine thiols in milk and cheese. They noted that both thiol and disulphide quenched the fluorescence in N sodium hydroxide solution. Mironov *et al.* (1971) applied the same principle to determine thiols and disulphides in proteins and peptides, using a reagent containing EDTA and ammonium nitrate to stabilize the thiols. Quenching at pH 7·4 was due to thiols; that at pH 13–14, to both thiols and disulphides. Senter *et al.* (1973) also applied a modification of Vakaleris and Pofahl's method to determine thiol groups in whey from milk processed by turbulent flow, ultra-high temperature procedures.

Haviř *et al.* (1965) visualized thiols (e.g. thiophenol, cysteine, 6-mercapto-purine, thioglucose) on paper and thin-layer chromatograms through the suppression of fluorescence of TMF; they considered it superior to nitro-prusside.

An identification method due to Wronski (1968b) may be mentioned here. He measured the quenching effect (expressed as the ratio of fluorescence of TMF + sample to fluorescence without sample) of ca. 20 thiols, at 465 nm and 510 nm and at various pH values. The values tabulated by him lend themselves well to identification of the thiols.

14.2.7. *Back-Titration with Acid.*

Unused mercuric acetate has been determined in this way. E.g. Kundu and Das (1959) treated *n*-octanethiol with excess mercuric acetate in methanol, also this thiol and benzyl mercaptan with phenylmercuric acetate (a better reagent in the presence of olefines with which it does not react). They back-titrated the unused mercuric salt with hydrochloric acid in butanol, using thymol blue or diphenylcarbohydrazide indicators:

$$Hg(OCOCH_3)_2 + 2HCl \rightarrow HgCl_2 + 2CH_3COOH$$

Direct titration with phenylmercuric acetate proved too slow.

Berger and Magnuson (1964) estimated mercaptosilanes by reaction in toluene with mercuric acetate in methanol 0.05% in acetic acid. They slowly titrated the stirred solution with hydrochloric acid in butanol, using thymol blue as indicator.

It is surprising that there are no other examples of this simple procedure.

14.2.8. *Back Titration with Nickel(II).*

Busev *et al.* (1968) determined thiols from catalytic cracking of aliphatic sulphides (also individual thiols, such as methane- and butanethiol) by treating with measured excess of *p*-dimethylaminophenylmercuric acetate and back-titrating with nickel diethylthiophosphate to diphenylcarbazone indicator.

14.2.9. *Back Titration with EDTA*

Cernenco and Crisan (1974) precipitated dodecanethiol with excess mercury(II) and back-titrated unused reagent with EDTA to xylenol orange.

14.3. Analytical utilization of a reaction product

Nearly all the analytical methods classified under this heading are based on a mercury-containing reaction product. The small number of other methods are given afterwards.

14.3.1. *Mercury-Containing Product*

The procedures mostly depend on separation of the product from excess reagent and perhaps from non-thiol material. This is achieved in several ways: precipitation of a poorly soluble derivative (usually from aqueous solution); removal of a soluble derivative (in an organic solvent); retention on the structure of the sample (e.g. tissue), excess reagent being washed away; retention on an artificial surface as in chromatographic-type procedures. Mercury-containing carriers are used, which retain thiols from solutions passed through them.

The derivative properties of low solubility, light absorption, fluorescence, melting point, radioactivity, polarographic activity and atomic absorption spectrum are then drawn on, together with chemical treatment which regenerates the original thiol or reagent, either of which can be evaluated analytically. The headings in this paragraph are convenient for dealing with the subject matter.

(a) *Solubility.* Some qualitative tests for thiols depend on their low solubility. Lecher and Siefken (1926) commented on the cloudiness yielded by thiols

D

with mercuric chloride and that this test is more sensitive than the red colour with nitrous acid; 1 drop of thiol in 50 ml of ethanol gave turbidity with a mercuric chloride/alcohol reagent. André and Kogane–Charles (1947) detected thiols in rapeseed oil through the cloudiness yielded with 5% ethereal mercuric chloride. Shemyakin and Berdnikov (1965) mention the white precipitate with mercuric nitrate among tests with metal cations for methyl thiouracil. Takai and Asami (1972) identified methanethiol in paddy soils by passing nitrogen through the heated sample into 4% mercuric cyanide after separating hydrogen sulphide with lead acetate. The positive response was a white precipitate.

Segal and Starkey (1953) determined methanethiol in mixtures by passing through 4% mercuric cyanide and evaluating the mercaptide gravimetrically after filtering and drying in vacuo over calcium chloride. Gravimetric procedures for mercury derivatives appear very rare.

Del Vecchio and Argenziano (1946) separated penicillamine from ammonia and amines by precipitation at pH 7·5 with mercuric chloride, recovering the thiol by treatment with hydrogen sulphide.

(b) *Light absorption.* Coloured mercurials were introduced especially in histochemical work. The first compound of this type was the red 1-(4-chloromercuriphenylazo)-2-naphthol of Bennett and Yphantis (1948). This was shown capable of attachment to biological tissue. Bennett (1951) used it as an optical tracer, as did Mescon and Flesch (1952) who immersed skin sections for 1–3 h in an 80% alcoholic solution of the reagent; Roberts (1960) for protein bound thiols in Coleus wound meristems; and Mauri *et al.* (1954) for thiols in blood cells.

The last named authors also introduced a water-soluble reagent, 1-hydroxy-2-(4-chloromercuriphenylazo)-8-aminonaphthalene-3,6-disulphonic   acid, neutralized with sodium bicarbonate and also red.

Other compounds with the phenylazo group used in histochemistry include: 1-(4-acetoxymercuriphenylazo)-2-naphthol, used in 0·003% amyl acetate solutions by Schrauwen (1963) for visualizing thiols on paper electropherograms (down to ca. 1 µg/cm^2) as orange–red spots, darkening on exposure to hydrogen chloride; 4-phenylazo-2-chloromercuriphenol and 4-(p-nitrophenylazo)-2-chloromercuriphenol by Chang and Liener (1964) which yields a 1:1 complex as a coloured label for cysteine; 4-(p-dimethylaminobenzeneazo)phenylmercuric acetate by Engel and Zerlotti (1964) for histochemical demonstration of protein-bound thiol groups, staining directly in an alcoholic glycine-buffered solution of the reagent and obtaining good contrast by observing in monochromatic light of 430 nm. Replacement of the phenylazo group by toluylazo, anisidylazo, phenetidylazo or diphenylaminoazo groups gives reagents yielding darker and better visible colours according to Szydlowska and Junikiewicz (1972).

Colour reactions with mercury compounds not containing an azo group have been used by Wronski (1960e) who detected thiols through the violet colour with mercuriated phenolphthalein.

Colorimetric or spectrophotometric methods for thiols are based on light absorbance of mercury-containing derivatives. Direct titration procedures, e.g. that of Boyer (1954) have been mentioned above under "Direct Titration", Section 2.14.1. Among non-titrimetric methods may be quoted that of Nakamura et al. (1968) who determined volatile thiols in saké by passing nitrogen through the acidified sample into ice-cooled p-chloromercuri-benzoate reagent and recording the increase in absorbance, due to the reaction product, at 250 nm; and that of Ambrosino et al. (1969) for thiols in certain proteins through absorbance data at 255 nm of the reaction product also with p-chloromercuribenzoate.

More frequently used are methods based on the coloured reagents for histochemical work, notably that of Bennett and Yphantis (1948). Flesch and Kun (1950) determined thiols in tissue homogenates by treating with an amyl acetate solution of the 1-(4-chloromercuriphenylazo)-2-naphthol reagent. The red precipitate was centrifuged, dissolved in concentrated hydrochloric or sulphuric acid and estimated colorimetrically. They found proportionality between the amount of precipitate and of thiol. Flesch and Kun also based a photometric procedure on the diminution of reagent amount; this is given in Section 2.14.2 above. Cannefax and Freedman (1955) criticized Flesch and Kun's method, pointing out that although glutathione and cysteine (and thiourea) yielded precipitates in phosphate buffer of pH 7·1, neither BAL nor mercaptoacetate did so and even interfered in the precipitation of glutathione. Rausch and Ritter (1955) studied the method, as did Malinský and Černoch (1959), who recommended methanol and ethanol as solvents. Munakata and Niinami (1963) applied the method to determine cysteine residues in wool and mercaptoacetate present. They extracted the reaction product of the dye with the mercaptoacetate into dimethylformamide (in which the product with cysteine is not soluble) and evaluated the extract colorimetrically. The wool was then hydrolysed with N sodium hydroxide for 20 min at 60°C, liberating the original dye from the derivative with cysteine. It was extracted with dichloroethane and likewise estimated colorimetrically. Bahr (1966) investigated the reaction in quantitative determination of thiols. Sakai (1968) determined micro amounts of thiol in proteins by reaction in acetone–phosphate buffer, pH 7·1 (1 + 1), removing residual traces of the reagent from the centrifuged precipitate with acetone. He then decomposed the product with a trace of hydrochloric acid in acetone and measured the absorbance of the solution at 470 nm. Later (1972) Sakai reviewed the procedure.

Zak et al. (1965) employed 4-(4-hydroxybenzeneazo)-phenylmercuric acetate in glycine-containing buffer of pH 8·6 for determining protein thiol. They removed unused dye on Dowex 1-X8 in the chloride form. After filtering through glass fibre, the filtrate was brought to pH 10·8 and evaluated spectrophotometrically at 434 nm.

Del Vecchio and Argenziano (1946) detected penicillamine through the orange colour with Nessler reagent.

(c) *Fluorescence*. The physical property of fluorescence has been utilized in direct titration methods and in those based on unused (fluorescent) reagent. Thiols are more active as quenchers. The only reference which can be quoted here, even with only meagre justification, is that of Wronski (1968b). He measured the fluorescence changes at 505 nm after reacting thiols with tetrakis (acetoxymercuri) fluorescein in acetate or phosphate buffer. The change was characteristic of each thiol in the 460–500 nm excitation range, serving for identification.

(d) *Melting point*. Wertheim (1929) suggested mercury(II) and lead(II) derivatives for the identification of thiols. The former were prepared by dissolving a few drops of the thiol in 2 ml of ethanol and then adding an excess of 10% mercuric cyanide. Wertheim quotes only ten melting points, and, except for methanethiol, recommends other derivatives. Challenger and Greenwood (1949) identified propanethiol and allyl mercaptan through the melting points of their mercury(II) salts.

(e) *Radioactivity*. Radioactive mercury reagents have been used in a few cases. Stratton and Frieden (1967) reacted with 3-chloromercuri-2-methoxy-1-propylurea containing [203]Hg. After electrophoretic separation, the product was estimated autoradiographically. A [203]Hg-reagent was used also by Cullen and McGuinness (1971) for soluble protein thiol groups and the labelled protein was monitored continuously with a liquid scintillation counter; for example, methylmercuric nitrate was used and separation was performed on Sephadex G-10F, eluting with dilute citric acid of pH 2.

Erwin and Pedersen (1968) used [14]C-*p*-chloromercuribenzoate and removed excess reagent by gel filtration. Aliquots of the eluate of the product were evaporated to dryness and counted with a gas-flow counter. This same labelled reagent was used by Krakow and Goolsby (1971) for protein thiol groups. They retained the reaction product on a nitrocellulose membrane filter of low affinity for the reagent. The radioactivity of the product was then measured and enabled 5–20 µg amounts of thiol to be estimated.

Herber (1962) estimated already radioactive methanethiol (with ^{35}S or ^{14}C) by precipitating with aqueous mercuric cyanide, drying the precipitate to constant weight and radioassaying it. An isotope dilution method was employed by Matsushima (1962) to determine thiophenol in coal tar acid. He added the ^{35}S-isotope, then isolating the compound by precipitation with mercuric chloride.

(f) *Polarography.* Lamprecht and Katzlmeier (1961) estimated the thiol groups in crystallized enzyme by treatment with *p*-chloromercuribenzoate and evaluating the product polarographically through the catalytic wave at -1.6 to -1.7 V in ammoniacal cobalt solutions.

(g) *Atomic absorption spectrum.* Suzuki *et al.* (1969) treated protein-bound thiols with mercuric chloride or *p*-chloromercuribenzoate and subsequently determined the protein-bound mercury by atomic absorption spectrometry. This is a highly sensitive method and enabled amounts of mercury down to 0.1 μg to be estimated; it also demands a sample size which is only ca. 2% of that required for other spectrophotometric procedures.

(h) *Chemical treatment of the product.* Three types of chemical treatment can be distinguished:

(1.) *Reaction of a grouping in the derivative.* Titration with acid under certain conditions is one example falling under this heading, and the equation may be written,

$$RS—Hg—OCOCH_3 + H^+ \rightarrow RSHg + CH_3COOH$$

Alicino (1960) titrated some sulphur compounds, including thiouracil, in acetic acid in the presence of mercuric acetate with perchloric acid and using the indicators crystal violet (purple to green) and quinaldine red (pink to colourless). Mercuric acetate is not titrated under these conditions. Lamathe (1966, 1967) determined thiols in petroleum products in this way. She added 2% mercuric acetate to the sample in toluene and then an excess of 0.1 N perchloric acid in acetic acid. Unused acid was back-titrated by standard sodium carbonate. Lamathe gives the above equation. She also performed a blank titration with mercuric acetate to correct for basic nitrogen compounds in the sample.

Yakovlev and Sokolovskii (1955) used 2-naphthol-3-mercuric chloride as a histochemical reagent for tissue thiols. To the treated specimens they then added tetrazotized 4,4'-diamino-3,3'-dimethoxybenzidine (*p,p'*-di-anisidine) which coupled with the naphthol, yielding a red or violet localizing azo dye.

The mercury in the derivative has been principal point of attack. Thus Spragg (1958) visualized the mercury derivatives of thiols, e.g. cysteine, glutathione, on paper chromatograms by spraying with 0.03% dithizone in 0.1 N ammonium hydroxide to yield salmon pink spots after gentle warming. One or two exceptions were found which did not react, e.g. thioethanol and mercaptoacetic acid; he prepared the derivatives from 0.1% mercuric chloride in 40% ethanol at pH 4 in acetic buffer. White and Wolfe (1962) likewise prepared mercaptides from thiols with *p*-chloro-mercuribenzoate and *p*-chloromercuriphenylsulphonate and visualized them on paper chromatograms using dithizone solution in carbon tetra-chloride (among several reagents tested); it gave gold to peach-coloured spots. Howard and Baldry (1969) separated thiols from thioethers by adding an excess of phenylmercuric acetate; they extracted the thiol derivatives with ethyl acetate, concentrated the extract and submitted it to thin-layer chromatography, visualizing with 0·1% dithizone in acetic acid as orange zones.

Wronski (1966a) determined the equivalent weights of thiols by first titrating with p-chloromercuribenzoate to thiofluorescein indicator in order to estimate the amount of thiol. After acidifying the solution to methyl orange with phosphoric acid, he centrifuged the precipitate, ultimately weighed it and then 'treated it with bromine. Finally the mercury content of the weighed precipitate was determined by titration with mercaptoacetate to thiofluorescein.

(2.) *Regeneration of thiol (or reagent).* Regeneration has been accomplished with acid or with another thiol in an exchange reaction. Examples of the subsequent treatment are given below.

Segal and Starkey (1953) separated methanethiol from mixtures by passing through 4% aqueous mercuric cyanide. As well as estimating the product gravimetrically (mentioned above, p. 86), they regenerated the thiol with hydrochloric acid and passed it into standard iodine reagent; they back-titrated unused iodine with thiosulphate. Koch and Paul (1963) also separated thiols from other sulphur compounds in gases by precipitating with 4% mercuric cyanide, and using an iodometric final stage.

There are several examples of treating the separated mercaptide with the acidic p-N-dimethylaminoaniline/iron(III) reagent to give a blue product, generally evaluated at ca. 500 nm. More details of the colorimetric procedure are on p. 201 under the heading "p-N-Dimethylaminoaniline". Thus Brychta and Rudolf (1956) bubbled natural gas through neutral mercuric cyanide or chloride and then added the reagent to the absorbent solution. Sliwinski and Doty (1958) investigated the methanethiol produced in meat irradiated with γ-rays. The vapours were swept into 5% mercuric acetate at 0–4°C and the colorimetric reagents were then added. Ocker and Rotsch (1959) used Sliwinski and Doty's method to determine thiols in the volatile components of the aroma of bread and baked goods. Moore et al. (1960) used 5% mercuric acetate as absorbent for determining thiols in air, following with dilution and addition of reagent. Ikeya (1964) swept volatile thiols from beer at 75°C into mercuric acetate absorbent. Sainsbury and Maw (1967) applied his method but with a trap of 5% mercuric acetate/2% zinc acetate, cooled in ice. Steffen (1968) estimated volatile thiols (and hydrogen sulphide) in beer by passing carbon dioxide through the acidified sample and into cadmium hydroxide to remove the sulphide, then into mercuric acetate at 0–4°C. Mercuric acetate was used also by Uvarova and Siridchenko (1969) as an absorbent in the determination of thiols, e.g. ethanethiol, in gas, using the same colorimetric procedure.

Recently, Wronski and Kudzin (1974) extracted alkanethiols (C_4, C_6, C_8, C_{12}) from hydrocarbon solvents (hexane, benzene or a 1:1 mixture) with a solution of 0·2 M sodium 2-hydroxymercuri-3-nitrobenzoate in 2 N potassium hydroxide-triethanolamine-2-ethoxyethanol-ethylene glycol (1 + 1 + 2 + 2). The thiols were regenerated by treatment with sodium sulphide in dilute potassium carbonate containing EDTA, at pH 10; they titrated the thiols in the usual way with o-hydroxymercuribenzoate.

Regeneration of thiol, followed by gas chromatography, has been used also. Carson et al. (1960) suggested analysis of volatile thiol mixtures (C_1–C_4) through conversion to mercaptides, then thiol group exchange at 250°C with toluene-3,4-dithiol and submission of the liberated thiols to GLC. Feldstein et al. (1965) commented on some difficulties in the GLC stage, namely that the FID response of thiols does not follow the carbon number rule and that combustion of sulphur compounds yields a decreased detector response. Okita (1967, 1970) absorbed methane- and ethanethiols from air in 5% mercuric cyanide solution and also on glass fibre impregnated with it. After hydrolysis of the mercaptides with hydrochloric acid, he collected the regenerated thiols in cold acetone and separated them by gas chromatography.

Folkard and Joyce (1963) used mercuric cyanide to collect volatile thiols from cultures of marine organisms of reaction mixtures. After decomposing the mercaptides with acid, they converted the thiols to sulphides with 1-chloro,2,4-dinitrobenzene and separated and identified these with paper chromatography.

Solid mercury-containing carriers are used in some separations of thiols. The original thiols are regenerated from them by elution with other, suitable thiols. Miles et al. (1954) prepared a mercury- containing polymer by mixing ethanolic mercuric acetate with a phenol/formaldehyde resin. The yellow precipitate forming within 5 min selectively took up thiols from aqueous solutions. They could be eluted with 2-mercaptoethanol or hydrogen sulphide. Quantitative yields were obtained in tests on cysteine and glutathione. Besch and co-workers (1957) treated Dowex 2-X in the chloride form with p-chloromercuribenzoate and used the resin product to take up thiols from a rat-liver extract. A similarly prepared resin was used by Smith and Rodnight (1957) to remove thiols from blood; they used BAL (2,3-dimercapto-1-propanol) as eluent, excess of which was removed with ethyl acetate. They tested the procedure on mixtures of glutathione, cysteine and cysteinylglycine. McCormack et al. (1960) also used a resin prepared this way for selective removal of thiols from solutions or tissue homogenates by reverse dialysis. Eldjarn and Jellum (1963) anchored a bifunctional mercurial (3,6-bis(acetoxymercurimethyl)dioxan) to thiolated Sephadex through a mercaptide bond with one of the mercury atoms. They used this product to separate thiol- from non-thiol proteins through formation of mercaptides with the remaining available mercury atom. The proteins could then be eluted with cysteine. A further example is Liener's (1967) water-insoluble organomercurial "EMMA", obtained by reacting a linear ethylene-maleic anhydride copolymer with p-(acetoxymercuri)-aniline acetate (ultimately converted to chloride with sodium chloride) in the presence of hexamethylenediamine as cross-linking agent. Over a wide pH range it bound, on average, 1·4 mmol g^{-1} of low molecular thiols. Proteins known to contain thiol groups were also selectively bound and could be recovered virtually by subsequent elution with cysteine.

A last example under this heading concerns regeneration of reagent rather than thiol. Sakai (1968) reacted micro protein thiol amounts with 1-(4-chloromercuriphenylazo)-2-naphthol to yield a precipitate. This product was isolated and decomposed with hydrochloric acid to the original reagent. Sakai estimated it photometrically at 470 nm.

(3.) *Blockage, elimination or inactivation of thiol groups.* This may be in their determination by difference or merely to prevent interference. An early example is the method of Lugg (1930) for the determination of thiols with Folin's tungstophosphate reagent (p. 18). To counter criticism that reducing compounds other than thiols were interfering, Lugg carried out a parallel determination of reducing power of the sample in the presence of mercuric chlorides to prevent thiols participating. This furnished a blank to correct for other reducing matter. This procedure was adopted sometimes in slightly modified form, in subsequent years, e.g. by Kassell and Brand (1938) to determine cysteine.

Trop *et al.* (1968) applied the principle to qualitative analysis in a study of the paper chromatography of thiols, α-hydroxy- and α-ketoacids. They deactivated the thiols with mercuric chloride so as to permit their distinction from the other compound classes in several visualization procedures.

In the direct titration in non-aqueous medium (glacial acetic acid) of many basic compounds containing a thiol group, this group can be blocked by adding mercuric acetate and hence does not interfere as an acidic group. Bayer and Posgay (1961) used this principle in titrations of bases with perchloric acid in acetic acid to gentian violet as indicator.

Mercury(II) has been used also to block thiol groups and prevent their interference in other determinations. These cases do not strictly belong here, but the recent work of Stutzenberger (1973) may be quoted as an illustration. He eliminated the influence of dithiols in the assay of deoxynucleotides with diphenylamine by precipitation with mercuric chloride.

14.3.2. *Analytical Utilization of a Non-Mercury-Containing Product*

Examples of the application of this principle are rare and practically limited to quantitative determinations of hydrochloric acid yielded in the reaction:

$$RSH + HgCl_2 \rightarrow RS{-}HgCl + HCl$$

Sampey and Reid (1932), for example, shook the thiol sample in benzene solution with 1% aqueous mercuric chloride for 3 min and then titrated the aqueous layer with sodium hydroxide to methyl red or orange. Ratkovics and Szepesváry (1958) treated thiols in methanol–benzene (4 + 1) with saturated mercuric chloride and subsequently titrated with methanolic potassium hydroxide to bromocresol green indicator or using the "high frequency" method. Prilezhaeva *et al.* (1962) shook sulphur-containing solid fuel samples for 100 h at room temperature with 12% ethanolic mercuric chloride. They then filtered, complexed the excess of mercury(II) in the filtrate with sodium chloride, and titrated the hydrochloric acid, equivalent to the thiol content, with sodium hydroxide to a methyl red or potentiometric end-point. In a recent example, Borodataya *et al.* (1971) estimated small amounts of ethanethiol in gas by passing through 3% mercuric chloride solution. At pH < 1·8, the acid is quantitatively formed in the reaction and yields a linear curve relating pH and concentration of thiol. Amounts down to 11 µg could be estimated.

Segal and Starkey (1953) tried several methods for determining methanethiol. In one of these, methanethiol was treated with 4% aqueous mercuric cyanide, according to:

$$2CH_3SH + Hg(CN)_2 \rightarrow Hg(SCH_3)_2 + 2CN^-$$

They filtered the mercaptide and determined the amount of cyanide by adding an excess of standard mercuric nitrate and back-titrating the unused reagent with thiocyanate to iron(III) indicator. Feigl *et al.* (1969) give a spot test for thiols, based on reaction with mercuric cyanide and dilute sulphuric acid. On warming to 80–90°C, hydrogen cyanide is evolved and is detected at the mouth of the reaction test tube through the blue yielded with a test paper moistened with copper ethyl acetoacetate/tetrabase solution in chloroform.

Another example of determination *via* a non-mercury-containing reaction product is given by Ozawa and Egashira (1962). They used the mercury complex of Ruhemann's purple, the end-product of reaction of ninhydrin with primary amines, 2-(1,3-dixoindan-2-ylimino)-1,3-indandione). This they did by adding a slight excess of mercuric chloride to an aqueous solution of the sodium salt of the dye. One reaction with thiols (which is quantitative at pH 4·7), Ruhemann's purple is set free and can be determined spectrophotometrically at 570 nm. They were able to estimate 5–50 µmol amounts of thiol.

D*

15. MOLYBDENUM(VI)

(See Chapter 1, "Oxidation" (Section 30, "Molybdate").

16. NICKEL(II)

The reaction of nickel(II) with thiols, which may be more complex than a straightforward formation of mercaptide, has been studied e.g. kinetically by Davies *et al.* (1968) in terms of the reaction with L-cysteine. Analytical applications have been few; some are given below:

Hartmann (1930) distinguished glutathione from cysteine by the fact that the former but not the latter forms a compound with nickel salts, e.g. sulphate, in carbon monoxide (the reverse is found with iron salts).

Shemyakin and Berdnikov (1965) proposed several tests for identifying methyl thiouracil, based on colour reactions with metal ions; e.g. nickel(II) nitrate yields a light green precipitate.

Thiol acids react with nickel(II) to yield stable complexes with absorption maximum in the ultraviolet. Sanderson *et al.* (1972) recently studied the kinetics of the exchange reaction between nickel(II) sulphate and the two thiol acids, cysteine and thiolactic acid, at pH 9 and hence in the presence of citrate to maintain nickel in solution. From absorbance measurements at 275 nm on the reaction mixture they were able to analyse mixtures of these two compounds.

Bruening and Bruening (1972) observed that thiols, sulphides and thiophene reversibly form complexes with transition metal compounds, e.g. nickel. They used nickel cyclohexylbutyrate, among other compounds, as a stationary phase in gas chromatography; the best results were obtained by adding 0·05–0·5% to 2·5 or 5% SF 96 silicone oil on Chromosorb G columns containing traces of allylthiol. They found improved separations of these sulphur compounds.

Two very recent direct titrations using nickel(II) can be quoted:

Bahadur and Sen (1973) titrated L-cysteine (also L-histidine), in the presence of other amino acids, at pH 10 (bicarbonate, carbonate) with nickel(II) sulphate or chloride. They used catechol violet as indicator, taking greenish blue as the end-point colour. Ranganayaki and Srivastava (1973) published a closely similar method but with 1-(2-pyridylazo)-2-naphthol(PAN) as indicator, changing from purple-violet to yellow.

17. PALLADIUM(II)

Palladium salts yield a yellowish colour with numerous sulphur compounds, including thiols. These probably react at least initially according to the equation,

$$RSH + Pd^{2+} \rightarrow RS - \overset{+}{Pd}(II) + H^+$$

Palladium mercaptides, $Pd(SR)_2$ and even oxidation products may be yielded too.

The discovery of the colour reaction between palladium(II) and thiols dates back to 1914 (Mylius and Mazzucchelli), and it has found limited analytical use in detection and also quantitative work, e.g. Reith (1934) detected methanethiol in air through the orange-brown coloration or precipitate given with a palladium(II) chloride/dilute sulphuric acid reagent. On dissolving the precipitate in 25% hydrochloric acid, the smell of the thiol was noticeable. Reith stated that semiquantitative thiol determinations would be possible using colour standards. Toennies and Kolb (1951) used an acid palladium chloride/potassium iodide/acetone reagent to visualize reducing compounds on PCs, with which it gives a bleached spot. They quote cystine (misprint evidently for cysteine). In a recent variation of the test, Frei *et al.* (1973) used a mixed calcein/palladium chloride spray reagent for detecting 10–100 ng amounts of thiols and other sulphur compounds on thin-layer chromatograms. They prepared the reagent from equal volumes of 10^{-3} M calcein and 0·005 M palladium chloride in 0·1 M hydrochloric acid, then adjusting the pH to 7·2 with phosphate and subsequently diluting with water and acetone. The palladium evidently reacts with the thiol and liberates the calcein, which fluoresces.

Some quantitative colorimetric procedures may be mentioned: Peurifoy *et al.* (1964) detected thiol odorant in liquefied petroleum gas through the yellow colour obtained by passing the gas through a tube containing silica gel impregnated with palladium(II) chloride, hydrochloric and sulphuric acids. The length of the yellow zone gave a rough value of the thiol concentration. Peurifoy and O'Neal (1965) later patented this device. Åkerfeldt and Lövgren (1964) based a spectrophotometric method for several sulphur-containing compound classes (disulphides, sulphides, sulphinic acids, thiols) on treatment of up to 1 μmol of sample with a reagent containing 20 μmols of ammonium tetrachloropalladinate(II) per ml of M hydrochloric acid. They measured the absorbance between 350 and 415 nm after 5 min in the cause of thiols which react faster than the other sulphur compounds. Kvapil (1965) determined 2-mercapto-4-hydroxy-5-methoxypyrimidine by neutralizing a 100–200 mg sample in N sodium hydrozide to phenolphthalein with sulphuric acid and adding sodium acetate and 1 ml of 0·5% palladium chloride. He evaluated the red-orange colour at 450 nm. The product, a 1:1 compound, is stable at pH 5–8.

A titration method for thiols was developed by Haglund and Lindgren (1965). They add 10 to 30 μmol of sample to a solution of *p*-nitroso-N-dimethylaniline indicator in ethanol/hydrochloric acid and removed oxygen with a current of nitrogen. They titrated this with an ammonium tetrachloropalladinate(II) reagent, taking as end-point the formation of a red complex between reagent and the indicator; this was best done by photometric titration at 490 nm. Titration was possible in the presence of the slower reacting sulphides. Their examples included cysteine and cysteamine.

A very recent method for glutathione in erythrocytes makes use of the blue-purple complex of palladium(II) with chlorpromazine (565 nm absorption maximum). Glutathione displaces the chlorpromazine from the complex and the diminution in absorbance is proportional to thiol amount (Lee and Tan, 1974).

18. PLATINUM(IV)

See Chapter 1, "Oxidation" (Section 44).

19. SILVER(I)

Silver(I) is the most used reagent for determining thiols. The initial reaction is formation of mercaptide, poorly water-soluble:

$$RSH + Ag^+ \rightarrow RSAg + H^+$$

The silver mercaptides have a tendency to carry down additional silver reagent (nitrate), which can falsify results based on silver consumption or the silver-containing product. Nevertheless, the silver method has been extensively employed to determine thiols, notably in petroleum products and biological material.

Procedures are divided into the usual three groups: direct titration; use of excess reagent and determination of the unused; and utilization of a reaction product. As with mercury(II), there are many more direct titration procedures than others.

There are few examples of the use of silver(I) in qualitative tests or identification.

19.1. Direct titration

Nearly 200 references to direct titration of thiols are given in highly summarized form in Table VI below. Almost all are potentiometric or amperometric titrations using a burette silver nitrate reagent. About ten are coulometric procedures, the earliest published already in 1954; it is surprising that this elegant principle has not been used more. The other end-point indications are two conductometric, one radiometric, one "high frequency" and four colorimetric, altogether less than 5% of the total. The pH of the reaction medium generally lies between ca. 5 (acetate buffer) and 9 (ammonium buffer). Some authors titrated in nitrogen to prevent atmospheric oxidation.

19.2. Determination of unused reagent

About 40 references are given to the use of excess silver(I) reagent in analytical problems similar to those tackled by direct titration, namely, analysis of biological samples, petroleum and allied materials; lower alkanethiols, thiophenols and thiol amino acids have been the customary test compounds here also.

On account of the variety of materials investigated, it is difficult to say anything general about reaction conditions. Reaction times have nearly always been short. The reagent has been silver nitrate, or electrolytically generated silver ion in one case. Unused reagent has been determined titrimetrically or by other methods. This offers a subdivision of the information:

19.2.1. *Back-Titration of Unused Reagent*

The standard procedure is evidently with thiocyanate ion to ferric ion as indicator. The halides (chloride, bromide and iodide) have also found use and also the standard thiols dodecanethiol and glutathione; instrumental (potentiometric, amperometric) end-point indication was employed with these. Two radiometric titrations of ^{110}Ag are quoted, one with thiocyanate, one with ammonium rhodanine. In a novel example, unused silver was reacted with tetracyanonickelate, $Ni(CN)_4^{2-}$ to yield an equivalent of Ni^{2-} which was then titrated with EDTA to murexide.

Examples are shown in Table VII.

19.2.2. *Other Determinations of Unused Reagent*

This principle has been only rarely applied. Andreen *et al.* (1963) determined thiols (and hydrogen sulphide) as odorants in natural gas through the conductivity change of a silver nitrate solution through which the sample was passed. Selyuzhitskii *et al.* (1965) added thiol samples to excess silver nitrate in a buffer of 0·01 M ammonium hydroxide and 0·8 M ammonium nitrate and related thiol amount to the diminution in diffusion current of the reagent. Ladenson and Purdy (1973) determined protein thiol (also glutathione) in a similar way; however, they generated excess silver ion electrolytically and then measured the fall in amperometric current on adding the sample. Gupta and Boltz (1971) estimated milligram amounts of several compounds, including 2-mercaptobenzothiazole, by precipitating with excess silver, heating for 1 min at 50–60°C, centrifuging and determining the unused reagent *via* atomic absorption spectrometry.

Table VI. Direct titrations with silver(I) (with silver nitrate unless otherwise given)

Sample	Conditions	End-point indication	References
C_2–C_8 thiols	Sample in ethanol, reagent in isopropanol	Potentiometric	Tamele and Ryland (1936)
Mercaptobenzothiazole	In 80% acetone	Potentiometric	Spacu (1939)
C_1–C_8 alkanethiols	In N alkali, 0·05 N in NH_4OH	Potentiometric	Tamele et al. (1941)
Ethanethiol	In acetate buffer/ethanol or ammoniacal medium; test of a new reference electrode	Potentiometric	Lykken and Tuemmler (1942)
Petroleum products	(modification of Tamele and Ryland's procedure, 1936)	Potentiometric	Davies and Armstrong (1943)
Dodecanethiol (in ethanol)	NH_4OH/NH_4NO_3	Amperometric	Kolthoff and Harris (1946)
Primary thiols	In absolute ethanol, + conc NH_4OH	Amperometric	Laitinen et al. (1946)
Thiols in presence of disulphides, also from them by reduction	Ammoniacal solution containing alcohol from the reduction	Amperometric	Kolthoff et al. (1946)
Gasoline	In presence of alkali (e.g. Na cresylate) and amine (e.g. pyridine) + solubilizer such as C_4 or C_5 alcohol	Nitroprusside, to disappearance of purple	Mapstone (1948, 1952)
Protein, denatured with guanidine; also cysteine, glutathione, mercaptoacetate	Sample in 95% ethanol, + NH_4OH + NH_4NO_3	Amperometric	Benesch and Benesch (1948)
Low result with Kolthoff/Harris method (1946) attributed to oxidation of the thiols			Frank et al. (1948)
Dodecanethiol, from determination of acrylonitrile	In ethanol or isopropanol, + NH_4OH	Amperometric (according to Kolthoff and Harris)	Beesing et al. (1949)

Material / test	Conditions, medium	Method	Reference
Cysteine, also from cystine	Ammoniacal medium	Amperometric (according to Kolthoff and Harris)	Hata (1949, 1951)
Bovine serum albumin; tested on glutathione (µg amounts)	NH_4OH/NH_4Cl	Amperometric	Rosenberg et al. (1950)
Blood and tissues (glutathione)	In $\leqslant 80\%$ ethanol + NH_4OH/NH_4NO_3 + EDTA to complex metals	Amperometric	Benesch and Benesch (1950)
Serum protein	Modification of Benesch and Benesch (1948)		Weissman et al. (1950)
	N_2 atmosphere and oxygen-free reagents	Amperometric	Strafford et al. (1950)
Cysteine: cystine after reaction with SO_3^{2-} or Na/Hg	Ammoniacal buffer	Amperometric	Kolthoff and Stricks (1950)
Mercaptobenzothiazole	In ethanol + NH_4OH + NH_4NO_3 + KNO_3	Amperometric	Liberti and Cervone (1950)
Serum	+ NH_4OH + NH_4NO_3; in 97% ethanol (based on Benesch and Benesch, 1948)	Amperometric	Schoenbach et al. (1950)
Proteins; tested on cysteine and glutathione	Modification of Benesch and Benesch (1948)		Iwaki (1951)
Mercaptobenzothiazole	10^{-2}–10^{-4} M range	Amperometric	Liberti (1951)
Kraft pulp digester blow gas	Essentially method of Lykken and Tuemmler (1942)		Felicetta et al. (1952, 1953)
Disulphides, after reduction with Zn	Method of Kolthoff and Harris (1946)		Earle (1953)
Mercaptoacetic acid	Tested several methods, including that of Kolthoff and Harris (1946)		Walker (1953)
Dodecanethiol in acrylate polymers	In acetone, + NH_4OH + NH_4NO_3; method of Kolthoff and Harris (1946)		Haslam et al. (1953)

Table VI (cont.)

Sample	Conditions	End-point indication	References
Dodecanethiol from determination of acrylonitrile	In isopropanol + 1% acetic acid	Potentiometric	Janz and Duncan (1953)
Milk after heat treatment	Method of Kolthoff and Harris (1946)		Zweig and Block (1953)
Urine	New, faster-rotating Pt electrode for amperometric titration		Herbert and Denson (1954)
	Iwaki's method (1951) claimed to give high results through reaction of other compounds, e.g. o-aminobenzoic acid		Yoshida (1954)
Whole blood	In ethanol + $(NH_4)_2CO_3$; nitrogen atmosphere	Amperometric	Del Pianto and Silvestroni (1954)
Petroleum stocks	In ethanol/benzene/NH_4OH/NH_4NO_3; coulometric Ag	Amperometric	Leisey (1954)
Protein disulphide, treated with SO_3^{2-}	+ NH_4OH + NH_4NO_3 + EDTA in 90% ethanol	Amperometric	Carter (1954)
Petroleum fractions	Method of Tamele and Ryland (1936)	Amperometric	McCoy and Weiss (1954)
Gluten	pH 4·4 to 8·7, best 4 to 5; dispersion in acetic acid, + KCl + urea	Amperometric	Matsumoto and Shimada (1955)
E.g. cysteine, homocysteine, mercaptoacetic acid	pH 1·8 to 2·2 or 6·5 to 7 (phosphate buffers)	Potentiometric	Cecil and McPhee (1955)
Crystalline proteins; tested on t-dodecanethiol, glutathione	Tris/HNO_3/KCl, pH 7·4	Amperometric	Benesch et al. (1955)
Hydrocarbons	In acetone/NH_4OH/NH_4NO_3	Amperometric	Grimes et al. (1955)
Gluten	Dispersion in acetic acid +40% urea + NH_4OH/NH_4NO_3	Amperometric	Matsumoto (1955)

Substance	Medium/conditions	Method	Reference
Haemoglobin	Ammoniacal buffer	Amperometric	Ingram (1955)
Protein	Ammoniacal buffer	Amperometric	Heide (1955)
Cysteine, also cystine + SO_3^{2-}	NH_4OH/NH_4NO_3	Amperometric (special rotating and vibrating electrodes)	Ray Sarkar and Sivaraman (1956)
Methanethiol standard	Ammoniacal solution	Potentiometric	Brychta and Rudolf (1956)
Blood coagulant factors, e.g. thrombin, prothrombin, fibrinogen	Tris buffer, pH 7·4	Amperometric	Carter (1956)
Blood	+ satd $(NH_4)_2CO_3$ + ethanol	Amperometric	Del Pianto and Sabatini (1956)
Mercaptobenzothiazole	Aqueous alcoholic or aqueous acetone solution	Conductometric	Lorenz and Echte (1956)
Wheat flour	Tris buffer + urea	Amperometric	Kong et al. (1957)
Gasoline	Acetate buffer/methanol/benzene to remove interference of free S	Potentiometric	Karchmer (1957)
Sickle-cell haemoglobin	NH_4OH/NH_4NO_3	Amperometric	Ingram (1957)
Normal and sickle-cell haemoglobin	Kolthoff/Harris method (1946) at 0° and 38°C	Amperometric	Murayama (1957, 1958)
E.g. petroleum distillates	Separated by GLC after extraction with alkali; issuing gases passed into cell containing 75% ethanol; coulometric Ag^+		Liberti and Cartoni (1957); Liberti et al. (1957); Liberti (1957)
Bovine serum albumin	In 4 M guanidine·HCl, + ammoniacal buffer (pH 9) or phosphate buffer (pH 7, poorer end-points); nitrogen atmosphere	Amperometric	Kolthoff et al. (1957)

Table VI (cont.)

Sample	Conditions	End-point indication	References
Combustion gases	Absorbed in methanol at $-75°C$	Potentiometric	Colombo et al. (1957)
Plant tissues	Ammoniacal buffer; nitrogen atmosphere	Amperometric	Nagai et al. (1957)
From reduction of disulphides with $NaBH_4$	In presence of NH_4OH/NH_4NO_3	Potentiometric	Stahl and Siggia (1957)
E.g. t-dodecanethiol, gluta-thione, cysteine, mercaptoacetic acid	(High results found with thiol groups having neighbouring amine or carboxylic acid groups)	Amperometric	Sluyterman (1957)
Mercaptobenzothiazole	Ammoniacal or alcoholic solution + KNO_3	Amperometric	Čihalík and Kudrnovská-Pavlíková (1957)
	Tris, ammonium buffers (study of effect of pH; optimum pH varies with the thiol)	Amperometric	Burton (1958)
Tissue, micro- and ultramicro	NH_4OH/NH_4NO_3	Amperometric	Yakovlev and Torchinskii (1958)
Hydrocarbon mixtures	Compared three methods, including potentiometric and amperometric titration with Ag^+		Romováček and Bednář (1958)
Haemoglobin	NH_4OH/NH_4NO_3 buffer	Amperometric, with Hg-covered rotating Pt electrode; also with other electrodes	Hommes and Huisman (1958)
Blood, tissue	Tris buffer, pH 7·4	Amperometric	Bhattacharya (1958)
From reduction of petroleum disulphides	Eessentially method of Tamele and Ryland (1936)		Hubbard et al. (1958)

Sample/Source	Conditions	Method	Reference
Petroleum naphtha	Acidic or basic solvents (methanol–benzene–acetate buffer, or methanol–benzene–ammonium buffer); nitrogen atmosphere	Potentiometric	Karchmer (1958)
From reduction of disulphides	In methanol/benzene, + sodium acetate	Potentiometric	Karchmer and Walker (1958)
Flour gluten	At pH 8 or 10, e.g. in tris buffer	Amperometric	Bloksma (1959)
From disulphides in natural rubber, after reduction with $LiAlH_4$	Reduction medium, + NH_4NO_3/ethanol	Amperometric	Studebaker and Nabers (1959)
From disulphides in beer, reduced electrolytically	Reduction medium, + tris + HNO_3 + KCl + gelatine	Amperometric	Brenner et al. (1959)
Grain proteolysates	Medium of NH_4OH/NH_4NO_3	Amperometric	Kotlyar (1959)
Flour, dough extracts	Tris buffer; nitrogen atmosphere	Amperometric	Matsumoto and Hlynka (1959)
E.g. cysteine; 2-mercapto-ethanesulphonate; 2,3-di-mercaptopropane sulphonate	In NH_4OH/NH_4NO_3 + $Ca(NO_3)_2$ to hinder reaction of Ag^+ with reaction products which otherwise gives high results	Amperometric	Belitser and Lobachevskaya (1959)
Meat undergoing thermal denaturation	Compared ferricyanide method with amperometric Ag^+ titration (latter gave 35% higher values!)		Ullmann (1959)
Hydrocarbons	In ammoniacal isopropanol	Ammonium dithizonate → deep red	Kunkel et al. (1959)
Non-protein and protein thiols in rat blood and tissues	Tris buffer, pH 7·4	Amperometric	Bhattacharya (1959)

Table VI (cont.)

Sample	Conditions	End-point indication	References
Biological thiols, e.g. glutathione, cysteine, dodecanethiol; bovine serum albumin	Tris buffer, pH 7·4	Amperometric	Harada et al. (1959)
Protein, also disulphides after reaction with SO_3^{2-}	In 8 M urea, + Na_2CO_3	Amperometric (stationary Pt electrode, solution stirred)	Carter (1959)
Proteins, non-proteins	Tris buffer, pH 7·4 + KCl + gelatin	Amperometric	Singh (1959–60)
Thiol/hydrogen sulphide mixtures	Aqueous samples in N NaOH, 0·05 N in NH_4OH; petroleum samples in H_2O–ethanol (1 + 39), + Na acetate	Potentiometric; curves indicate presence of H_2S and enable both to be estimated	Tamele et al. (1960)
Mercaptoacids	Found high results, due to reaction of the –COOH group		Carr and Bit-Alkhas (1960)
Liquid fuels	Acetone solution, + NH_4OH/NH_4NO_3 in ethanol	Amperometric	Romováček and Holub (1960)
E.g. glutathione, cysteine, mercaptoethanol (study of reactions with Ag^+ and p-chloromercuribenzoate)	Tris/KCl/HNO_3, of pH 7·5	Amperometric	Hoch and Vallee (1960)
Hydrocarbons	Benzene/95% ethanol/NH_4OH/NH_4NO_3; coulometric, automatic procedure	Amperometric	Leisey (1960, 1961)

Sample	Notes	Technique	Reference
E.g. cysteine, homocysteine methyl ester, 2-mercaptoethylamine, penicillamine, toluenethiol, mercaptoacetate	Examined influence of thiol structure; high values with some believed due to reaction of Ag^+ with $-SRNH_2$ or $-SRNH_3^+$	Amperometric	Aibara et al. (1960)
Thiols among other compound classes	Titrated with $^{110}Ag^+$	Radiometric	Bebesel and Sirbu (1960)
Liquid hydrocarbons	In ethanol, + Na acetate	Potentiometric	Brejcha and Sima (1960)
Mercaptobenzothiazole in technical preparations	In acetone, reagent in ethanol	Potentiometric (method of Tamele and Ryland, 1936)	Czerwiński and Vieweger (1960)
Glutathione, mercaptoethanol, mercaptoacetate (cysteine gave poor results)	NH_4OH/NH_4NO_3; nitrogen atmosphere	Amperometric	Kolthoff and Eisenstädter (1961b)
Rabbit serum; rat diaphragm; also disulphides + SO_3^{2-}	NH_4OH/NH_4NO_3	Amperometric	Schwartz et al. (1961)
Tested on n-hexanethiol	Acetone/25% aqueous NH_4OH; reagent in isopropanol; in acetone/acetic acid then:	"High frequency"; potentiometric	van Meurs (1961)
Ethanethiol in odorized natural gas	Passed through ethanol	Amperometric	Sadykhov and Korobtsova (1961)
Aqueous extracts from irradiated cottonseed proteins	NH_4OH/NH_4NO_3	Amperometric	Ibragimov et al. (1961)
Sulphate turpentine (methanethiol)	In ethanol–benzene (3 + 1) in presence of ammonia	Potentiometric	Prokshin (1962)
Milk proteins; cysteine	Na acetate, pH 10·25	Amperometric	Yoshino et al. (1962)
Blood	NH_4OH/NH_4NO_3	Amperometric	Sokolovskii (1962)

Table VI (cont.)

Sample	Conditions	End-point indication	References
Milk and dairy products; tested on cysteine, mercapto-acetate (both good) and glutathione (poor; oxidation?)	NH_4OH/NH_4NO_3, pH 9·8	Amperometric	Kiermeier and Hamed (1962)
5-Mercapto-1-phenyltetrazole		Amperometric	Stevancevic (1962)
6-Mercaptopurine in presence of other purine bases	In 0·2 N H_2SO_4 (other purines react only in alkaline solution)	Potentiometric	Linek et al. (1963)
E.g. glutathione, cysteine, 2-mercaptoethylamine (cysteamine)	Tris buffer, pH 7·4 + KCl; reverse titration to accelerate and dispel oxidation dangers	Amperometric	Börresen (1963)
Cysteine + glutathione; latter alone after reaction of former with glyoxalate		Amperometric	Gadal (1963)
E.g. glutathione (cysteine less good)	Tris buffer, pH 7·4	Amperometric	Nedić and Berkeš (1963)
Thiols in petroleum products after separation as dinitro-phenylsulphides and re-generation with iPrOH (see p. 134)	Ammoniacal solution		Dahmen et al. (1963)
Sour natural gases, after GLC separation	Passed into acetic acid–water (7 + 3); coulometric titration	Potentiometric	Fredericks and Harlow (1964)
Hydrocarbons	In methanol/isopropanol/benzene/water/Na acetate; nitrogen atmosphere	Potentiometric	Hammerich and Gonder-mann (1964)

Sample	Conditions/method	Detection	Reference
E.g. glutathione	Modified Kolthoff/Harris (1946) method		Shol'ts (1964)
Thiols from disulphides by reduction with Bu_3P	10% aqueous methanol, + a little $HClO_4$	Amperometric	Humphrey and Potter (1965)
Bovine serum albumin	Critical study of amperometric titration with Ag^+ (also with $Hg(II)$ reagents)		Kolthoff et al. (1965)
Gasoline boiling range stocks after GLC separation	Automatic coulometric titration	Potentiometric	Brand and Keyworth (1965)
Protein disulphides, treated with SO_3^{2-} + 8 M urea	Modified method of Carter (1959)	Amperometric	Okulov (1965)
Reduced glutathione in red blood cells	Tris buffer, pH 7·3, 13 mM in KCl	Amperometric	Grimes (1965)
Alkane- and arenethiols, β-mercaptopropionic acid	Aqueous or alcoholic sample + ethanol + a little $HClO_4$	p-N-dimethylamino-benzylidene-rhodanine (yellow → red or purple)	Wronski (1965b)
Blood serum	NH_4OH/NH_4NO_3 (essentially Kolthoff/Harris, 1946)	Amperometric	Murzakaev (1966)
Cysteine, glutathione; proteins	Imidazole buffer, pH 6 to 7	Amperometric	Sluyterman (1966)
Thiol used in emulsion polymerization of chloroprene	Toluene acetone (3 + 4) + sample in alcohol + NH_4OH/NH_4NO_3 in water	Amperometric	Melkonyan et al. (1966)
Aqueous effluents from oil-processing industries	In N NaOH and 0·05 N NH_4OH	Potentiometric (2 inflections, first S^{2-}, then RS^-)	Vajta et al. (1966)
Milk	Tris buffer, pH 7; at 35°C (based on Cecil and McPhee, 1955)	Potentiometric	Kiermeier and Petz (1966)

Table VI (cont.)

Sample	Conditions	End-point indication	References
	Tris buffer, near neutral pH; (modification of Rosenberg et al. (1950))	Amperometric	Rothfus (1966)
n-Dodecanethiol; suspension of denatured fungal cells	Ethanolic, ammoniacal solution	Amperometric (with stationary electrode in stirred soln.)	Richmond and Somers (1966)
Gases	Coulometric, in ammoniacal medium		Dworak and Davis (1967)
Fuels	(Compared with Cu^{2+} titration which was less satisfactory)	Potentiometric	Stekhun et al. (1967) also Bol'shakov et al. (1967)
Yeast cells	(Compared with photometric method with p-chloromercuribenzoate)	Amperometric	Wei (1967)
Wheat gluten protein	Tris buffer, pH 7·5, + KNO_3 + EDTA + 8 M urea	Amperometric	Okada and Yonezawa (1967)
Cereals	Ground grain + HCl, HNO_3, Tris, EDTA in succession; at room temp.	Amperometric	Janicki and Wierzbowski (1967)
Serum	In ammoniacal buffer, NH_4OH/NH_4NO_3	Amperometric in rapidly stirred soln. (spiral Pt electrode)	Shmakotina and Uksusnikov (1968)
Thiols from study of organo-metallic compounds cleaving disulphides	Isopropanol solution, + NH_4OH/NH_4NO_3; nitrogen atmosphere	Amperometric	Uraneck et al. (1968)
Thioamino acids (e.g. cysteine, glutathione, N-acetylcysteine) in presence of ascorbic acid	Citrate/HNO_3 buffer, pH 2·3	Potentiometric at constant current of 0·7 mV	Santi and Peillon (1968)

Material	Conditions	Method	Reference
Blood	Tris and phosphate buffer mixtures	Amperometric	Bokarev et al. (1968)
	Compared several methods, including Sokolovskii's (1962)		Korochanskaya (1968)
Crude oil	Propanol/benzene/ammoniacal medium; also in heptane/propanol/ethanol $(5 + 4 + 1)$, $+ NH_4OH/NH_4NO_3$ medium	Potentiometric amperometric	Obolentsev et al. (1968)
Motor fuels	Ammoniacal solution	Potentiometric	Podlipskii et al. (1969)
Cracking gasoline; also from disulphides by reduction with Zn	Ammoniacal medium; 5–20°C	Potentiometric	Gulyaeva and Blokh (1969)
Certain proteins	At pH 7·4 (also with Hg(II) reagent)	Amperometric	Ambrosino et al. (1969)
E.g. cysteine, albumin	At pH 9·5; inert atmosphere (compared with Hg(II) procedures)	Potentiometric	Toribara and Koval (1970)
Animal tissue and meat products (tested influence of many salts); glutathione as standard	Tris buffer, pH 7·4; low values in presence of $CuSO_4$ improved by adding EDTA	Amperometric	Hofmann (1970)
E.g. lower alkanethiols	Semi-automatic coulometric, based on Kolthoff/Harris, 1946; ethanolic solutions	Amperometric	Devay and Garai (1970)
Coarse-grained wheat	Supernatant from treatment with 4 M urea and centrifuging, + tris buffer, pH 7; nitrogen atmosphere	Potentiometric	Kuehbauch and Wuensch (1971)
E.g. glutathione, cysteine, mercaptopropionic acid	Tris/HNO_3 buffer, pH 7	Amperometric	Hofmann (1971a)
E.g. glutathione, 2-mercaptoethanol, cysteine	(Little influence of pH between 2·5 and 9; buffers giving complexes with Ag^+ best avoided)	Potentiometric with ion-specific electrode	Gruen and Harrap (1971)

Table VI (cont.)

Sample	Conditions	End-point indication	References
Waste waters	Alkaline, ammoniacal medium	Potentiometric	Stankevich (1971)
Straight-run gasoline	Ammoniacal solution	Potentiometric	Shcherbina et al. (1971)
Thiol impurities in dimethyl sulphoxide	Ammoniacal medium	Potentiometric	Khazova and Bogomolov (1971)
Biological substrates	NH_4OH/NH_4NO_3	Amperometric	Selyuzhitskii et al. (1971)
Blood glutathione	Ammoniacal medium: coulometric titration (study of conditions)	Biamperometric	Ladenson and Purdy (1971b)
Thiol products from protein disulphide + $NaBH_4$	Solution from reduction (containing urea, EDTA) + acetone + HNO_3	Potentiometric with ion-selective electrode	Harrap and Gruen (1971)
	Semi-automatic, coulometric titration	Potentiometric	Devay et al. (1972)
E.g. thiophenol, petroleum products	Tested apparatus of Devay et al. (1972)		Garai et al. (1972)
	At pH 7·4 (tris + HNO_3) (compared with five other methods); gave correct results with only few compounds (e.g. glutathione, o-aminothiophenol)	Amperometric	Wenck et al. (1972)
Tested on dodecanethiol	Aqueous-alcoholic ammoniacal solution; coulometric titration	Biamperometric	Švajgl and Holle (1972)
Milk	Tris buffer, pH 7·4	Amperometric	Hofmann (1972)
Thiols, also sulphides, disulphides	Coulometric titration (review of methods)	Amperometric (with rotating Cu, Ag electrodes)	Sement et al. (1972)

Thiol/air mixtures	Bubbled through aqueous alcoholic solution containing Na acetate	Potentiometric; time of bubbling (at constant rate) to potential jump	Kirchner (1972)
Trialkoxysilanethiols, $(RO)_3SiSH$ ($R = C_3$ to C_5, cyclohexyl); also amine salts of the thiols	In benzene–isopropanol (1 + 3)	Potentiometric	Wojnowski and Kwiatkowska-Sienkiewicz (1973)
Protein; tested on glutathione; also disulphides in presence of SO_3^{2-}	Tris, ammonium or imidazole buffer	Biamperometric	Ladenson and Purdy (1973)
Thiols, hydrogen sulphide	Aqueous solution at high pH (1–2) or in ethanol–benzene at pH 108; reverse titration best with volatile compounds (H_2S, CH_3SH, C_2H_5SH)		Peter and Rosset (1974)

Table VII. Back titration of unused silver(I)

Sample and Conditions	Back Titrant	References
Naphthas; tested on C_2–C_4 alkanethiols	SCN^- to red with Fe(III) alum	Borgstrom and Reid (1929)
2-Mercaptobenzothiazole; in NH_4OH	SCN^- or Cl^- to dichlorofluorescein	Ushakov and Galanov (1934)
Hydrocarbon solvents; tested on C_3–C_5 alkanethiols, thiophenols, benzyl mercaptan; sample + methanol as solubilizer; well shaken	SCN^- to Fe(III) alum	Malisoff and Anding, jr (1935)
Analysis scheme for mixtures of sulphur compounds; 1 min shaking	SCN^- to Fe(III)	Bell and Agruss (1941)

Table VII (*cont.*)

Sample and Conditions	Back Titrant	References
Refinery caustic scrubbing solutions	SCN^-	Koons (1941)
Natural gas	+ excess NaCl and back titrated with Ag^+	(Anonymous) (1945)
Gasoline	Method of Borgstrom and Reid (1929)	Mapstone (1946a)
Cresylic acid; sample in kerosene-amyl alcohol (5 + 1)	Method of Borgstrom and Reid (1929)	Mapstone (1946b)
E.g. glutathione; used excess to give the soluble complex $AgSR.AgNO_3$	KBr, potentiometric at pH 2	Cecil (1950)
Methylthiouracil; in NH_4OH, 2·5 h heated at 100°, then filtered from Ag_2S	SCN^- (filtrate)	Wojahn (1951)
Methylthiouracil, among many sulphur compounds; in NH_4OH solution; filtered	SCN^- (filtrate acidified with HNO_3)	Middeldorf (1951)
Methanethiol	SCN^- to Fe(III)	Segal and Starkey (1953)
Gaseous thiols; in acetone/NH_4OH/NH_4NO_3	Dodecanethiol potentiometric	Grimes *et al.* (1955)
Gases; passed into reagent	SCN^-	Hammar (1955)
6-Mercaptopurine; in ammoniacal solution	SCN^- to Fe(III)	Blažek and Stejskal (1956)
Beer; 30 min reaction	Glutathione in extrapolation procedure to $t = 0$	Brenner *et al.* (1957)
E.g. in benzene; in presence of HNO_3; shaken 5 min, then filtered	SCN^- (filtrate) to Fe(III)	Constantinescu and Constantinescu (1957)
Hydrocarbon mixtures; nitrogen atmosphere; compared 3 methods	Dodecanethiol/isopropanol amperometric	Romováček and Bednář (1958)

6-(4-Carboxybutylthio)-purine; in dilute HNO_3; filtered	SCN^- (filtrate)	Jančík et al. (1960)
Thiophenol in tar acids; + aqueous methanol + excess $^{110}AgNO_3$	Radiometrically; with ammonium rhodanine or SCN^-	Matsushima (1962)
6-Mercaptopurine; in dil. NH_4OH, several min; filtered	Filtrate + $K_2Ni(CN)_4$, then liberated Ni^{2+} titrated with EDTA to murexide	Hennart (1962)
Mercaptobenzothiazole in rubber; ethanolic extract	NaCl	Gordon et al. (1962)
Homogenates from plant tissue	Glutathione amperometric	Laurinavicius (1963)
Thiols in sulphur-containing solid fuels; shaken 100 h	KCl amperometric at 0°C	Prilezhaeva et al. (1963)
Microdetermination on plant extracts; + methanol + NH_4OH/NH_4NO_3	+ equivalent of glutathione and titrated with Ag^+	Spanyar et al. (1964)
Myofibrils and muscle tissue; suspended in pH 7·4, tris, buffer; 1 h reaction	+ excess glutathione, then back-titrated with Ag^+	Hamm and Hofmann (1966)
Slowly reacting protein thiols	+ excess glutathione; then back-titrated with Ag^+	Hofmann and Hamm (1967)
Muscle tissue; + tris buffer, pH 7·4; 1 h reaction in dark flask	KI amperometric	Hofmann (1971b)
pH 7·4, 1 h reaction	+ equivalent of glutathione, then back-titrated amperometric with Ag^+	Hofmann (1971a)

19.3. Analytical utilization of a reaction product

As under mercury(II), distinction may be made between silver-containing and other reaction products:

19.3.1. Silver-containing Product

Coloured and/or insoluble silver-containing products serve for detection of thiols. Thus André and Kogane-Charles (1947) detected thiols in rapeseed oil through the darkening of a saturated alcoholic silver nitrate solution with subsequent formation of a precipitate. Karr (1954) proposed several tests for thiols in petroleum samples; in one, 6 drops of sample are agitated for 1 min with 1 drop of 2% aqueous silver nitrate, the positive outcome being a white or pale yellow precipitate in the lower, aqueous layer. Saxe (1967) detected thiols in the Weiss ring-oven technique through the yellow ring given by spraying with 0·1 N silver nitrate and exposure to ultraviolet light.

Gastovo and Pileri (1954) visualized 6-mercaptopurine on chromatograms with ammoniacal silver nitrate reagent, which yielded orange–red. Reguera and Asimov (1950) also used a composite reagent for chromatographic visualization, spraying with 2% silver nitrate and then 0·5% sodium dichromate. Immersion in 0·5 N nitric acid leached out silver dichromate not bound to the (purine) complexes studied. Semenza and Lucchelli (1955) modified this procedure slightly, using it for 2-amino-6-mercaptopurine among other compounds.

Quantitative gravimetric determination of the silver mercaptide appears rare. Examples are the work of Ushakov and Galanov (1934) for mercaptobenzothiazole; and of Laitinen et al. (1946) for primary thiols precipitated from ammoniacal ethanol. The final stage was drying for 2 h at 60–70°C; tertiary thiols lend themselves less well to this procedure since the silver salts are soluble in 95% ethanol.

Formation of a silver derivative has more usually been followed by regeneration of the thiol which is then determined by other standard methods. For example, Hartner and Schleiss (1936) precipitated glutathione from biological material as its silver (also cadmium) derivative; they then added acid, bromide and bromate to the precipitate and ultimately estimated unused bromate. Porter et al. (1963) studied polysulphides by reduction with lithium aluminium hydride and estimation of the thiols formed. In one of their procedures, they swept the products into a silver nitrate/pyridine solution at −10°C; the thiols were regenerated from the precipitate by treatment with sulphuric acid/thiourea, extracted with light petroleum and subjected to gas chromatography. Brand and Keyworth (1964) precipitated thiols in petroleum products with silver nitrate in formamide at −65°C, subsequently recovering them and carrying out gas chromatography also. Muenze (1969) determined thiol concentrations down to 5×10^{-10} M in blood serum by passing through

KPS-200 ion exchanger containing ^{110}Ag, previously introduced by passing through silver nitrate containing the isotope. The products were eluted with ammonium hydroxide and evaluated with a scintillation counter.

19.3.2. *Product not Containing Silver*

The best known procedure here is based on titration of the acid yielded in the reactions,

$$RSH + Ag^+ \rightarrow RSAg + H^+$$

It has the advantage of being independent of formation of double salt, $RSAg.AgNO_3$. Mapstone (1946a) introduced the method for thiol determination in gasoline. Canbäck (1947) and Berggren and Kirsten (1951) determined 4-propyl-2-thiouracil by adding aqueous silver nitrate and bromothymol blue to the cooled acetone solution and titrating with standard sodium hydroxide. Thiols were determined by Pellerin and Gautier (1961) by adding silver nitrate/pyridine to the sample in the same solvent, water, ethanol or acetone and, after 15 min at room temperature, titrating with alcoholic sodium hydroxide potentiometrically or to thymol blue or thymolphthalein. Saville (1961) preferred this method to determination via silver consumption. He determined saturated or unsaturated thiols in aqueous pyridine after reaction for 5 min by titrating with sodium hydroxide to phenolphthalein. Porter *et al.* (1963) used this method among others to determine thiols derived from the reduction of polysulphides.

Two methods classifiable here depend on the use of a reagent other than silver nitrate. Kunkel *et al.* (1959) determined small amounts of thiols in hydrocarbons by adding silver dithizonate in carbon tetrachloride. The liberated dithizone they evaluated colorimetrically at 615 nm after 5 min. Jones (1963) patented a device for continuously monitoring thiol amounts in hydrocarbons by bringing the sample into contact with an aqueous solution of a relatively poorly soluble metal salt. Silver chloride was an example, yielding the still less soluble silver mercaptide and liberating chloride ions. These were estimated potentiometrically.

In studies of thiol-disulphide exchange Fava *et al.* (1957) assayed *n*-butane-, *t*-butane-, *n*-hexane- and benzenethiol containing ^{35}S by precipitating the silver salt and ultimately converting it into barium sulphate; this was estimated radiometrically.

20. VANADIUM(V)

See Chapter 1, "Oxidation" (Section 54, "Vanadate")).

21. ZINC(II)

Perhaps surprisingly, only four references to the use of zinc reagents can be quoted: Colovos and Freiser (1969) titrated 2,3-dimercapto-1-propanol (BAL) complexometrically with zinc chloride in a carbonate/bicarbonate buffer of pH 10, using eriochrome black T as indicator (blue → violet). A less selective procedure for cysteine and cystine comes from Taniguchi (1971) who determined these in protein by treatment with a 0·6 % aqueous suspension of zinc hydroxide at pH 9·5 in nitrogen atmosphere. After 2–4 d at 100°C the sulphide yielded was converted to hydrogen sulphide by acidification and this was determined through the "methylene blue" reaction with p-dimethylaminoaniline/ferric ion. In this procedure the zinc presumably plays a role secondary to that of the hydroxyl ions.

The thiol sites of serum albumin were investigated by Saroff and Mark (1953) with the help of amperometric titration with several reagents. These included zinc acetate, at pH 6·5. The results showed that zinc binds the albumin at 8 similar sites. This is not an analytical procedure of quantitative determination but might serve as a basis for such.

Bruening and Bruening (1972) found that thiols and sulphides underwent reversible complexation with compounds of transition metals, e.g. zinc cyclohexylbutyrate and zince hexylmercaptide. They used these compounds as 0·05–0·5 % additives to 2·5 or 5 % SF 96 silicone oil/Chromosorb G columns to improve gas chromatographic separation of the sulphur compounds.

22. ORGANOMETALLIC REAGENTS

Such reagents can be as conveniently classified here as anywhere else. There are few examples of their use.

The classical active hydrogen determination of Tschugaeff–Zerewitinoff was tested on some thiols (propane-, isobutane-, isopentane-, benzenethiols and benzyl mercaptan) by Zerewitinoff (1908). Sensibly quantitative yields of methane were found with the methylmagnesium iodide reagent in diamyl ether at 25°:

$$RSH + CH_3MgI \rightarrow CH_4 + RSMgI$$

The method is unspecific because many other functional groups react similarly, e.g. hydroxyl, amino, acetylenic terminal hydrogen atoms.

Terent'ev and Shor (1947) found that methylzinc iodide had no advantage over the Grignard reagent, and in fact reacted more slowly.

Metal alkyls also do not react quantitatively, as found in attempts to introduce bismuth triethyl (see Section 2.2., "Bismuth(III)"), lead tetra-ethyl (see Chapter 1, Section 29, "Lead(IV)") and zinc diethyl (Haurowitz, 1929).

Lithium aluminium hydride was introduced in the forties as a possible substitute for methylmagnesium iodide in determining active hydrogen. Krynitski *et al.* (1948) included octanethiol among the compounds studied. They obtained a quantitative yield of hydrogen within 5–10 min in diethyl ether at 0°C:

$$4RSH + LiAlH_4 \rightarrow 4H_2 + RSLi + (RS)_3Al$$

The method is as unspecific as the original procedure with methylmagnesium iodide.

REFERENCES

Abbott, C. F. (1953). *J. Pharm. Pharmacol.* **5**, 53.
Aibara, K., Herreid, E. O. and Wilson, H. K. (1960). *J. Dairy Sci.* **43**, 1736.
Åkerfeldt, S. and Lövgren, G. (1964). *Anal. Biochem.* **8**, 223.
Alfredo, A., Aguillo, E., Morini, C. and Agoff, J. (1973). *Galaxia,* No. 46, 4, 8; *Chem. Abs.* **78**, 125721.
Alicino, J. F. (1960). *Microchem. J.* **4**, 551.
Allison, A. C. and Cecil, R. (1958). *Biochem. J.* **69**, 27.
Ambrosino, C., Vancheri, L., Lausarot, P. M. and Papa, G. (1969). *Ric. Sci.* **39**, 924.
André, E. and Kogane-Charles, M. (1947). *Ann. Agron.* **17**, 393.
Andreasch, R. (1879). *Chem. Ber.* **12**, 1390.
Andreasch, R. (1889). *Monatsh.* **10**, 73.
Andreen, B. H., Kniebes, D. V. and Tarman, P. B. (1963). *Inst. Gas Technol., Tech. Rept.* No. 7; *Chem. Abs.* **62**, 2645; see also Andreen, B. H. and Kniebes, D. V. (1962). *Proc. Operating Sect., Am. Gas Assoc.* CEP-62-13; *Chem. Abs.* **60**, 7843.
Anonymous (1945). *Amer. Gas J.* **162**, No. 6, 47, 60.
Anson, M. L. (1940). *J. Biol. Chem.* **135**, 797; also (1941). *J. Gen. Physiol.* **24**, 399.
Asquith, R. S., Miro, P. and Garcia, J. J. (1968). *Dominguez. Text. Res. J.* **38**, 1057; *Chem. Abs.* **70**, 12501.
Bahadur, K. and Sen, P. (1973). *Analysis,* **2**, 126.
Bahr, G. F. (1966). *Introd. Quant. Cytochem.* 469; *Chem. Abs.* **68**, 719.
Bald, E. (1972). *Przegl. Pap.* **28**, 142; *Anal. Abs.* **25**, 961.
Bardos, T. J. and Kalman, T. I. (1966). *J. Pharm. Sci.* **55**, 606.
Barrett, S., Croft, A. G. and Hartley, A. W. (1971). *J. Sci. Food Agric.* **22**, 173.
Barron, E. S. G. and Singer, T. P. (1945). *J. Biol. Chem.* **157**, 221.
Bayer, I. and Posgay, G. (1961). *Acta Pharm. Hung.* **31**, Suppl. 43; also *Pharm. Zentralh.* **100**, 65.
Beach, E. F. and Teague, D. M. (1942). *J. Biol. Chem.* **142**, 277.
Bebesel, P. and Sirbu, I. (1960). *Rev. Chim. (Bucharest)*, **11**, 288.
Beesing, D. W., Tyler, W. P., Kurtz, D. M. and Harrison, S. A. (1949). *Anal. Chem.* **21**, 1073.
Belcher, R., Gawargious, Y. A. and Macdonald, A. M. G. (1965). *Anal. Chim. Acta,* **33**, 210.
Belitser, V. O. and Lobachevskaya, O. V. (1959). *Ukr. Biokhim. Zh.* **31**, 579.
Bell, R. T. and Agruss, M. S. (1941). *Ind. Eng. Chem., Anal. Ed.* **13**, 297.
Benesch, R. and Benesch, R. (1948). *Arch. Biochem.* **19**, 35.

E

Benesch, R. E. and Benesch, R. (1950). *Arch. Biochem.* **28**, 43.
Benesch, R. and Benesch, R. E. (1951). *J. Amer. Chem. Soc.* **73**, 3391.
Benesch, R. and Benesch, R. E. (1952). *Arch. Biochem. Biophys.* **38**, 425.
Benesch, R. E., Lardy, H. A. and Benesch, R. (1955). *J. Biol. Chem.* **216**, 663.
Bennett, H. S. (1951). *Anat. Record,* **110**, 231; *Chem. Abs.* **46**, 8703.
Bennett, H. S. and Yphantis, D. A. (1948). *J. Amer. Chem. Soc.* **70**, 3522.
Berg, R., Cox, R. P. Barness, J. G. and Huset, C. A. (1970). *U.S. Patent No.* 3,493,484 of Feb. 3.
Berger, A. and Magnuson, J. A. (1964). *Anal. Chem.* **36**, 1156.
Berggren, A. and Kirsten, W. (1951). *Farm. Revy,* **50**, 245.
Besch, D. K., Goldzieher, J. W. and McCormack, S. (1957). *Science,* **126**, 650.
Bhatia, M. S., Bajaj, K. L., Singh, S. and Bhatia, I. S. (1972). *Analyst (London),* **97**, 890.
Bhattacharya, S. K. (1958). *Biochem. J.* **69**, 43; also (1959). *Nature (London),* **183**, 1327.
Binet, L. and Weller, G. (1934a). *Bull. Soc. Chim. Biol.* **16**, 1284.
Binet, L. and Weller, G. (1934b). *Compt. Rend.* **198**, 1185.
Binet, L. and Weller, G. (1935). *Compt. Rend. Soc. Biol.* **119**, 939.
Binet, L. and Weller, G. (1936). *Bull. Soc. Chim. Biol.* **18**, 358.
Binet, L. and Weller, G. (1938). *Bull. Soc. Chim. Biol.* **20**, 123.
Blažek, J. and Stejskal, Z. (1956). *Česk. Farm.* **5**, 29.
Bloksma, A. H. (1959). *Cereal Chem.* **36**, 357.
Börresen, H. C. (1963). *Anal. Chem.* **35**, 1096.
Bokarev, K. S., Chertok, N. O. and Saval'ev, A. N. (1968). *Fizial. Rast.* **15**, 926; *Chem. Abs.* **70**, 17511.
Bol'shakov, G. F., Stekhun, A. I., Chalykh, N. D. and Pokhitun, L. E. (1967). *Neftepererab. Neftekhim.* **2**; *Chem. Abs.* **67**, 101623.
Borgstrom, P. and Reid, E. E. (1929). *Ind. Eng. Chem., Anal. Ed.* **1**, 186.
Borodataya, V. A., Sokolina, L. F., Afanasev, Yu. M. and Gladkii, V. A. (1971). *Zav. Lab.* **37**, 1313.
Boyd, G. A. (1933). *Oil & Gas J.* **32**, No. 8,16, 31.
Boyer, P. D. (1954). *J. Amer. Chem. Soc.* **76**, 4331.
Brand, V. T. and Keyworth, D. A. (1964). General Papers, Vol. 9, No. 3, 45, August, Division of Petroleum Chemistry, American C.S. (quoted in *Anal. Chem.* **37**, 1424).
Brand, V. T. and Keyworth, D. A. (1965). *Anal. Chem.* **37**, 1424.
Brand, E., Cahill, G. F. and Kassell, B. (1940). *J. Biol. Chem.* **133**, 431.
Brejcha, A. and Sima, J. (1960). *Sb. Praci Výzkumu Chem. Využiti Uhlí, Dehtu Ropy,* No. 1, 166; *Chem. Abs.* **58**, 5427.
Brenner, M. W. and Khan, A. A. (1974). *J. Inst. Brew.* **80**, 544.
Brenner, M. W. and Laufer, L. (1972). *Proc. Am. Soc. Brew. Chem.* **98**, 103; *Anal. Abs.* **26**, 3543.
Brenner, M. W., Jakob, G. and Owades, J. L. (1957). *J. Inst. Brewing,* **63**, 408.
Brenner, M. W., Owades, J. L., Schapiro, G. J. and Laufer, S. (1959). *Proc. Europ. Brew. Conv. Rome,* 239; *Anal. Abs.* **7**, 3996.
Bruening, W. and Bruening, I. M. de R. (1972). *Bol. Tec. PETROBRAS,* **15**, 319; *Chem. Abs.* **79**, 73328.
Brychta, M. and Rudolf, J. (1956). *Paliva,* **36**, 307.
Burley, R. W. (1956). *Textile Res. J.* **26**, 332.
Burton, H. (1958). *Biochim. Biophys. Acta,* **29**, 193.
Busev, A. I. and Teternikov, L. I. (1969). *Zh. Anal. Khim.* **24**, 918.

Busev, A. I. and Teternikov, L. I. (1971). *Anal. Letters,* **4**, 53.
Busev, R. I., Shelemina, N. V., Teternikov, L. I. and Danilova, T. A. (1968). *Anal. Letters,* **1**, 763.
Busev, A. I., Teternikov, L. I. and Maslennikova, Z. V. (1971). *Zh. Anal. Khim.* **26**, 1852.
Bydalek, T. J. and Poldoski, J. E. (1968). *Anal. Chem.* **40**, 1878.
Calcutt, G. (1960). *Biochim. Biophys. Acta,* **44**, 364.
Calcutt, G. and Doxey, D. (1959). *Exptl. Cell Research,* **17**, 542.
Calzolari, C. and Donda, A. (1954). *Univ. Studi. Trieste, Fac. Sci. Inst. Chim.* No. 3.
Canbäck, T. (1947). *Farm. Revy,* **29**, 465.
Cannan, R. K. and Richardson, G. M. (1929). *Biochem. J.* **23**, 1242.
Cannefax, G. R. and Freedman, L. D. (1955). *Proc. Soc. Exptl. Biol. Med.* **89**, 337.
Carr, E. M. and Bit-Alkhas, M. (1960). *Anal. Biochem.* **1**, 158.
Carson, J. F., Weston, W. J. and Ralls, J. W. (1960). *Nature (London),* **186**, 801.
Carter, J. R. (1954). *Science,* **120**, 895.
Carter, J. R. (1956). *Proc. Soc. Exptl. Biol. Med.* **91**, 406.
Carter, J. R. (1959). *J. Biol. Chem.* **234**, 1705.
Cecil, R. (1950). *Biochem. J.* **47**, 572.
Cecil, R. (1955). *Biochim. Biophys. Acta,* **18**, 154.
Cecil, R. and McPhee, J. R. (1955). *Biochem. J.* **59**, 234.
Cecil, R. and McPhee, J. R. (1957). *Biochem. J.* **66**, 538.
Ceresa, F. and Guala, P. (1939). *Giorn. Accad. Med. Torino,* **102**, 193; *Chem. Abs.* **35**, 6995.
Cernenco, A. and Crisan, T. (1974). *Materiale Plast.* **11**, 239, *Anal. Abs.* **29**, 2C 79.
Černoch, M. (1966). *Coll. Czech. Chem. Commun.* **31**, 782.
Challenger, F. and Greenwood, D. (1949). *Biochem. J.* **44**, 87.
Chang, S. F. and Liener, L. E. (1964). *Nature (London),* **203**, 1065.
Chromý, V. and Svoboda, V. (1963). *Talanta,* **10**, 1109.
Číhalík, J. and Kudrnovská-Pavlíková, E. (1957). *Chem. Listy,* **51**, 76.
Claësson, P. (1881). *Chem. Ber.* **14**, 411.
Clark, R. E. D. (1957). *Analyst (London),* **82**, 178.
Clark, R. E. D. and Neville, R. G. (1959). *J. Org. Chem.* **24**, 110.
Colombo, P., Corbetta, D., Pirotta, A. and Sartori, A. (1957). *Tappi,* **40**, 490.
Colovos, G. and Freiser, H. (1969). *Talanta,* **16**, 1605.
Constaninescu, M. and Constaninescu, T. (1957). *Lucr. Inst. Petrol Gaze (Bucharest),* **3**, 353; *Anal. Abs.* **5**, 4192.
Contopoulos, A. N. and Anderson, H. H. (1950). *J. Lab. Clin. Med.* **36**, 929.
Csagoly, E. (1957). *Acta Pharm. Hung.* **27**, 267.
Cullen, M. C. and McGuinness, E. T. (1971). *Anal. Biochem.* **42**, 455.
Cynajek, F. and Szlanga, J. (1954). *Farm. Polska,* **10**, 17.
Czerwiński, W. and Vieweger, H. (1960). *Chem. Anal. (Warsaw),* **5**, 1011.
Dahmen, E. A. M. F., Dijkstra, R. and Verjaal, A. J. (1963). *Erdöl & Kohle,* **16**, 768.
Davies, E. R. H. and Armstrong, J. W. (1943). *J. Inst. Petroleum,* **29**, 323.
Davies, G., Kustin, K. and Pasternack, R. F. (1968). *Trans. Far. Soc.* **64**, 1006.
Del Pianto, E. and Sabatini, A. (1956). *Ric. Sci.* **26**, 482.
Del Pianto, E. and Silvestroni, P. (1954). *Ric. Sci.* **24**, 67.
Del Vecchio, G. and Argenziano, R. (1946). *Boll. Soc. Ital. Biol. Sper.* **22**, 1189.
Devay, J. and Garai, R. (1970). *Hung. Sci. Instrum.* No. 19, 39; *Chem. Abs.* **74**, 119858.
Devay, J., Garai, T., Havas, J. and Juhasz, B. (1972). *Hung. Sci. Instrum.* **22**, 11; *Chem. Abs.* **77**, 42869.

Doornbos, D. A. (1967). *Pharm. Weekbl.* **102**, 1095.
Dubský, J. V. and Šindelář, V. (1938). *Mikrochim. Acta,* **3**, 258.
Dworak, D. D. and Davis, E. N. (1967). *U.S. Patent No.* 3,338,812 of Aug. 29.
Earle, T. E. (1953). *Anal. Chem.* **25**, 769.
Edelhoch, H., Katchalski, E., Maybury, R. H., Hughes, jr., W. L. and Edsall, J. L. (1953). *J. Amer. Chem. Soc.* **75**, 5058.
Ehrlich, E. (1967). *Bull. Soc. Chim. Biol.* **49**, 889.
Eldjarn, L. and Jellum, E. (1963). *Acta Chem. Scand.* **17**, 2610.
Engel, M. B. and Zerlotti, E. (1964). *J. Histochem. Cytochem.* **12**, 156.
Erwin, J. G. and Pedersen, P. L. (1968). *Anal. Biochem.* **25**, 477.
Faragher, W. F., Morrell, J. C. and Monroe, G. S. (1927). *Ind. Eng. Chem.* **18**, 1281.
Fava, A., Iliceto, A. and Camera, E. (1957). *J. Amer. Chem. Soc.* **79**, 833.
Fecko, J. and Zaborniak, F. (1968). *Chem. Anal. (Warsaw),* **13**, 659.
Feigl, F. Golstein, D. and Libergott, E. K. (1969). *Anal. Chim. Acta,* **47**, 553.
Feldstein, M., Balestrieri, S. and Levaggi, D. A. (1965). *J. Air Pollution Control Assoc.* **15**, 215; *Chem. Abs.* **63**, 7556.
Felicetta, V. F., Peniston, Q. P. and McCarthy, J. L. (1952); *Can. Pulp Paper Ind.* **5**, No. 12, 16, 18, 20, 22, 24, 26–7, 30, 41; *Chem. Abs.* **47**, 5115; also (1953). *Tappi,* **36**, 425.
Fernandez Diez, M. J., Osuga, D. T. and Feeney, R. E. (1964). *Arch. Biochem. Biophys.* **107**, 449.
Flesch, P. and Kun, E. (1950). *Proc. Soc. Explt. Biol. Med.* **74**, 249.
Folkard, A. E. and Joyce, A. E. (1963). *J. Sci. Food Agric.* **14**, 510.
Forbes, W. F. and Hamlin, C. R. (1968). *Can. J. Chem.* **46**, 3033.
Fraenkel-Conrat, H. (1955). *J. Biol. Chem.* **217**, 373.
Fraenkel-Conrat, J., Cook, B. B. and Morgan, A. F. (1952). *Arch. Biochem. Biophys.* **35**, 157.
Frank, R. L., Smith, P. V., Woodward, F. E., Reynolds, W. B. and Canterine, P. J. (1948). *J. Polymer Sci.* **3**, 39.
Frater, R. and Hird, F. J. R. (1965). *Biochem. J.* **96**, 895.
Fredericks, E. M. and Harlow, G. A. (1964). *Anal. Chem.* **38**, 263.
Frei, R. W., Machellan, B. L. and MacNeil, J. D. (1973). *Anal. Chim. Acta,* **66**, 139.
Freytag, H. (1953). *Z. Anal. Chem.* **138**, 259.
Fridovich, I. and Handler, P. (1957). *Anal. Chem.* **29**, 1219.
Fritz, J. S. and Palmer, T. A. (1961). *Anal. Chem.* **33**, 98.
Gabrielyantz, S. M. and Artem'eva, O. A. (1934). *Groznenskii Neftyanik,* **4**, 41; *Chem. Abs.* **29**, 2725.
Gadal, P. (1963). *Compt. Rend.* **256**, 4311.
Garai, T., Almasi, E. and Devay, J. (1972). *Hung. Sci. Instrum.* 15; *Chem. Abs.* **77**, 56190.
Gastovo, F. and Pileri, A. (1954). *Rass. Med. Sper.* **1**, 122.
Gershkovich, E. E. (1971). *Gig. Tr. Prof. Zabol.* **15**, 58; *Chem. Abs.* **76**, 17449.
Ghiglione, C. and Bozzi-Tichadou, M. (1954). *Bull. Soc. Chim. Biol.* **36**, 659.
Gilman, H. and Nelson, J. F. (1937). *J. Amer. Chem. Soc.* **59**, 935.
Gordon, B. E., Melamed, E. A. and Belova, N. A. (1962). *Kauchuk i Rezina,* **21**, No. 8,53; *Chem. Abs.* **58**, 1611.
Graff, S., Maculla, E. and Graff, A. M. (1937). *J. Biol. Chem.* **121**, 11.
Grafnetterová, J. (1960). *Časopis Lékářů Českých,* **99**, 182.
Gregg, D. C., Bouffard, P. E. and Barton, R. (1961). *Anal. Chem.* **33**, 269.
Grimes, A. J. (1965). *Nature (London),* **205**, 94.

Grimes, M. D., Puckett, J. E., Newby, B. J. and Heinrich, B. J. (1955). *Anal. Chem.* **27**, 152.

Gruen, L. C. and Harrap, B. S. (1971). *Anal. Biochem.* **42**, 377.

Gulyaeva, L. I. and Blokh, N. V. (1969). *Neftekhimiya*, **9**, 308; *Chem. Abs.* **71**, 23445.

Gupta, H. K. L. and Boltz, D. F. (1971). *Microch. J.* **16**, 571.

Hack, A. G. and Allan, A. L. (1968). *Rev. Invest. Agropecuar., Ser.* **2**, 5, 77; *Chem. Abs.* **70**, 95576.

Haglund, H. and Lindgren, I. (1965). *Talanta*, **12**, 499.

Hakewill, H. and Rueck, E. M. (1946). *Proc. Amer. Gas Assoc.* **28**, 529.

Hamm, R. and Hofmann, K. (1966). *Z. Lebensm.-Untersuch. u. Forsch.* **130**, 133.

Hamm, R. and Hofmann, K. (1967). *Z. Lebensm.-Untersuch. u. Forsch.* **136**, 7.

Hammar, C. G. B. (1955). *Svensk Kem. Tidskr.* **67**, 307.

Hammerich, T. and Gondermann, H. (1964). *Erdöl & Kohle*, **17**, 20.

Harada, Y., Ito, K. and Udaka, K. (1959). *Mie Med. J.* **8**, 343; *Chem. Abs.* **53**, 17228.

Harding, C. I., Hendrickson, E. R., Sundaresan, B. B. and Walker, C. V. (1964). *Am. Chem. Soc., Div. Waste Water Chem.*, Preprints, **4**, 119.

Harrap, B. S. and Gruen, L. C. (1971). *Anal. Biochem.* **42**, 398.

Harris, L. (1922). *Biochem. J.* **16**, 746.

Hartmann, H. (1930). *Biochem. Z.* **223**, 489.

Hartner, F. and Schleiss, E. (1936). *Mikrochemie*, **20**, 63.

Haslam, J., Grossman, S., Squirrel, D. C. M. and Loveday, S. F. (1953). *Analyst (London)*, **78**, 92.

Hata, T. (1949). *Bull. Res. Inst. Food Sci., Kyoto Univ.* No. 2, 13; *Chem. Abs.* **46**, 3597.

Hata, T. (1951a). *Bull. Res. Inst. Food Sci., Kyoto Univ.* No. 4, 45.

Hata, T. (1951b). *Mem. Res. Inst. Food Sci., Kyoto Univ.* No 1, 19.

Haurowitz, F. (1929). *Mikrochem.* **7**, 88.

Haviř, J. Vřešťál, J. and Chromý, V. (1965). *Chem. Listy*, **59**, 431.

Heide, K. (1955). *Behringwerke Mitt.* No. 30, 97; *Chem. Abs.* **52**, 14743.

Henderson, L. M., Agruss, M. S. and Ayers, jr., G. W. (1940). *Ind. Eng. Chem., Anal. Ed.* **12**, 1.

Hennart, C. (1962). *Talanta*, **9**, 97.

Herber, R. H. (1962). *Anal. Chem.* **34**, 340.

Herbert, F. J. and Denson, J. R. (1954). *Anal. Chem.* **26**, 440.

Hladký, Z. (1965). *Z. Chem.* **5**, 424.

Hladký, Z. and Vřešťál, J. (1969). *Coll. Czech. Chem. Commun.* **34**, 984.

Hoch, F. L. and Vallee, B. L. (1960). *Biochim. Biophys. Acta*, **91**, 1.

Hofmann, K. (1970). *Z. Anal. Chem.* **250**, 256.

Hofmann, K. (1971a). *Z. Anal. Chem.* **256**, 187.

Hofmann, K. (1971b). *Z. Lebensm.-Untersuch. u. Forsch.* **147**, 68.

Hofmann, K. (1972). *Nahrung*, **16**, 197.

Hofmann, K. and Hamm, R. (1967). *Z. Anal. Chem.* **231**, 199.

Holt, W. L. and Mattson, L. N. (1949). *Anal. Chem.* **21**, 1389.

Holzapfel, H. and Stottmeister, U. (1967). *Z. Anal. Chem.* **232**, 331.

Hommes, F. A. and Huisman, T. H. J. (1958). *Biochem. J.* **68**, 312.

Hopkins, F. G. (1929). *J. Biol. Chem.* **84**, 269.

Horowitz, M. and Klotz, I. M. (1956). *Arch. Biochem. Biophys.* **63**, 77.

Howard, G. E. and Baldry, J. (1969). *Analyst (London)*, **94**, 589.

Hubbard, R. L., Haines, W. E. and Ball, J. S. (1958). *Anal. Chem.* **30**, 91.

Hughes, W. L. (1949). *Cold Spring Harbor Symposium Quant. Biol.* **14**, 79.

Human, J. P. E. (1958). *Textile Res. J.* **28**, 647.

Humphrey, R. E. and Potter, J. L. (1965). *Anal. Chem.* **37**, 164.

Ibragimov, A. P., Mukhamedov, A. and Tulganov, A. (1961). *Tr. Tashkentsk. Konf. po Mirnomy Ispol'z At. Energii, Akad. Nauk Uz. SSR* **1**, 306; *Chem. Abs.* **56**, 14590.

Ikeya, T. (1964). *Jozo Kagaku Kenkyu Hokoku,* **10**, 23; *Chem. Abs.* **64**, 8895.

Ingram, V. M. (1955). *Biochem. J.* **59**, 653.

Ingram, V. M. (1957). *Biochem. J.* **65**, 760.

Isles, T. E. and Jocelyn, P. C. (1963). *Biochem. J.* **88**, 84.

Iwaki, G. (1951). *J. Physiol. Soc. Japan,* **13**, 356; *Chem. Abs.* **46**, 1606.

Jančík, F., Kakáč, B. and Buděšínský, B. (1960). *Česk. Farm.* **9**, 329.

Janicki, J. and Wierzbowski, J. (1967). *Rocz. Wyzsz. Szk. Roln. Poznaniu* **35**, 99; *Chem. Abs.* **72**, 65460.

Jansen, E. F. (1948). *J. Biol. Chem.* **176**, 657.

Jansen, E. F. and Jang, R. (1952). *Arch. Biochem. Biophys.* **40**, 358.

Janz, G. J. and Duncan, N. E. (1953). *Anal. Chem.* **25**, 1410.

Jaselskis, B. and Schlough, S. D. (1974). *Anal. Chem.* **46**, 915.

Jones, R. H. (1963). *U.S. Patent* No. 3,114,699 of Dec. 17.

Joniau, M., Bloemmen, J. and Lontie, R. (1970). *Biochim. Biophys. Acta* **214**, 468.

Joyet-Lavergne, P. (1938). *Compt. Rend. Soc. Biol.* **128**, 59.

Kapoor, R. C. (1959). *Z. Anal. Chem.* **166**, 1.

Karchmer, J. H. (1957). *Anal. Chem.* **29**, 425.

Karchmer, J. H. (1958). *Anal. Chem.* **30**, 80.

Karchmer, J. H. and Walker, M. T. (1958). *Anal. Chem.* **30**, 75.

Karr, jr. C. (1954). *Anal. Chem.* **26**, 528.

Kassell, B. and Brand, E. (1938). *J. Biol. Chem.* **125**, 115.

Khazova, I. V. and Bogomolov, B. D. (1971). *Izv. Vyssh. Ucheb. Zaved. Les. Zh.* **14**, 167; *Chem. Abs.* **75**, 115578.

Khusmitdinova, Z. S. (1965). *Metody Analiza Radioaktivn. Preparatov, Sb. Statei,* 148; *Chem. Abs.* **63**, 11985.

Kiermeier, F. and Hamed, M. G. E. (1962). *Nahrung,* **6**, 638.

Kiermeier, F. and Petz, E. (1966). *Z. Lebensm.-Untersuch. u. Forsch.* **131**, 75.

Kirchner, A. (1972). *Chem.-Ztg.* **96**, 527.

Klotz, I. M. and Carver, B. R. (1961). *Arch. Biochem. Biophys.* **95**, 540.

Koch, H. and Paul, D. (1963). *Brennstoff-Chem.* **44**, 231.

Koeppe, O. J., Boyer, P. D. and Stulberg, M. P. (1956). *J. Biol. Chem.* **219**, 569.

Körbl, J. and Vaníček, J. (1969). *Czech. Patent* No. 133,383 of July 15.

Kolthoff, I. M. and Eisenstädler, J. (1961a). *Anal. Chim. Acta,* **24**, 280.

Kolthoff, I. M. and Eisenstadler, J. (1961b). *Anal. Chim. Acta,* **24**, 83.

Kolthoff, I. M. and Harris, W. E. (1946). *Ind. Eng. Chem., Anal. Ed.* **18**, 161.

Kolthoff, I. M. and Stricks, W. (1950). *J. Amer. Chem. Soc.* **72**, 1952.

Kolthoff, I. M., May, D. R., Morgan, P., Laitinen, H. A. and O'Brien, N. (1946). *Ind. Eng. Chem., Anal. Ed.* **18**, 442.

Kolthoff, I. M., Stricks, W. and Morren, L. (1954). *Anal. Chem.* **26**, 366.

Kolthoff, I. M., Anastasi, A., Stricks, W., Tan, B. H. and Desmukh, G. S. (1957). *J. Amer. Chem. Soc.* **79**, 5102.

Kolthoff, I. M., Shore, W. S., Tan, B. H. and Matsuoka, M. (1965). *Anal. Biochem.* **12**, 497.

Kong, R. W., Mecham, D. K. and Pence, J. W. (1957). *Cereal Chem.* **34**, 201.

Koons, R. D. (1941). *Refiner Natural Gasoline Mfr.* **20**, 393.

Korobeinik, F. G. (1968). *Lab. Delo,* 692; *Anal. Abs.* **18**, 1896.
Korochanskaya, S. P. (1968). *Lab. Delo,* 755; *Chem. Abs.* **70**, 54754.
Kotlyar, G. I. (1959). *Biokhimiya,* **24**, 15.
Krakow, J. S. and Goolsby, S. P. (1971). *Biochem. Biophys. Res. Commun.* **44**, 453.
Krynitski, J. A., Johnson, J. E. and Carhart, H. W. (1948). *J. Amer. Chem. Soc.* **70**, 486.
Kuehbauch, W. and Wuensch, A. (1971). *Z. Lebensm.-Untersuch. u. Forsch.* **146**, 9.
Kul'berg, L. M. and Presman, Sh. E. (1940). *Lab. Prakt. (USSR).* **15**, No. 4, 15; *Chem. Abs.* **34**, 7951.
Kundu, K. K. and Das, M. N. (1959). *Anal. Chem.* **31**, 1358.
Kunkel, R. K., Buckley, J. E. and Gorin, G. (1959). *Anal. Chem.* **31**, 1098.
Kvapil, J. (1965). *Czech. Patent* No. 114,448 of Apr. 15.
Ladenson, J. H. and Purdy, W. C. (1971a). *Clin. Chem.* **17**, 908.
Ladenson, J. H. and Purdy, W. C. (1971b). *Anal. Chim. Acta,* **57**, 465.
Ladenson, J. H. and Purdy, W. C. (1973). *Anal. Chim. Acta,* **64**, 259.
Laitinen, H. A., O'Brien, A. S. and Nelson, S. S. (1946). *Ind. Eng. Chem., Anal. Ed.* **18**, 47.
Lamathe, J. (1966). *Compt. Rend.* **263**, 872.
Lamathe, J. (1967). *Chim. Anal. (Paris),* **49**, 119.
Lamprecht, W. and Katzlmeier, H. (1961). *Z. Anal. Chem.* **181**, 216.
Laurinavicius, R. (1963). *Lietuvos TSR Mokslu Akad. Darbai Ser. C.* 111; *Chem. Abs.* **60**, 12350.
Leach, S. J. (1959). *Biochim. Biophys. Acta,* **33**, 264.
Leach, S. J. (1960). *Austr. J. Chem.* **13**, 520.
Lecher, H. and Siefken, W. (1926). *Chem. Ber.* **59**, 2597, 2600.
Lee, K.-T. and Tan, I.-K. (1974). *Clin. Chim. Acta,* **53**, 153.
Leisey, F. A. (1954). *Anal. Chem.* **26**, 1607.
Leisey, F. A. (1960). *U.S. Patent* No. 2,928,774 of Mar. 15.
Leisey, F. A. (1961). *U.S. Patent* No. 2,989,377 of Jun. 20.
Lennartz, T. A. and Middeldorf, R. (1949). *Süddeut. Apoth.-Z.* **89**, 593.
LeRosen, A. L., Moravek, R. T. and Carlton, J. K. (1952). *Anal. Chem.* **24**, 1335.
Leussing, D. L., Mislan, J. P. and Goll, R. J. (1960). *J. Phys. Chem.* **64**, 1070.
Liberti, A. (1951). *Ann. Chim. (Rome),* **41**, 363.
Liberti, A. (1957). *Anal. Chim. Acta,* **17**, 247.
Liberti, A. and Cartoni, G. P. (1957). *Chim. e Ind.* **39**, 821.
Liberti, A. and Cervone, E. (1950). *Atti Accad Nazl. Lincei, Rend. Classe Sci. Fis. Mat. e Nat.* **8**, 613; *Chem. Abs.* **45**, 69.
Liberti, A., Lepri, F., Ciavetta, L. and Cartoni, G. P. (1957). *Ric. Sci.* **27**; *Suppl. A, Polarografia,* **3**, 21; *Chem. Abs.* **52**, 6055.
Liener, I. E. (1967). *Arch. Biochem. Biophys.* **121**, 67.
Linek, K., Peciar, C. and Fedoronka, M. (1963). *Chem. Zvesti,* **17**, 510.
Lontie, R. and Beckers, G. (1956). *J. Indian Chem. Soc.* **33**, 285.
Lorenz, O. and Echte, E. (1956). *Kautsch. u. Gummi,* **9**, WT 300.
Lugg, J. W. H. (1930). *Biochem. J.* **26**, 2144, 2160.
Lur'e, Ya. Ya., Alferova, L. A. and Titova, G. A. (1963). *Zav. Lab.* **29**, 412.
Lykken, L. and Tuemmler, F. D. (1942). *Ind. Eng. Chem., Anal. Ed.* **14**, 67.
McAllister, R. A. (1952). *J. Pharm. Pharmacol.* **4**, 311.
McCormack, S., Goldzieher, J. W. and Besch, P. K. (1960). *Biochim. Biophys. Acta,* **38**, 293.

McCoy, R. N. and Weiss, F. T. (1954). *Anal. Chem.* **26**, 1928.

MacDonnell, R. L., Silva, R.B. and Feeney, R. E. (1951). *Arch. Biochem. Biophys.* **32**, 288.

Maclaren, J. A., Leach, S. J. and Swan, J. M. (1960). *Proc. Intern. Wool Textile Res.*, 2nd Conf. No. 12, 655; *Chem. Abs.* **56**, 655.

McMurray, C. H. and Trentham, D. R. (1969). *Biochem. J.* **115**, 913.

Mairesse-Ducarmois, C. A., Vandenbalck, J. L. and Patriarche, G. J. (1973). *J. Pharm. Belg.* **28**, 300.

Malinský, J. and Černoch, M. (1959). *Acta Univ. Palackianae Olomucensis,* **18**, 91; *Chem. Abs.* **54**, 22807.

Malisoff, W. M. and Anding, jr., C. E. (1935). *Ind. Eng. Chem., Anal. Ed.* **7**, 86.

Mapstone, G. E. (1946a). *Austr. Chem. Inst., J. & Proc.* **13**, 232, 373.

Mapstone, G. E. (1946b). *Austr. Chem. Inst., J. & Proc.* **13**, 269.

Mapstone, G. E. (1948). *Austr. Chem. Inst., J & Proc.* **15**, 236.

Mapstone, G. E. (1952). *Petroleum Processing,* **7**, 1655.

Marras, G. and Dall'Olio, G. (1953). *Boll. Soc. Ital. Biol. Sper.* **29**, 1719.

Matoušek, L. and Laučíková, O. (1953). *Chem. Listy,* **47**, 1062.

Matsumoto, H. (1955). *J. Ferment. Technol.* **33**, 223; *Chem. Abs.* **49**, 15107.

Matsumoto, H. and Hlynka, I. (1959). *Cereal Chem.* **36**, 513.

Matsumoto, H. and Shimada, M. (1955). *J. Ferment. Technol.* **33**, 290; *Chem. Abs.* **49**, 16249.

Matsushima, I. (1962). *Coal Tar (Japan),* **14**, 646; *Chem. Abs.* **60**, 2334.

Mauri, C., Vaccari, F. and Kaderavek, G. P. (1954). *Haematologica (Pavia),* **38**, 263; *Chem. Abs.* **48**, 13790.

Medes, G. (1936). *Biochem. J.* **30**, 1293.

Melkonyan, L. G., Bagdasaryan, R. V. and Bunyatyants, Zh. V. (1966). *Armyansk. Khim. Zh.* **19**, 402; *Chem. Abs.* **66**, 3475.

Mescon, H. and Flesch, P. (1952). *J. Invest. Dermatol.* **18**, 261.

van Meurs, N. (1961). *J. Electroanal. Chem.* **2**, 17.

Michaelis, L. and Barron, E. S. G. (1929). *J. Biol. Chem.* **83**, 191.

Middeldorf, R. (1951). *Arzneimittelforsch.* **1**, 311.

Mildner, P., Mihanovic, B. and Wintersteiger-Cvoriscec, D. (1972). *Croat. Chim. Acta,* **44**, 407; *Chem. Abs.* **78**, 81693.

Miles, H. J., Stadtman, E. R. and Kielley, W. W. (1954). *J. Amer. Chem. Soc.* **76**, 4041.

Mironov, G. P., Kaminskii, Yu. G. and Kondrashova, M. N. (1971). *Vop. Med.-Khim.* **17**, 83; *Chem. Abs.* **74**, 136219.

Misra, G. J. and Tandon, J. P. (1970). *Z. Naturforsch.* **25**, 30.

Mizuguchi, J., Takahashi, F. and Saito, Y. (1962). *Nippon Kagaku Zasshi,* **83**, 957.

Moncorps, C. and Schmid, R. (1932). *Z. Physiol. Chem.* **205**, 141.

Montagnani, A., Santoianni, P. and di Ieso, F. (1959). *Boll. Soc. Ital. Biol. Sper.* **35**, 903.

Moore, H., Helweg, H. L. and Graul, R. J. (1960). *Am. Ind. Hyg. Assoc. J.* **21**, 466; *Chem. Abs.* **55**, 6746.

Mrowetz, G. and Klostermeyer, H. (1972). *Z. Lebensm. u. Forsch.* **149**, 134.

Muenze, R. (1969). *Wiss. Z. Karl-Marx Univ. Leipzig, Math.-Naturw. Reihe,* **18**, 621; *Chem. Abs.* **72**, 87007.

Muftic, M. (1970). *Anal. Biochem.* **36**, 539.

Munakata, H. and Niinami, Y. (1963). *Sen-i Gakkaishi* **19**, 392; *Chem. Abs.* **62**, 13307.

Murayama, M. (1957). *J. Biol. Chem.* **228**, 231.

Murayama, M. (1958). *J. Biol. Chem.* **233**, 594.
Murzakaev, F. G. (1966). *Lab. Delo,* 148.
Mussini, A. (1958). *Boll. Soc. Ital. Biol. Sper.* **34**, 1419.
Mylius, F. and Mazzucchelli, A. (1914). *Z. Anorg. Chem.* **89**, 1.
Nagai, Y., Kono, T. and Funahashi, S. (1957). *Bull. Agr. Chem. Soc. Japan,* **21**, 121; *Chem. Abs.* **51**, 10682.
Nakamura, Y. and Nemoto, Y. (1961). *Sen-i Gakkaishi,* **17**, 428; *Chem. Abs.* **55**, 17019.
Nakamura, K., Yao, T. and Sato, S. (1968). *Nippon Jozo Kyokai Zasshi,* **63**, 673; *Chem. Abs.* **69**, 75583.
Nametkin, S. S., Putsillo, V. G. and Shcheglova, E. P. (1943). *Bull. Acad. Sci. URSS, Classe Sci. Tech.* No. $\frac{1}{2}$ 10; *Chem. Abs.* **39**, 2863.
Namikoshi, K. and Yamakawa, S. (1973). *Bunseki Kagaku,* **22**, 776.
Nedić, M. and Berkeš, I. (1963). *Acta Pharm. Jugoslav.* **13**, 13.
Obolentsev, R. D., Salimgareeva, I. M., Zaluzhnaya, V. I. and Baikova, A. Ya. (1968). *Khim. Seraorg. Soedin., Soderzh. Neftyakh Nefteprod.* **8**, 280; *Chem. Abs.* **71**, 126779.
Ocker, H. D. and Rotsch, A. (1959). *Brot u. Gebäck,* **13**, 165.
Oehme, F. (1960). *Erdöl u. Kohle,* **13**, 394.
Oelsner, W. and Huebner, G. (1964). *Chem. Tech.* (*Berlin*), **16**, 432.
Oganesyan, S. S. and Dzhbanibekova, V. A. (1958). *Dokl. Akad. Nauk Armanyan SSR,* **27**, 227; *Chem. Abs.* **53**, 15172.
Oganesyan, S. S. and Zaminyan, T. S. (1960). *Vopr. Radiobiol. Akad. Nauk Arm. SSR, Sektor Radiobiol., Sb. Trudov,* **1**, 107; *Chem. Abs.* **56**, 15761.
Ohara, M., Kurata, J. and Haga, T. (1947). *Igaku to Seibutsugaku,* **11**, 344; *Chem. Abs.* **47**, 9396.
Okada, K. and Yonezawa, D. (1967). *Nippon Nogei Kagaku Zasshi,* **41**, 329; *Chem. Abs.* **67**, 115855.
Okita, T. (1967). *Koshu Eiseiin Kenkyu Hokoku,* **16**, 59; *Chem. Abs.* **68**, 117009.
Okita, T. (1970). *Atmos. Environ.* **4**, 93; *Chem. Abs.* **72**, 82678.
Okulov, V. I. (1965). *Lab. Delo,* 591.
Ozawa, K. and Egashira, S. (1962). *Bunseki Kagaku,* **11**, 509.
Palmer, T. A. (1962). *Diss. Abs.* **22**, 3366.
Papp, J. (1971). *Svensk Papperstidn.* **74**, 310; also *Cellul. Chem. Technol.* **5**, 147; *Anal. Abs.* **22**, 2404; **23**, 453.
Papp, J. and Havas, J. (1970). *Proc. Anal. Chem. Conf.,* 3rd 2, 41; *Chem. Abs.* **74**, 49424.
Pellerin, E. and Gautier, J. A. (1961). *Ann. Pharm. Franç.* **19**, 81.
Pepe, F. A. and Singer, S. J. (1956). *J. Amer. Chem. Soc.* **78**, 4583.
Peter, F. and Rosset, R. (1974). *Anal. Chim. Acta,* **70**, 149.
Peurifoy, P. V. and O'Neal, M. J. (1965). *U.S. Patent* No. 3,208,828 of Sep. 28.
Peurifoy, P. V., O'Neal, M. J. and Dvoretzky, I. (1964). *Anal. Chem.* **36**, 1853.
Pihar, O. (1953). *Chem. Listy,* **47**, 1647, 1652.
Pirie, N. W. (1931). *Biochem. J.* **25**, 614.
Polaczek, L. and Kuszczak, H. (1966). *Chem. Anal.* (*Warsaw*), **11**, 873.
Podlipskii, L. A., Ropyanaya, M. A. and Kruglova, L. I. (1969). *Khim. Tekhnol. Topl. Masel,* **14**, 61; *Chem. Abs.* **70**, 69842.
Porter, M., Saville, B. and Watson, A. A. (1963). *J. Chem. Soc.* (*London*), 346.
Prilezhaeva, E. N., Fedorovskaya, N. P., Miesserova, L. V., Domanina, O. N. and Khaskina, I. M. (1963). *Tr. Inst. Goryuch. Iskop., Akad. Nauk. SSSR,* **21**, 159; *Chem. Abs.* **60**, 6663.

E*

Proctor, C. D. (1956). *Trans. Illinois State Acad. Sci.* **49**, 80; *Chem. Abs.* **51**, 16205.
Prokshin, G. F. (1962). *Izv. Vysshikh Ucheb. Zaved. Lesn. Zh.* **5**, No. 5, 161; *Chem. Abs.* **58**, 7033.
Przybylowicz, E. D. and Rogers, L. B. (1958). *Anal. Chim. Acta*, **18**, 596.
Pszonicka, M. and Skwara, W. (1970). *Chem. Anal. (Warsaw)*, **15**, 175.
Ranganayaki, S. and Srivastava, B. L. (1973). *Microchem. J.* **18**, 699.
Ratkovics, F. and Szepesváry, P. (1958). *Magy. Kem. Folyoirat*, **64**, 472.
Rausch, L. and Ritter, S. (1955). *Klin. Wochschr.* **33**, 1009.
Ray Sarkar, B. C. and Sivaraman, R. (1956). *Analyst (London)*, **81**, 668.
Reguera, R. M. and Asimov, I. (1950). *J. Amer. Chem. Soc.* **72**, 5781.
Reith, J. F. (1934). *Rec. Trav. Chim.* **53**, 18.
Richmond, D. V. and Somers, E. (1966). *Chem. Ind. (London)*, 18.
Riesz, C. H. and Wohlberg, C. (1943). *Am. Gas Assoc. Proc.* **25**, 259.
Robert, B. and Robert, L. (1956). *Ann. Biol. Clin. (Paris)*, **14**, 587.
Roberts, L. W. (1960). *Am. J. Botany*, **47**, 110.
Romováček, J. and Bednář, J. (1958). *Paliva*, **38**, No. 1, 9.
Romováček, J. and Holub, A. (1960). *Sb. Vysoké Skoly Chem.-Technol. v Praze, Oddil Fak. Technol. Paliv Vody*, **4**, 73; *Chem. Abs.* **62**, 14395.
Rosenberg, S., Perrone, J. C. and Kirk, P. L. (1950). *Anal. Chem.* **22**, 1186.
Rossouw, S. D. (1940). *Onderstepoort J. Vet. Sci. Animal Ind.* **14**, 461; *Chem. Abstr.* **35**, 2443.
Rossouw, S. and Wilken-Jorden, T. J. (1935). *Biochem. J.* **29**, 219.
Rothfus, J. A. (1966). *Anal. Biochem.* **16**, 167.
Rublev, V. V., Shishina, N. V. and Molodtsova, V. I. (1974). *Izv. Vyssh. Ucheb. Zaved., Khim. Tekhnol.* **17**, 472; *Chem. Abs.* **81**, 9509.
Sadykhov, I. D. and Korobtsova, M. I. (1961). *Azerbaidzhan. Khim. Zh.* No. 3, 131; *Chem. Abs.* **56**, 7575.
Sainsbury, D. M. and Maw, G. A. (1967). *J. Inst. Brew.* **73**, 293.
Sakai, H. (1968). *Anal. Biochem.* **26**, 269.
Sakai, H. (1972). *Tampakushitsu, Kakusan Koso*, **17**, 211; *Chem. Abs.* **77**, 44899.
Sakamoto, M., Yamada, Y. and Tonami, H. (1969). *J. Appl. Polym. Sci.* **13**, 1845.
Sampey, J. R. and Reid, E. E. (1932). *J. Amer. Chem. Soc.* **54**, 3404.
Sanderson, D., Bittikofer, J. A. and Purdue, H. L. (1972). *Anal. Chem.* **44**, 1934.
Santi, N. and Peillon, E. (1968). *Ann. Pharm. Franç.* **26**, 177.
Saroff, H. A. and Mark, H. J. (1953). *J. Amer. Chem. Soc.* **75**, 1420.
Sasago, K., Wilson, H. K. and Herreid, E. O. (1963). *J. Dairy Sci.* **46**, 1348.
Saville, B. (1961). *Analyst (London)*, **86**, 29.
Saxe, M. H. (1967). *Anal. Chem.* **39**, 1676.
Schoenbach, E. B., Armistead, E. B. and Weissman, N. (1950). *Proc. Soc. Exp. Biol. Med.* **73**, 44.
Schrauwen, J. A. M. (1963). *J. Chromatogr.* **10**, 113.
Schultz, J. and Vars, H. M. (1947). *J. Biol. Chem.* **167**, 715.
Schwartz, A., Pora, E. A., Kis, Z., Madar, I. and Fabian, N. (1961). *Comun. Acad. Rep. Populare Romîne*, **11**, 45; *Chem. Abs.* **55**, 21224.
Segal, W. and Starkey, R. L. (1953). *Anal. Chem.* **25**, 1645.
Selig, W. (1973). *Mikrochim. Acta*, 453.
Selyuzhitskii, G. V., Smol'skii, A. V. and Shvartsman, I. E. (1965). *Tr. Leningrad. Sanit. Gig. Med. Inst.* **81**, 97; *Chem. Abs.* **66**, 93560.
Selyuzhitskii, G. V., Timofeev, V. P. and Nikon, A. M. (1971). *Gig. Sanit.* **36**, 73; *Chem. Abs.* **75**, 137360.

Semco, M. B. (1950). *U.S. Patent* No. 2,529,886 of Nov. 14.

Sement, E., Rousselet, F., Girard, M. L. and Chemla, M. (1972). *Ann. Pharm. Franç.* **30**, 691.

Semenza, G. and Lucchelli, P. (1955). *Boll. Soc. Ital. Biol. Sper.* **31**, 842.

Sen, P. and Bahadur, K. (1974). *Talanta,* **21**, 968.

Senter, S. D., Stone, W. K. and Thomas, W. C. (1973). *J. Dairy Sci.* **56**, 1331.

Serrano Bergés, L. and Fernández, T. (1966). *Anales R. Soc. Esp. Fís. Quim.* **62**, 807.

Shah, R. G. and Gandhi, R. S. (1970). *Text. Res. J.* **40**, 857.

Shakh, Ts. I. and Kagan, F. Yu. (1962). *Farmatsevt. Zh. (Kiev),* **17**, 12; *Chem. Abs.* **59**, 6200.

Shaw, J. A. (1940). *Ind. Eng. Chem., Anal. Ed.* **12**, 668.

Shcherbina, E., Efimova, T. A., Tenenbaum, A. E., Mikhal'skaya, L. I. and Astakhov, V. A. (1971). *Zh. Prik. Khim.* **44**, 1589.

Shemyakin, F. M. and Berdnikov, A. I. (1965). *Aptech. Delo,* **14**, 63.

Shinohara, K. and Kilpatrick, M. (1934). *J. Biol. Chem.* **105**, 241.

Shmakotina, Z. V. and Uksusnikov, V. I. (1968). *Lab. Delo,* 243.

Shol'ts, Kh. F. (1964). *Biokhimiya,* **29**, 577.

Simpson, R. B. and Saroff, H. A. (1958). *J. Amer. Chem. Soc.* **80**, 2129.

Singer, S. J., Fothergill, J. E. and Sheinoff, J. R. (1960). *J. Amer. Chem. Soc.* **82**, 565.

Singh, D. (1959–60). *J. Sci. Res. Banares Hindu Univ.* **10**, 6; *Chem. Abs.* **55**, 8152.

Singh, D. and Varma, A. (1963–4). *J. Sci. Res. Banares Hindu Univ.* **14**, 33; *Anal. Abs.* **12**, 3936.

Sliwinski, R. A. and Doty, D. M. (1958). *J. Agr. Food Chem.* **6**, 41.

Sluyterman, L. A. Ae. (1957). *Biochim. Biophys. Acta,* **25**, 402.

Sluyterman, L. A. Ae. (1966). *Anal. Biochem.* **14**, 317.

Smith, M. J. and Rodnight, R. (1957). *Biochem. J.* **72**, 1P.

Sokolovskii, V. V. (1962). *Lab. Delo,* **8**, 3.

Spacu, P. G. (1939). *Bull. Sect. Sci. Acad. Roumaine,* **22**, 142; *Chem. Abs.* **34**, 1938.

Spanyar, P., Kevei, J. and Blazovich, M. (1964). *Kozp. Elelmiszeripari Kutatointez. Kozlem.* No. 1, 21; *Chem. Abs.* **67**, 656; **68**, 737.

Spragg, S. P. (1958). *Nature (London),* **182**, 1314.

Spray, G. H. (1947). *Biochem. J.* **41**, 360.

Stahl, C. R. and Siggia, S. (1957). *Anal. Chem.* **29**, 154.

Stankevich, B. E. (1971). *USSR Patent* No. 299,795 of Mar. 26.

Staszewski, S. and Zygmunt, B. (1973). *Chem. Anal. (Warsaw),* **18**, 85.

Stebletsova, V. D. and Evstifeev, M. M. (1971). *Metody Khim. Anal. Stokov Vod Predpr. Khim. Prom.* 33; *Chem. Abs.* **78**, 47533.

Steffen, P. (1968). *Nahrung,* **12**, 701.

Stekhun, A. I., Starikova, A. I., Bondarenko, L. P. and Nosal, T. P. (1967). *Khim. Tekhnol. Topl. Masel,* **12**, 56; *Chem. Abs.* **67**, 75093.

Stepień, M. and Gaczyński, R. (1961). *Chem. Anal. (Warsaw),* **6**, 1045.

Stevancevic, D. B. (1962). *Glasnik Hem. Drustva Beograd.* **27**, 367; *Chem. Abs.* **61**, 43.

Strafford, N., Cropper, F. R. and Hamer, A. (1950). *Analyst (London),* **75**, 55.

Stratton, L. P. and Frieden, E. (1967). *Nature (London),* **26**, 932.

Stricks, W. and Chakravarti, S. K. (1961). *Anal. Chem.* **33**, 194.

Studebaker, M. L. and Nabers, L. G. (1959). *Rubber Chem. & Technol.* **32**, 941; *Chem. Abs.* **56**, 2541.

Stutzenberger, F. (1973). *Anal. Biochem.* **56**, 294.

Sullivan, B., Dahle, L. and Larson, R. (1961). *Cereal Chem.* **38**, 272.

Suzuki, M., Coombs, T. L. and Vallee, B. L. (1969). *J. Chromatogr.* **43**, 11.

Švajgl, O. and Holle, B. (1972). *Chem. Prům.* **22**, 286; also *Sb. Pr. Výzk. Chem. Využiti Úhlí, Dehtu Ropy,* No. 12,127; *Chem. Abs.* **78**, 126553.
Swan, J. M. (1965). *Austr. J. Chem.* **18**, 411.
Swenson, A. D. and Boyer, P. D. (1957). *J. Amer. Chem. Soc.* **79**, 2174.
Szydlowska, H. and Junikiewicz, E. (1972). *Folia Histochem. Cytochem.* **10**, 279; *Chem. Abs.* **77**, 161706.
Takai, Y. and Asami, T. (1962). *Soil Sci., Plant Nutr. (Tokio)*, **8**, 132; *Chem. Abs.* **57**, 14206.
Tamele, M. W. and Ryland, L. B. (1936). *Ind. Eng. Chem., Anal. Ed.* **8**, 16.
Tamele, M. W., Ryland, L. B. and Irvine, V. A. (1941). *Ind. Eng. Chem., Anal. Ed.* **13**, 518.
Tamele, M. W., Ryland, L. B. and McCoy, R. N. (1960). *Anal. Chem.* **32**, 1007.
Taniguchi, N. (1971). *Anal. Biochem.* **40**, 200.
Tattrie, B. L. and Connell, G. E. (1967). *Can. J. Biochem.* **45**, 551.
Terent'ev, A. P. and Shor, N. I. (1947). *Zh. Obshch. Khim.* **17**, 2075.
Thibert, R. J. and Ke, P. J. (1971). *Mikrochim. Acta,* 531.
Toennies, G. and Kolb, J. J. (1951). *Anal. Chem.* **23**, 823.
Torchinskii, Yu. M. (1959). *Ukr. Biokh. Zh.* **31**, 589; *Chem. Abs.* **54**, 13204.
Toribara, T. Y. and Koval, L. (1970). *Talanta,* **17**, 1003.
Trop, M., Sprecher, M. and Pinsky, A. (1968). *J. Chromatogr.* **32**, 426.
Tsen, G. C. and Anderson, J. A. (1963). *Cereal Chem.* **40**, 314.
Tupeeva, R. B. (1967). *Gig. Sanit.* **32**, 62.
Ullmann, J. (1959). *Spisy Přirodoved. Fak. Univ. v Brné,* No. 400, 45; *Chem. Abs.* **54**, 15747.
Uraneck, C. A., Burleigh, J. E. and Cleary, J. W. (1968). *Anal. Chem.* **40**, 327.
Ushakov, M. I. and Galanov, A. S. (1934). *Z. Anal. Chem.* **99**, 185.
Uvarova, E. I. and Siridchenko, A. K. (1969). *USSR Patent* No. 246,152 of Jun. 11.

Vajta, L., Szebenyi, I., Horvath, M. and Vermes, E. (1966). *Period. Polytech. Chem. Eng. (Budapest),* **10**, 309; *Chem. Abs.* **67**, 36223.
Vakaleris, D. G. and Pofahl, T. R. (1968). *J. Dairy Sci.* **51**, 1166.
Veibel, S. and Wronski, M. (1966). *Anal. Chem.* **38**, 910.
Vickery, H. B. and White, A. (1932–3). *J. Biol. Chem.* **99**, 701.
Walker, G. T. (1953). *Mfg. Chemist,* **24**, 376, 429.
Wei, Y.-S. (1967). *Bull. Inst. Chem., Acad. Sinica* No. 13, 8; *Chem. Abs.* **67**, 79519.
Weissman, N., Schoenbach, E. B. and Armistead, E. B. (1950). *J. Biol. Chem.* **187**, 153.
Wenck, H., Schwabe, E., Schneider, F. and Flohé, L. (1972). *Z. Anal. Chem.* **258**, 267.
Werner, R. C. and Levy, M. (1958). *J. Amer. Chem. Soc.* **80**, 5735.
Wertheim, E. (1929). *J. Amer. Chem. Soc.* **51**, 3661.
White, B. J. and Wolfe, R. G. (1962). *J. Chromatogr.* **7**, 516.
Wojahn, H. (1951). *Arch. Pharm.* **284**, 243.
Wojnowska, M. and Wojnowski, W. (1973). *Chem. Anal. (Warsaw)*, **18**, 1117.
Wojnowski, W. and Kwiatkowska-Sienkiewicz, K. (1973). *Z. Anorg. Allgem. Chem.* **396**, 333.
Wolfram, L. J. and Lennhoff, M. (1967). *Text. Res. J.* **37**, 145.
Wronski, M. (1958a). *Analyst (London)*, **83**, 314.
Wronski, M. (1958b). *Zesz. Nauk Univ. Lódzkiego,* Ser. II, No. 4, 181; *Chem. Abs.* **54**, 3051.

Wronski, M. (1960a). *Analyst* (*London*), **85**, 527.
Wronski, M. (1960b). *Anal. Chem.* **32**, 133.
Wronski, M. (1960c). *Chem. Anal.* (*Warsaw*), **5**, 295.
Wronski, M. (1960d). *Chem. Anal.* (*Warsaw*), **5**, 511.
Wronski, M. (1960e). *Z. Anal. Chem.* **175**, 432.
Wronski, M. (1961a). *Z. Anal. Chem.* **180**, 185.
Wronski, M. (1961b). *Z. Anal. Chem.* **184**, 193.
Wronski, M. (1961c). *Chem. Anal.* (*Warsaw*), **6**, 859.
Wronski, M. (1961d). *Acta Chim. Acad. Sci. Hung.* **28**, 303.
Wronski, M. (1962). *Z. Anal. Chem.* **192**, 294.
Wronski, M. (1963a). *Chem. Anal.* (*Warsaw*), **8**, 467.
Wronski, M. (1963b). *Analyst* (*London*), **88**, 562.
Wronski, M. (1964a). *Z. Anal. Chem.* **206**, 352.
Wronski, M. (1964b). *Analyst* (*London*), **89**, 800.
Wronski, M. (1965a). *Analyst* (*London*), **90**, 697.
Wronski, M. (1965b). *Talanta*, **12**, 593.
Wronski, M. (1966a). *Z. Anal. Chem.* **217**, 265.
Wronski, M. (1966b). *Chem. Anal.* (*Warsaw*), **11**, 1153; *Analyst* (*London*), **91**, 745.
Wronski, M. (1967). *Biochem. J.* **104**, 978.
Wronski, M. (1968a). *Faserforsch. Textiltech.* **19**, 40.
Wronski, M. (1968b). *Talanta*, **15**, 241.
Wronski, M. (1970). *Mikrochim. Acta*, 955.
Wronski, M. (1971a). *Z. Anal. Chem.* **253**, 24.
Wronski, M. (1971b). *Chem. Anal.* (*Warsaw*), **16**, 439.
Wronski, M. (1972). *Chem. Anal.* (*Warsaw*), **17**, 1183.
Wronski, M. and Burkart, P. (1958). *Faserforsch u. Textiltech.* **9**, 36.
Wronski, M. and Goworek, W. (1968). *Chem. Anal.* (*Warsaw*), **13**, 197.
Wronski, M. and Kudzin, Z. (1974). *Chem. Anal.* (*Warsaw*), **19**, 453.
Yakovlev, V. A. and Sokolovskii, V. V. (1955). *Dokl. Akad. Nauk SSSR*, **101**, 321.
Yakovlev, V. A. and Torchinskii, Yu. M. (1958). *Biokhimiya*, **23**, 755.
Yoshida, K. (1954). *Japan J. Nation's Health*, **23**, 362; *Chem. Abs.* **50**, 5821.
Yoshida, K. and Kurihara, M. (1952). *Bunseki Kagaku*, **1**, 89.
Yoshino, U., Wilson, H. K. and Herreid, E. O. (1962). *J. Dairy Sci.* **45**, 1459.
Yurow, H. W. and Sass, S. (1971). *Anal. Chim. Acta*, **56**, 297.
Zahler, W. L. and Cleland, W. W. (1968). *J. Biol. Chem.* **243**, 716.
Zahn, H., Gerthsen, T. and Meichelbeck, H. (1962). *Melliand. Textilber*, **43**, 1179.
Zak, R., Curry, W. M. and Dowben, R. M. (1965). *Anal. Biochem.* **10**, 135.
Zerewitinoff, T. (1908). *Chem. Ber.* **41**, 2236.
Zijp, J. W. H. (1956). *Rec. Trav. Chim.* **75**, 1060.
Zittle, C. A. and O'Dell, R. A. (1941). *J. Biol. Chem.* **139**, 753.
Zweig, G. and Block, R. J. (1953). *J. Dairy Sci.* **36**, 427.

3. REACTIONS OF ACYLATION, ARYLATION AND ALKYLATION USING HALIDES AND SIMILAR REAGENTS

These reactions may be formulated generally:

$$RSH + Hal\text{—}R' \rightarrow RS\text{—}R' + H^+ + Hal^-$$

R' contains an activating group, such as carbonyl or nitro.

The reaction products may furnish evidence of a thiol in an application of the reaction as a test; be useful derivatives for identification, sometimes after further treatment; or be evaluated in a quantitative determination. The reaction may be used also to remove thiol groups and to prevent their interference in other analytical procedures.

Quantitative determination through reagent consumption is extremely rare, mainly on account of the instability of most of the reagents.

A useful classification of analytical information is according to the reagent type, with subheadings:

$$O$$
$$\|$$

1. Reaction with acid halides, R—C—Hal

$$O$$
$$\|$$

2. Analogous reaction with acid anhydrides, $(R\text{—}C\text{—})_2O$

3. Reaction with activated aromatic halides, Ar—Hal

$$O$$
$$\|\quad|$$

4. Reaction with β-halocarbonyl compounds, R—C—C—Hal
$$|$$

$$|$$

5. Reaction with β-aromatically substituted halides, Ar—C—Hal
$$|$$

6. Miscellaneous halides

1. REACTION WITH ACID HALIDES

The identification of hydroxy-compounds and amines by conversion into esters or amides with acid chlorides and subsequent determination of melting points is well known. This has been applied also to thiols, probably first by Wertheim (1929), using 3,5-dinitrobenzoyl chloride. The melting points of the lower aliphatic thiol esters are inconveniently low and similar, but higher alkanethiols and arenethiols could be better identified:

$$RSH + R'COCl \rightarrow RS'COR' + HCl$$

Saunders (1934) identified cysteine and thiolactic acid also through their derivatives with 3,5-dinitrobenzoyl chloride. He dissolved the amino acid in excess alkali and added the calculated amount of reagent. After shaking for 2 min the derivative was yielded and could be precipitated by acidification.

El S. Amin (1958) characterized thiols by reacting with *p*-nitrophenylazo-benzoyl chloride,

$$O_2N-\!\!\left\langle\!\!\bigcirc\!\!\right\rangle\!\!-N=N-\!\!\left\langle\!\!\bigcirc\!\!\right\rangle\!\!-COCl$$

in pyridine/benzene for 1–2 days at 5°C, or at room temperature, or at 50°C, depending on the volatility. After adding water and extracting with benzene–ether (1 + 1), the organic layer was ultimately chromatographed on alumina and the thiol esters were eluted with benzene.

N,N-Diphenylcarbamoyl chloride, $(C_6H_5)_2N\!-\!COCl$, is another acid chloride used (by Hiskey *et al.*, 1961) to prepare N,N-diphenylthiocarbamates for identification of thiols. The sample in ethanol was treated with sodium and the product was added to ethanolic reagent. After warming for 5 min, the end-product was filtered and crystallized.

The standard quantitative differential method for determining alcohols and amines was applied to thiophenol by Olson and Feldman (1937). They mixed the sample in toluene with ca. 50% excess of acetyl chloride in pyridine at 0°C and allowed these to react for 20 min at 60°C. They then added water and titrated with standard alkali to phenolphthalein. A control with the reagent alone was carried out; the difference between the two titrations corresponds to thiol:

$$ArSH + CH_3COCl \rightarrow ArS-COCH_3 + HCl$$
$$H_2O + CH_3COCl \rightarrow CH_3COOH + HCl$$

Umbreit and Houtman (1967) determined some thiols (also alcohols and amines) by reaction with 3,5-dinitrobenzoyl chloride. They then decomposed the excess of reagent and evaluated the red colour yielded by the ester with alkali at 555 nm.

Some investigators have converted thiols to thiolo esters with an acid chloride as a preliminary step in a chromatographic procedure. Thus Gasparič and Borecký (1961) treated thiols (also hydroxy- and amino-compounds) with 3,5-dinitrobenzoyl chloride and then tested the separation of the thiolo esters by paper chromatography, using 20 solvent systems. Schwartz and Brewington (1968) used the 2,6-dinitrophenylhydrazone of pyruvic acid chloride to prepare such derivatives:

$$RSH \ + \ CH_3-\underset{\underset{N-NH-}{\overset{\|}{O}}}{C}-COCl \text{(2,4-dinitrophenyl)} \longrightarrow CH_3-\underset{\underset{N-NH-}{\overset{\|}{O}}}{C}-COSR \text{(2,4-dinitrophenyl)}$$

They separated these coloured products on silica gel G, using hexane–benzene–diethylamine $(3 + 2 + 5)$ and also by column chromatography. Thiols in food aromas were converted by Gascó and Barrera (1972) into thiolobenzoate esters by reaction with benzoyl chloride/pyridine; they separated the esters by GLC on 5% SE-20 silicone gum rubber on Chromosorb G-HP at 125°C. Recently, Korolczuk et al. (1974) likewise prepared thiolobenzoate esters of thiols containing up to 8 carbon atoms, then separated them by GLC on various supports impregnated with 4% SE-20 at temperatures programmed from 150–250°C at 10° per min. They found that the thiolobenzoates were superior to other tested derivatives (dinitro-phenylthioethers). To prepare the esters, they used both the Schotten-Baumann procedure with benzoyl chloride in aqueous sodium hydroxide, and also the method in pyridine solvent; the former was preferred.

2. REACTION WITH ACID ANHYDRIDES

Derivatives for identifying thiols, analogous to those with hydroxy-compounds and amines, can be prepared with anhydrides as well as with acid chlorides:

$$RSH + (R'CO)_2O \rightarrow RSCOR' + R'COOH$$

For example, Wertheim (1929) recommended 3-nitrophthalic anhydride to prepare these thiolo esters.

There are few quantitative applications of acylation with anhydrides. Toennies and Kolb (1942) determined non-basic hydrogen in amino acids, e.g. the thiol group in cysteine, using an acetic anhydride/acetic acid reagent containing perchloric acid to yield a perchlorate salt with the amino group and prevent its acetylation. After 2 h at room temperature, an aliquot of the reaction mixture was added to a measured excess of anthranilic acid to react with the perchloric acid and unreacted anhydride. They titrated the unused anthranilic acid from this last reaction after 3 h, using perchloric acid–acetic acid and crystal violet as indicator. Schenk and Fritz (1960) determined alcohols, amines and also thiols (e.g. propanethiol) with an acetic anhydride–ethyl acetate–perchloric acid reagent employing a reaction time of 5 min. They then added aqueous pyridine and titrated with sodium hydroxide to cresol red–thymol blue mixed indicator. They performed a control determination without sample; the difference in titre from that using the sample gives the amount of thiolo ester and hence of thiol. Berger and Magnuson (1964) applied Schenk and Fritz's method to mercapto-silanes, and Magnuson and Knaub (1965) used a similar method to determine mercaptogermanes; they merely replaced the ethyl acetate solvent by 1,2-dichloroethane and used phenolphthalein indicator.

Esters of the N-acylated amino acids may be gas chromatographed, and this has found considerable application in recent years, particularly with the lower alkyl esters of N-acetyl- and N-trifluoroacetyl-derivatives, prepared using acetic or trifluoracetic anhydrides, respectively. Although cysteine is usually among the amino acid examples, it is evidently N-acylated, so that this technique is only mentioned here without entering into details.

A colour reaction for the thiol group is from Palomo Coll and Palomo Coll (1963). A colour with a red element (rose, scarlet, orange) is given when the sample is treated with acetic acid, acetic anhydride and drops of a concentrated nitric acid/acetic acid/urea reagent. It is possible that nitrous acid is formed and reacts to give RSNO; the rôle of the anhydride is then not clear.

3. REACTION WITH ACTIVATED AROMATIC HALIDES

Nearly all the procedures given under this heading were carried out with 1-chloro- or 1-fluoro-2,4-dinitrobenzene. These reagents introduce chromophoric nitro groups which facilitate detection and quantitative determination.

Bost et al. (1932) first used 1-chloro-2,4-dinitrobenzene to identify thiols. They mixed the sample in absolute ethanol with an equivalent amount of dilute sodium hydroxide and with the reagent in ethanol. Most thiols reacted sensibly instantaneously but a good yield was ensured by refluxing for 10 min. The 2.4-dinitrophenyl sulphides separated on cooling:

Since their melting points lie rather near together, they are best oxidized to sulphones in acetic acid with a ca. 50% excess of aqueous permanganate. The sulphones mostly separate and can be crystallized from absolute ethanol. This procedure has been adopted in standard handbooks of organic qualitative analysis. The oxidation to sulphones is not always necessary and performed. Thus Grogan *et al.* (1955) characterized polymethylene dithiols by reaction with 1-chloro-2,4-dinitrobenzene (and sulphenyl chlorides, see below) to yield bis-sulphides. Carson and Wong (1957) applied the method to thiols, including also those derived from disulphides by reduction with lithium aluminium hydride; they subsequently chromatographed the sulphides on alumina containing 1% of fluorescent zinc sulphide, on which the compounds were recognizable in u.v. light as dark shadows against a yellow fluorescing background. In another example, Rittner *et al.* (1962) identified the sulphides through their X-ray diffraction patterns.

Paper and thin-layer chromatography is used to separate the dinitrophenyl sulphides in analyses of thiol mixtures. Thus Day and Patton (1959) applied this to thiols (1–7 C-atoms) and mercaptoacids, using ascending paper chromatography with heptane as mobile phase and methanol as stationary phase; they detected down to 30 ng of sulphide in u.v. light of 360 nm wavelength. Folkard and Joyce (1963) studied volatile thiols (and disulphides) aspirated from cultures of marine organisms or reaction mixtures. They prepared the 2,4-dinitrophenyl sulphides from the chloro-reagent and separated them by chromatography on Whatman 3MM paper impregnated by liquid paraffin and using chloroform–methanol–water–liquid paraffin (15 + 20 + 9 + 6) as mobile phase.

Génévois (1964) reviewed analysis based on conversion into, and separation of, polynitro derivatives and mentioned the reaction of thiols with 1-fluoro-2,4-dinitrobenzene. According to Larroquère (1961), quoted by Génévois, other groups (such as amino groups) do not react at pH 6; he discussed the possibilities of chromatographic separation.

Brady and Hoskinson (1971) prepared the 2,4-dinitrophenyl derivatives of amino acids and of certain pyrimidines and purines and chromatographed on thin layers of unmodified and esterified keratin, using neutral and basic (with ammonia) solvents. Compounds such as cysteine yield a N,S-diaryl derivative (so that their R_f-values are then very low). The derivatives were detected in visible or u.v. light. In another recent work, Thamm (1972) detected mercapto acids (e.g. mercaptoacetic, thiolactic and 3-mercapto-propionic acids) in permanent wave lotions by converting to the 2,4-dinitro-phenyl sulphides and carrying out thin-layer chromatography on silica gel; they visualized with mercuric chloride or lead acetate.

Some examples of quantitative application of the reaction can be given: Dahmen *et al.* (1963) converted thiols in petroleum products to sulphides with 1-chloro-2,4-dinitrobenzene and potassium hydroxide in alcoholic milieu, separated the products on silica gel and evaluated them photo-metrically at 355 nm; they found that the molar absorptivities of sulphides from primary and secondary thiols were double those of sulphides derived from tertiary thiols. Obara *et al.* (1966) separated thiols as their 2,4-dinitro-phenyl sulphide derivatives (from the fluoro-reagent and sodium hydrogen carbonate in acetone, left overnight at ambient temperature) by thin-layer chromatography on silica gel with the mobile phase benzene–xylene–carbon tetrachloride (2 + 1 + 1). They located the zones in u.v. light, scraped them off and extracted with acetone. After evaporation to dryness and dissolution in ethanol, they were evaluated at ca. 330 nm or were dissolved in carbon tetrachloride and their IR-spectra recorded.

Details can be given of the method of Obtemperanskaya and Tikhonov (1966) for alkane- and arenethiols. A solution of 6 to 9 mg of sample in 100 ml of dioxan is prepared and an aliquot of several ml is mixed with 1 ml of dioxan solution containing 9 mg of 1-fluoro-2,4-dinitrobenzene and 200 mg of potassium fluoride. This is heated for 5 min at 50–60°C. Fifteen ml of 0·1 N sodium hydroxide are then added and unused reagent is destroyed by heating for 15 min at 90–95°C. The solution is extracted for 5 min with 15 ml of hexane, and 5 ml of the hexane layer are made up to 25 ml with acetone and 0·6 ml of 2 N sodium hydroxide. The absorbance of this solution is evaluated after 35 min at 640–660 nm or after 13 h at 550–560 nm for tertiary thiols. The colour reaction of the 2,4-dinitrophenyl sulphide is that of Janovský and later Obtemperanskaya and co-workers (1970) reviewed this principle of introducing a 2,4-dinitrophenyl group, extracting and then developing the colour according to the Janovský reaction with an active methylene group (acetone) in alkaline conditions.

Zuber *et al.* (1955) used paper electrophoresis to separate cysteine as S-(2,4-dinitrophenyl)cysteine from wool. The reaction was carried out on the wool sample with the fluoro-reagent in acetate buffer of pH 5 at 60°C for 5 h. The wool sample was then subjected to acid hydrolysis and after electrophoresis of the hydrolysate the cysteine derivative was located in u.v. radiation, the zone cut out, eluted with dilute hydrochloric acid and evalu-ated at 320–340 nm. Later, Zahn and co-workers (1961) used 1-fluoro-2,4-dinitrobenzene containing ^{14}C in various buffers from pH 2·52 up to 9.3. After hydrolysis and electrophoresis, they used a methane gas flow counter to evaluate the electropherogram. They studied the effect of pH, reagent concentration, temperature and time on the reaction with cysteine but found that it was not quantitative, even at lower pH.

Gascó and Barrera (1972) reacted thiols from materials such as fish with 1-chloro-2,4-dinitrobenzene and oxidized the sulphide products with permanganate to sulphones. They subjected sulphides and sulphones to GLC on 5% SE-20 on Chromosorb G-HP at 185°C. Korolczuk *et al.* (1974)

also prepared 2,4- and 2,5-dinitrophenyl sulphides of thiols containing up to 8 carbon atoms, and studied their separation in GLC on various carriers impregnated with 4% SE-30 at temperatures programmed from 200 to 300°C at 10°/min.

Tonialo *et al.* (1972) followed the progress of titration of peptide thiol groups with 2-fluoro-3-nitropyridine solution through measurements of circular dichroism at 367 nm.

Birkett *et al.* (1970) used 7-chloro-4-nitrobenzo-2-oxa-1,3-diazole (NBD chloride) in sodium citrate (pH 7) containing EDTA as a specific reagent for thiol groups in amino acids and proteins.

Obtemperanskaya and Egorova (1969, 1970) titrated alkane- and arene-thiols potentiometrically in dioxan–isopropanol (1 + 5) with tetraethyl-ammonium hydroxide in benzene–ethanol (7 + 1) in a nitrogen atmosphere. They found that accuracy was greatly improved by adding a 2 to 3-fold excess of certain polynitroaromatic halides, e.g. 1-chloro- or 1-bromo-2,4-dinitrobenzene or 1-chloro-2,4-dinitronaphthalene. Probably these react as above to liberate hydrogen chloride which is easily titrated.

4. REACTION WITH β-HALOCARBONYL COMPOUNDS

This reaction leads to the formation of β-ketothioethers:

$$RSH + Hal—\overset{|}{\underset{|}{C}}—CO—R' \rightarrow RS'—\overset{|}{\underset{|}{C}}—CO—R' + HHal$$

The reagents most used are the halosubstituted acetates, bromo- and iodo-acetate, together with iodoacetamide.

Work on this reaction dates back to the thirties. Dickens (1933) and Rapkine (1933) reported the inhibition of glycolysis by the reaction of iodoacetate and iodoacetamide with the thiol group in glutathione, and showed that iodoacetate may react not only with the thiol groups of simpler molecules such as cysteine and glutathione, but also with those of proteins. Michaelis and Schubert (1934) doubted whether the two reagents were really specific for the thiol group since they may also react with amino groups, e.g. in amino acids; however, the thiol group of cysteine reacts but the amino group does not when too large an excess of reagent is avoided. Barron (1951) also stated that iodoacetate is not specific for thiols and will react with the amino group of amino acids at physiological pH values. Smythe (1936) measured the rates of reaction of thiols with iodoacetate and iodoacetamide, following the reaction through the evolution of carbon dioxide from sodium hydrogen carbonate as a result of the acid formed. The amide was found to react faster than the salt (pH 6·1) and the order of reaction rates for the thiols was thioglucose > thiosalicylic acid > cysteine > glutathione > thioglycol.

4.1. Blocking of thiol groups

The halide reagents are nevertheless used for demonstration of thiol groups through their blocking action. Barron and Singer (1945) mentioned iodoacetate and iodoacetamide as reagents for detecting enzyme thiol groups. Board (1951) observed that the images given on Eastman NTB photographic emulsion by thiols such as cysteine, glutathione and dimercaptopropylurea were all inhibited by pretreatment with iodoacetate. More recently, Feigl and Caldas (1969) eliminated anions such as mercaptides, RS^-, by warming for 5 min with chloroacetate in alkaline solution.

This inactivation of thiols has found quantitative use. Mirsky and Anson (1935) blocked cysteine in their determination of cystine in presence of cysteine, using the Folin tungstophosphate method. Brdička (1938) used iodoacetate to eliminate thiol groups in polarographic determinations. Sanford and Humoller (1947) determined cystine in hydrolysed hair in the presence of cysteine by alkylating the latter with an excess of 1% sodium iodoacetate at pH 8·3 for 30 min at 100°C. They then employed the Sullivan method with 1,2-naphthoquinone-4-sulphonate to determine the residual cystine, and also to determine the sum of cysteine and cystine; the amount of the former was obtained by difference. Sandegren et al. (1950) used Brdička's polarographic method to determine protein disulphide + thiol and then the former alone after inactivating the thiol with iodoacetate. In work on the disulphides in rats, Bonting (1950) likewise determined cysteine + cystine in borate buffer of pH 10 using Brdička's procedure. After treatment with iodoacetate, he then estimated cystine alone. Bromoacetate was employed by Baraud (1966) in the determination of protein amino acids. He estimated cysteine through its iodine consumption at pH 0·7 to 3·25; the iodine consumption after first treating with bromoacetate was then measured to give a blank value.

4.2. Other reagents with identifiable groups

· Barrnett et al. (1955) reported preliminary histochemical work with 4-iodoacetamido-1-naphthol. Roberts (1956) described its use to detect protein-bound thiol groups in plant tissues. He used a reaction time of 2 h at 60°C in a mixture of ethanol and barbital buffer at pH 8·5. A coloured product for visualization was then obtained by coupling for 3–5 min with tetrazotized di-o-anisidine.

Fasold et al. (1964) introduced 2,2'-dicarboxy-4,4'-bis(iodoacetamido) azobenzene as a water-soluble, cleavable reagent for linking free thiol groups in proteins. It was claimed to react only with thiol groups. With glutathione it yielded a product separable from excess reagent by paper electrophoresis at pH 1·9 then reducible with dithionite to a ninhydrin-positive amine detectable by diazotization.

A sensitive test for 4-thiouracil and 4-thiouridine in amounts down to 100 nmol, proposed by Secrist *et al.* (1971), was based on reaction at 37°C and pH 8·3 to 8·4 with 4-bromomethyl-7-methoxy-2-oxo-2H-benzopyran,

a vinylogous β-halocarbonyl compound. The reaction product shows a white fluorescence in radiation of 400 nm. The water-insoluble reagent is adsorbed on to Celite to enable it to be used in aqueous solution.

4.3. Chromatography of derivatives

White and Wolfe (1962) carried out paper chromatography of the iodo-acetamide derivatives of thiols (e.g. cysteine, glutathione, thiomalic acid), visualizing with ninhydrin/acetone. Sanso and Rigoli (1970) separated cysteine and its methyl ester from other amino acids (including cystine) by reacting with iodoacetamide at pH 9 and then using thin-layer chromatography on silica gel GF_{254}; they too visualized with ninhydrin, which gave a yellow product.

4.4. Quantitative determination

Most quantitative applications of the reaction depend on the estimation of either the acid or the halide (iodide) yielded in the reaction. There are examples also of radioactive or spectrophotometric estimation of the organic reaction product.

4.4.1. Iodide Estimation

Rosner (1940) in studies of the reaction between denatured egg albumin with iodoacetate at pH 7·3 acidified the reaction mixture, added 3 % hydrogen peroxide and determined the iodine formed from the liberated iodide photometrically. Mazur *et al.* (1950) applied Rosner's method to determine thiol groups of ferritin or apoferritin through reaction with iodoacetamide at pH 7 or 7·4 (phosphate buffer). In their study of the "Bohr effect" with proteins, Benesch and Benesch (1961) treated haemoglobin thiol groups with iodoacetamide. They then precipitated the protein with trichloroacetic acid and also estimated iodide in the filtrate from this by adding 3 % hydrogen peroxide and sulphuric acid and evaluating the iodine at 465 nm.

Potentiometric titration of the iodide with silver ion was used by Stein and Guarnaccio (1960). They estimated thiols in reduced hair keratin by treating with solid iodoacetamide in pH 10 borate buffer for 45 min at room temperature with frequent stirring. They adjusted the pH of an aliquot to 7 before the final titration. Watts *et al.* (1961) similarly determined protein (and other) thiols by reaction with iodoacetate or iodoacetamide in borate buffer of pH 9, estimating the liberated iodide finally with a silver/silver iodide electrode.

Schreiber (1971) used the β-halocarbonyl reagent 1-(2-carboxyethyl)-4,5-dichloro-6-pyridazinone and its dibromo analogue to determine thiols. He

carried out the reaction at pH 10(chloro) or 9(bromo), then acidifying with sulphuric acid and titrating potentiometrically with silver nitrate. He was thus able to determine amounts down to ca. 20 μmol of compounds such as L-cysteine, glutathione, 2-mercaptoethanol, and mercaptoacetic acid.

Wenck *et al.* (1972) compared five methods for thiol determination and considered the best to be reaction with iodoacetate followed by potentiometric titration with silver ion using a silver/silver iodide electrode.

4.4.2. Acid Estimation

Fraenkel-Conrat *et al.* (1952) determined thiol groups in β-lactoglobulin by reaction with iodoacetamide at pH 8, then titrating with sodium hydroxide back to this value. In Benesch and Benesch's method (1957) for thiols, iodoacetamide in 2–5 fold excess is added at pH 5·5. The pH is then adjusted to 9(phenolphthalein) with alkali and, after 2 min at room temperature, the mixture is titrated with acid back to pH 5·5 (methyl red). The same authors (1961) titrated back to pH 7·3 in a study concerned with haemoglobin thiols. Brenner *et al.* (1960) determined disulphides in beer by electrolytic reduction to thiols and then applying Benesch and Benesch's method. Singh and Varma (1963–4) treated thiols for 8 min with iodoacetate at pH 9·2 to 10; they then added an excess of sodium hydroxide and back-titrated potentiometrically with hydrochloric acid.

4.4.3. Spectrophotometric Methods

Avi-Dor and Mager (1956) reacted thiols, e.g. cysteine and glutathione, with fluoropyruvate to yield products with absorption maxima at 265–275 nm, or higher if unsubstituted α- or β-amino groups are present, as in cysteine:

$$RSH + FCH_2CO—COO^- \rightarrow RSCH_2CO—COO^- + HF$$

They mixed aqueous solutions of sample and reagent and allowed the reaction to take place for 5 min. They then estimated cysteine from the absorbance at 300 nm ($\varepsilon = 5800$); glutathione interferes negligibly since its maximum is at 268 nm and the molar absorptivity only 670. In later work, Avi-Dor and Lipkin (1958) utilized the enhanced absorbances and bathochromic shifts brought about by borate. After reaction for 5 min in borate medium, they brought the pH to 2 with hydrochloric acid and evaluated glutathione through the absorbance at 285 nm. Cysteine and glutathione can be determined by measurements at 285 and 320 nm and the evaluation of simultaneous equations. Herrington et al. (1967) determined aminoethanethiols in biological materials by a method based on reaction with 3-fluoropyruvate at pH 9, and absorbance measurements in the 270 to 300 nm range.

4.4.4. Radiometric Methods

Free haptoglobin thiol was examined by Tattrie and Connell (1967) by reaction with ^{14}C-iodoacetate in 7·2 M guanidine hydrochloride. After acidic hydrolysis, they determined the ^{14}C-carboxymethylcysteine content radiometrically. Goren et al. (1968) used ^{14}C-bromoacetate to prepare e.g. ^{14}C-carboxymethylcysteine from protein thiols. After separation on an amino acid analyser (Beckman Spinco 120B) they passed the radioactive products through a flow cell for scintillation counting.

5. REACTION WITH β-AROMATICALLY SUBSTITUTED HALIDES

Bucher (1951) prepared derivatives for identifying thiouracils by warming for 1 min with benzyl chloride in the presence of alkali:

$$RSH + ClCH_2C_6H_5 \rightarrow RSCH_2C_6H_5 + HCl$$

Sawicki et al. (1958) reacted many aromatic compounds with benzal or piperonal chloride in trifluoroacetic acid/chloroform. The diarylmethane products showed characteristic colours and absorption maxima. Down to 0·1 μg of thiophenol in 3 ml solution could be detected after 5–10 min reaction time, as an orange product of absorption maximum 482 nm.

Kawahara (1968, 1969, 1971) detected thiols in water (after concentration with active charcoal and extracting with 5% alkali) through conversion to a sulphide with α-bromo,2,3,4,5,6-pentafluorotoluene by heating in acetone +

potassium carbonate. This product gives a high response in the electron capture detector and was thus separated and detected by GLC using a tritium foil detector of this type and a column of 5% FS 1265 + 3% DC 200(1 + 1) on Chromosorb P of 60–80 mesh at 200° with nitrogen as carrier gas.

6. MISCELLANEOUS HALIDES

Alkyl halides have been used for blocking thiol groups, e.g. ethyl bromide by Patterson et al. (1941) in the determination of cystine in hair. A reaction time of 18–20 h was needed, after which they hydrolysed in the usual way with 6 N hydrochloric acid, to which the ethylated product is stable. They then determined the cystine by Sullivan's method. The long reaction time is a disadvantage and these reagents have been displaced by the more reactive halides.

Bacq, Desreux, Fredericq, Fischer (1946, 1957) investigated the reaction of a number of halide reagents with protein thiol groups. Apart from some already mentioned (iodoacetate, iodoacetamide) they found that chloropicrin, CCl_3NO_2, was capable of blocking thiol groups. It appears, however, that this oxidizes to disulphide.

Sulphenyl halides, reacting to give disulphides, have proved to be useful reagents:

$$RSH + ArSCl \rightarrow RSSAr + HCl$$

Grogan et al. (1955) characterized polymethylenedithiols through their derivatives with 1-chloro-2,4-dinitrobenzene (see above), and also with 2-nitro- and 2,4-dinitrobenzenesulphenyl chlorides. They used the technique of Kharasch et al. (1953) for hydroxycompounds. Langford and Lawson (1957) surveyed characterization with the 2,4-dinitro-reagent, mentioning thiols. They added these to a solution of the reagent in carbon tetrachloride and warmed for a few min on the water bath. Solvent was removed to facilitate crystallization and the disulphide was crystallized from chloroform or ethanol. Böhme and Stachel (1957) mix a suspension of the reagent in cold ethanol with the thiol. The latter goes into solution on shaking and the disulphide finally separates. They recrystallized from ethanol or petroleum ether.

Obtemperanskaya and Kalinina (1971) applied the reaction quantitatively to determine primary and tertiary thiols, aliphatic and aromatic. Hydroxyl, amino or carboxyl groups do not interfere. A 5–10 mg sample is treated with 10 ml of 0·02 M 2-nitrobenzenesulphenyl chloride in acetic acid. After 2–3 min, 5 ml of 5% potassium iodide and 40 ml of water are added, resulting in the liberation of iodine from unreacted reagent:

$$2ArSCl + 2I^- \rightarrow I_2 + ArS—SAr + 2Cl^-$$

The iodine is titrated with 0·02 M thiosulphate.

Later, Obtemperanskaya *et al.* (1974) described a similar procedure using 2,4-dinitrobenzenesulphenyl chloride, with 1–2 min reaction time and amperometric final titration with thiosulphate. They were able to titrate alkane- and arenethiols in the presence of heterocyclic compounds containing the —N═C—SH group. Obtemperanskaya and Nguyen Kim Can (1974) separated the reaction product with thiols from unused dinitrosulphenyl chloride using TLC and then determined it spectrophotometrically at 470 nm after elution and subjection to the Janovský reaction with acetone/ alkali.

Bocco *et al.* (1970) used 2- or 4-nitrobenzenesulphenyl chlorides to determine cysteine (also tryptophan) in proteins, by reaction in aqueous acetic or

formic acid to form a —S—S—⟨ ⟩—NO$_2$ grouping. Treatment with

0·1 N sodium hydroxide then liberated the nitrobenzenethiol which they estimated through its absorbance at 450 nm.

Bradbury and Smyth (1973) reacted cysteine with 3-bromopropionate, Br—CH$_2$—CH$_2$—COO$^-$, to give *S*-carboxyethylcysteine. Unlike the corresponding carboxymethyl compound (from haloacetate), this does not undergo internal cyclization, and consequently retains its primary amino group so as to be detectable by conventional amino acid methods.

Krstěva *et al.* (1965) methylated protein thiols with dimethyl sulphate, then hydrolysed and remethylated (for total free thiols). Reduction in the presence of Raney nickel liberated methane which they determined by infrared spectrometry.

The Gibbs' reagent can be conveniently classified here although it is by no means sure that it reacts only with the thiol group. This reagent, 2,6-dichloro(or bromo) quinonechloroimide, is best known as a test reagent for phenols, with which it yields coloured indophenols:

McAllister (1950) proposed it as a test for methylthiouracil and thiouracil. In borate buffer of pH 8, these yield yellow products with the 2,6-dichloro-reagent; the products are soluble in chloroform, in contrast to those from other urinary materials such as uric acid or creatinine. McAllister and Howells (1952) applied the reaction to determine methyl- and propylthiouracil in tablets. They treated a solution in ammonium hydroxide with borate buffer of pH 8 and 0·4% solution of reagent in aldehyde-free absolute ethanol. After 45 min, they extracted the yellow product with chloroform and determined the absorbance of the solution.

The 2,6-dichloroquinonechloroimide reagent was used for PC visualization by Lederer and Silberman (1952) and Brueggeman and Schole (1967). The former sprayed the dried chromatogram of thiouracil and its 5-methyl and 5-propyl derivatives with alcoholic reagent and then 0·1 N sodium hydroxide to give intense orange-yellow spots on a purple background; the latter also worked with thiouracil derivatives from feeds, spraying with 4% reagent to obtain brown spots.

Kamiya (1959) used 2,6-dibromoquinonechloroimide in ethanol to test for enolizable thioketone groups or thiols. He treated the sample in ethanol + 1 drop of sodium hydroxide solution with 1 drop of reagent. Aliphatic thiols gave a yellow colour, aromatic and heterocyclic, orange. Since the aliphatic compounds possess no rings for a coupling reaction, formation of

at least in an initial reaction, appears likely.

Recently, Millingen (1974) identified vulcanization accelerators in TLC by iodoplatinate and 2,6-dibromoquinonechloroimide reagents. 2-Mercaptobenzothiazole yields an orange colour with the latter.

Stenerson (1967) employed the dibromo-reagent to determine many insecticides and also glutathione, cysteine hydrochloride and 2-mercaptoethanol. He mixed sample solution with reagent in acetic acid, absolute ethanol or acetic acid–diethyl ether (1 + 1) with N hydrochloric acid to adjust the pH to 1·5. After 3 min reaction, he read the absorbance at 435 nm. Glutathione required a 25 min reaction-time.

Ratliff et al. (1972) determined propylthiouracil in serum by reacting for 10 min at room temperature in pH 8 buffer (borate, sodium and potassium chlorides) with 2,6-dichlorobenzoquinonechloroimide. They then extracted the coloured product into chloroform and likewise evaluated at 435 nm.

REFERENCES

Amin, El. S. (1958). *J. Chem. Soc. (London)*, 4769.
Avi-Dor, Y. and Lipkin, R. (1958). *J. Biol. Chem.* **233**, 69.
Avi-Dor, Y. and Mager, J. (1956). *J. Biol. Chem.* **222**, 249.
Bacq, Z. M. and Fischer, A. (1946). *Bull. Soc. Chim. Biol.* **28**, 234.

Baraud, J. (1966). *Chim. Anal. (Paris)*, **48**, 179.
Barrnett, R. J., Tsou, K.-C. and Seligman, A. M. (1955). *J. Histochem. Cytochem.* **3**, 406.
Barron, E. S. G. (1951). *Advances in Enzymol.* **11**, 201.
Barron, E. S. G. and Singer, T. P. (1956). *J. Biol. Chem.* **157**, 221.
Benesch, R. and Benesch, R. (1957). *Biochim. Biophys. Acta*, **23**, 643.
Benesch, R. and Benesch, R. E. (1961). *J. Biol. Chem.* **236**, 405.
Berger, A. and Magnuson, J. A. (1964). *Anal. Chem.* **36**, 1156.
Birkett, D. J., Price, N. C., Radda, G. K. and Salmon, V. (1970). *FEBS Lett.* **6**, 346.
Board, F. A. (1951). *J. Cell. Comp. Physiol.* **38**, 377; *Chem. Abs.* **46**, 8176.
Bocco, E., Veronese, F. M., Fontana, A. and Benassi, C. A. (1970). *Eur. J. Biochem.* **3** 188.
Böhme, H. and Stachel, H. D. (1957). *Z. Anal. Chem.* **154**, 27.
Bonting, jr., S. L. (1950). *Biochim. Biophys. Acta*, **6**, 183.
Bost, R. W., Turner, J. O. and Norton, R. D. (1932). *J. Amer. Chem. Soc.* **54**, 1985; also (1933). *ibid.* **55**, 4956.
Bradbury, A. F. and Smyth, D. G. (1973). *Biochem. J.* **131**, 637.
Brady, P. R. and Hoskinson, R. M. (1971). *J. Chromatogr.* **54**, 65.
Brdička, R. (1938). *Acta Union Intern. Contra Cancerum*, **3**, 13.
Brenner, M. W., Owades, J. L., Schapiro, G. J. and Laufer, S. (1960). *Am. Brewer*, **93**, 28, 38.
Brueggeman, J. and Schole, J. (1967). *Landwirt. Forsch.* No. 21, 134; *Chem. Abs.* **68**, 2099.
Bucher, K. (1951). *Pharm. Helv. Acta*, **26**, 145.
Carson, J. F. and Wong, F. F. (1957). *J. Org. Chem.* **22**, 1725.
Dahmen, E. A. M. F., Dijkstra, R. and Verjaal, A. J. (1963). *Erdöl u. Kohle*, **16**, 768.
Day, E. A. and Patton, S. (1959). *Michrochem. J.* **3**, 137.
Desreux, V., Fredericq, E. and Fischer, P. (1946). *Bull. Soc. Chim. Biol.* **28**, 493.
Dickens, F. (1933). *Biochem. J.* **27**, 1141.
Fasold, H., Groeschel-Stewart, U. and Turba, F. (1964). *Biochem. Z.* **339**, 487.
Feigl, F. and Caldas, A. (1969). *Anal. Chim. Acta*, **47**, 555.
Folkard, A. E. and Joyce, A. E. (1963). *J. Sci. Food Agric.* **14**, 510.
Fraenkel-Conrat, J., Cook, B. B. and Morgan, A. F. (1952). *Arch. Biochem. Biophys.* **35**, 157.
Fredericq, E. and Desreux, V. (1947). *Bull. Soc. Chim. Biol.* **29**, 100.
Gascó, L. and Barrera, R. (1972). *Anal. Chim. Acta*, **61**, 253.
Gasparič, J. and Borecký, J. (1961). *J. Chromatogr.* **5**, 466.
Génévois, L. (1964). *Chim. Anal. (Paris)*, **46**, 539.
Goren, H. J., Glick, D. M. and Barnard, E. A. (1968). *Arch. Biochem. Biophys.* **126**, 607.
Grogan, C. H., Rice, L. M. and Reid, E. E. (1955). *J. Org. Chem.* **20**, 50.
Herrington, K. A., Pointer, K., Meister, A. and Friedman, O. M. (1967). *Cancer Res.* **27A**, 130.
Hiskey, R. G., Carroll, F. I., Smith, R. F. and Corbett, R. T. (1961). *J. Org. Chem.* **26**, 4756.
Kamiya, S. (1959). *Bunseki Kagaku*, **8**, 596.
Kawahara, F. K. (1968). *Anal. Chem.* **40**, 1009; also (1969). *Amer. Chem. Soc., Div. Water, Air Waste Chem., Gen. Pap.* **170**; *Chem. Abs.* **73**, 133835; and (1971). *Environ. Sci. Technol.* **5**, 235.
Kharasch, N., McQuarrie, D. P. and Buess, C. M. (1953). *J. Amer. Chem. Soc.* **75**, 2658.
Korolczuk, J., Daniewski, M. and Mielniczik, Z. (1974). *J. Chromatogr.* **100**, 165.
Krstěva, M., Alexiev, B., Ivanov, Č. and Yordanov, B. (1965). *Anal. Chim. Acta*, **32**, 465.

Langford, R. B. and Lawson, D. D. (1957). *J. Chem. Educ.* **34**, 510.
Lederer, M. and Silberman, H. (1952). *Anal. Chim. Acta,* **6**, 133.
McAllister, R. A. (1950). *Nature (London),* **166**, 789.
McAllister, R. A. and Howells, A. (1952). *J. Pharm. Pharmacol.* **46**, 259.
Magnuson, J. A. and Knaub, E. W. (1965). *Anal. Chem.* **37**, 1607.
Mazur, A., Litt, I. and Shorr, E. (1950). *J. Biol. Chem.* **187**, 485.
Michaelis, L. and Schubert, M. P. (1934). *J. Biol. Chem.* **106**, 331.
Millingen, M. B. (1974). *Anal. Chem.* **46**, 746.
Mirsky, A. E. and Anson, M. L. (1935). *J. Gen. Physiol.* **18**, 307.
Obara, Y., Ishikawa, Y. and Nishino, C. (1966). *Agric. Biol. Chem.* **30**, 164; *Anal. Abs.* **14**, 3408.
Obtemperanskaya, S. I. and Egorova, T. A. (1969). *Vest. Mosk. Gos. Univ., Ser. Khim.* 115; *Anal. Abs.* **19**, 400; also *Zh. Anal. Khim.* **24**, 1439; also (1970). *USSR Patent No.* 286,329 of Nov. 10.
Obtemperanskaya, S. I. and Kalinina, N. N. (1971). *Zh. Anal. Khim.* **26**, 1407.
Obtemperanskaya, S. I. and Nguyen Kim Can (1974). *Zh. Anal., Khim.* **29**, 2069.
Obtemperanskaya, S. I. and Tikhonov, Y. N. (1966). *Vestn. Mosk. Univ., Ser. II,* **21**, 95; *Chem. Abs.* **66**, 91439.
Obtemperanskaya, S. I., Tikhonov, Y. N., Moroz, N. S., Shinskaya, T. A., Egorova, A. G., Likhosherstova, V. N., Chalenko, R. G. and Tsypkina, I. Ya. (1970). *Probl. Anal. Khim.* **1**, 220; *Chem. Abs.* **75**, 58446.
Obtemperanskaya, S. I., Kalinina, N. N., Speranskaya, T. N. and Solov'eva, N. M. (1974). *Zh. Anal. Khim.* **29**, 949.
Olson, J. R. and Feldman, H. B. (1937). *J. Amer. Chem. Soc.* **59**, 2003.
Palomo Coll, A. L. and Palomo Coll, A. (1963). *Afinidad,* **20**, 150; *Chem. Abs.* **59**, 13356.
Patterson, W. I., Geiger, W. B., Mizell, L. R. and Harris, M. (1941). *J. Res. Nat. Bur. Stds.* **27**, 89.
Rapkine, L. (1933). *Compt. Rend. Soc. Biol.* **112**, 790.
Ratliff, C. R., Gilliland, P. F. and Hall, F. F. (1972). *Clin. Chem.* **18**, 1373.
Rittner, R., Tilley, G., Mayer jr., A. and Siggia, S. (1963). *Anal. Chem.* **34**, 237.
Roberts, L. W. (1956). *Science,* **124**, 628.
Rosner, L. (1940). *J. Biol. Chem.* **132**, 657.
Sandegren, E., Ekström, D. and Nielsen, N. (1950). *Acta Chem. Scand.* **4**, 1311.
Sanford, D. and Humoller, F. L. (1947). *Ind. Eng. Chem., Anal. Ed.* **19**, 404.
Sanso, G. and Rigoli, A. (1970). *Boll. Chim. Farm.* **109**, 266.
Saunders, B. C. (1934). *Biochem. J.* **28**, 580.
Sawicki, E., Miller, R., Stanley, T. and Hauser, T. (1958). *Anal. Chem.* **30**, 1130.
Schenk, G. H. and Fritz, J. S. (1960). *Anal. Chem.* **32**, 987.
Schreiber, W. (1971). *Z. Anal. Chem.* **254**, 345.
Schwartz, D. P. and Brewington, C. R. (1968). *Microchem. J.* **13**, 310.
Secrist, J. A., Barrio, J. R. and Leonard, N. J. (1971). *Biochem. Biophys. Res. Commun.* **45**, 1262.
Singh, D. and Varma, A. (1963–4). *J. Sci. Res. Banares Hindu Univ.* **14**, 19; *Anal. Abs.* **12**, 3925.
Smythe, C. V. (1936). *J. Biol. Chem.* **114**, 601.
Stein, H. H. and Guarnaccio, J. (1960). *Anal. Chim. Acta,* **23**, 89.
Stenerson, J. H. V. (1967). *Bull. Environ. Contam. Toxicol.* **2**, 364; *Chem. Abs.* **69**, 2041.
Tattrie, B. L. and Connell, G. E. (1967). *Can. J. Biochem.* **45**, 551.
Thamm, E. (1972). *Seifen, Öle, Fette, Wachse,* **98**, 258.
Toennies, G. and Kolb, J. J. (1942). *J. Biol. Chem.* **144**, 219.

Tonialo, C., Nisato, D., Biondi, L. and Signor, A. (1972). *J. Chem. Soc. (London)*, *Perkin Trans.* I, 1182.
Umbreit, G. R. and Houtman, R. L. (1967). *J. Pharm. Sci.* **56**, 349.
Watts, D. C., Rabin, B. R. and Cook, E. M. (1961). *Biochim. Biophys. Acta*, **48**, 380.
Wenck, H., Schwabe, E., Schneider, F. and Flohé, L. (1972). *Z. Anal. Chem.* **258**, 267.
Wertheim, E. (1929). *J. Amer. Chem. Soc.* **51**, 3661.
White, B. J. and Wolfe, R. G. (1962). *J. Chromatogr.* **7**, 516.
Zahn, H., Weigmann, H.-D. and Nischwitz, E. (1961). *Kolloid-Z.* **179**, 49.
Zuber, H., Traumann, K. and Zahn, H. (1955). *Z. Naturforsch.* **10b**, 457.

4. ADDITION TO UNSATURATED BONDS

Thiols add to the double bonds C=C, C=O and C=N according to the equation:

$$\text{RSH} + \overset{\displaystyle |}{\underset{}{\text{C}}}=\text{X} \rightarrow \overset{\displaystyle |}{\underset{\overset{|}{\text{RS}}}{\text{C}}}\!\!-\!\!\overset{}{\underset{\overset{|}{\text{H}}}{\text{X}}}$$

A convenient subdivision is according to these three groups.

1. ADDITION OF THIOLS TO THE —C=C— BOND

This reaction has been known for a long time. The earliest mention of addition of thiols to olefines is probably that of Posner (1905) who obtained good yields from olefines, such as trimethylethylene, styrene, pinene, etc., with benzyl mercaptan and thiophenol; yields were poor with sterically hindered olefines such as stilbene, with lower alkenes and with ethanethiol. A great deal of work was done on the reaction in subsequent years, much of it with preparative aims or in connection with reaction mechanisms.

The addition has found some analytical application but evidently only for determining or separating the olefine. A few examples may be mentioned: Thus Axberg and Holmberg (1933) determined iodine numbers of fats by treating with a measured excess of mercaptoacetic acid and estimating the unused thiol iodometrically. Hoog and Eichwald (1939) studied the reaction of mercaptoacetic acid with C_5 to C_8 monoolefines, adding propionic acid as a solubilizer. They found complete reaction often within an hour. Koenig and Swern (1957) referred to the rich patent literature on reactions of fats with thiols. They studied the addition of mercaptoacetic acid to long-chain mono-unsaturated compounds, such as oleic, elaidic and 10-undecenoic acids and methyl oleate and ricinoleate. Apart from the undecenoic acid, all yielded centrally branched addition products and unreacted chain compounds could be separated from them *via* urea complexes. Eshelman *et al.* (1960) separated unsaturated from saturated glycerides in fats by reaction

147

also with mercaptoacetic acid overnight at 50°C and ultimate extraction of the saturated glycerides with petroleum ether.

Possible application of this procedure to thiol determination is handicapped by the frequently slow reaction and modest yields. Carbon–carbon double bonds which are activated by neighbouring groups such as carbonyl or carbon–nitrogen multiple bonds, react much faster and usually quantitatively. Procedures with such reagents have found analytical application to thiols and are described below.

1.1. Addition to the —C=C—C=O group

Jacot–Guillarmod and Ceschini (1959) separated and identified volatile thiols (up to C_5) by reaction with an equivalent amount of acrolein at 35–40°C in the presence of 1% cupric acetate as catalyst. This yields 3-alkylthio-propionaldehydes:

$$CH_2=CH—CHO + RSH \rightarrow RS—CH_2—CH_2—CHO$$

They converted these aldehydes into their 2,4-dinitrophenylhydrazones for separation by ascending paper chromatography using a stationary phase of dimethylformamide and a mobile phase of cyclohexane saturated with it.

1.2. Addition to the —C=C—C=N group

Most examples which fall under this heading are based on reaction with acrylonitrile. During the last few years, vinyl heterocycles have been introduced also; these contain the grouping or are vinylogues.

1.2.1. *Addition to Acrylonitrile*

Acrylonitrile, a readily available compound, reacts with thiols quantitatively according to the equation:

$$CH_2=CH—C≡N + RSH \rightarrow RS—CH_2—CH_2—C≡N$$

It is generally added to mixtures containing thiols in order to remove them from the field of reaction with other subsequently introduced analytical reagents. The thiol amount is then obtained by difference. Thus Earle (1953) determined disulphides in the presence of thiols by treating the mixture for 10 min with excess acrylonitrile in the presence of potassium hydroxide. He then washed out excess reagent with water, reduced the disulphides with zinc to thiols and titrated these with silver ion. Wronski applied this to titrimetric analysis of mixtures of various sulphur compounds, for instance: thiols(ethanethiol, cysteine, mercaptoacetic acid) + sulphide, titrating aliquots with o-hydroxymercuribenzoate with and without treatment with acrylonitrile for 2–15 min (1960); thiourea + mercaptoacetic acid, likewise by mercurimetric titration with the same reagent, with and without treatment with alkaline acrylonitrile for 15 min (1961); thiocyanate ion/thiols (*e.g.* mercapto-acetic acid) by titrating the total with *o*-hydroxymercuribenzoate and then thiocyanate alone with silver ion after having reacted the thiol with acryloni-trile (1961); cyanide ion/thiols with similar mercurimetric titration of the sum and argentometric titration of remaining cyanide (1963). Wronski also differentiated alkanethiols, which react rapidly with acrylonitrile, from aro-matic or heterocyclic thiols. For example, he determined the sum of alkane-thiols and mercaptobenzothiazole by his standard titration with *o*-hydroxy-mercuribenzoate, then removed the former with the nitrile and re-titrated the thiazole (1961); he similarly masked alkanethiols in the presence of arenethiols by reaction for 1 min with acrylonitrile; mercurimetric titrations before and after this treatment yielded values for both thiol types (1964a).

Weil and Seibles (1961) reduced α- and β-lactoalbumin disulphides with 2-mercaptoethanol to the corresponding thiols and then converted these to *S*-cyanoethyl(cysteinyl) groups with acrylonitrile in 100% excess at pH 8 for 4 h at room temperature in a nitrogen atmosphere. The product was subsequently hydrolysed with 6N hydrochloric acid; this yielded *S*-carboxyethylcysteine which was isolated and determined chromatographic-ally. Kalan *et al* (1965) similarly determined disulphides in milk proteins by reducing to thiols with 2-mercaptoethanol, converting to *S*-cyanoethyl-cysteinyl groups with acrylonitrile, hydrolysing, and separating on Dowex 50X-12 ion exchanger.

Obtemperanskaya *et al.* (1957) determined thiols by reaction for up to 1 h with excess acrylonitrile/alkali. They then estimated unused nitrile by adding sodium sulphite and titrating the liberated alkali with hydrochloric acid to thymolphthalein/alizarin yellow:

$$CH_2{=}CH{-}CN + SO_3^{2-} + H_2O \rightarrow {}^-SO_3{-}CH_2{-}CH_2{-}CN + OH^-$$

Obtemperanskaya and Nguyen Dyk Hoc (1970) determined some alkane- and arenethiols by treating with excess acrylonitrile and estimating the unused amount by gas chromatography on 10% polyethyleneglycol 3000 on 60–80 mesh Chromosorb W at 100°C in helium carrier gas. Obtemperanskaya *et al.* (1973) determined heterocyclic compounds containing thiol groups, such as mercaptobenzothiazole, 6-mercaptopurine, 2-mercaptobenzimidazole) simi-larly and also estimated unreacted acrylonitrile by the sulphite method given above (Obtemperanskaya *et al.*, 1957).

Recently Greenhow and Loo (1974) determined alkane- and arenethiols in the presence of carboxylic acids and phenols by thermometric titration using potassium or tetrabutylammonium hydroxide. They added acrylonitrile to react with the thiols and to serve as indicator for the titration of the residual acids and phenols. The first excess of alkali titrant catalyses exothermic

F

polymerization of the nitrile, furnishing a thermometric end-point. Greenhow and Loo titrated total acidic compounds likewise, using added acetone as indicator; it dimerizes exothermically in the presence of alkali. The amount of thiol was given by the difference between the two titration values.

1.2.2. Vinylpyridines and -quinoline

Friedman and co-workers (1969, 1970; Friedman and Noma, 1970; Krull et al. 1971; Cavins et al., 1972) used these compounds to react quantitatively with thiol groups in proteins or reduced (disulphide groups) proteins:

$$RSH \quad + \quad CH_2{=}CH{-}\!\!\left\langle\!\!\bigcirc\!\!\right\rangle\!\!N \quad \longrightarrow \quad RS{-}CH_2{-}CH_2{-}\!\!\left\langle\!\!\bigcirc\!\!\right\rangle\!\!N$$

(or in 2-position)

Reaction is carried out at near neutral pH, e.g. at pH 7·5 from ammonium acetate, for several hours. The pyridine derivatives have an absorption maximum at 255 nm, the quinoline derivatives at 318 nm. Spectrophotometric determination is then possible at these wavelengths.

1.3. Addition to the —C=C—SO₂— group

Obtemperanskaya et al. (1974) recently used aryl vinyl, ethyl vinyl and 2-hydroxyethyl vinyl sulphones to determine aliphatic, aromatic and heterocyclic thiols via nucleophilic addition to the vinyl group. They used a threefold excess of sulphones and estimated the unused amount after 5 min reaction at 18–20°C by treatment with sodium sulphite to yield an equivalent amount of hydroxyl ion:

$$CH_2{=}CH{-}SO_2{-}R + H_2O + SO_3^{2-} \rightarrow \overset{(-)}{O_3}S{-}CH_2{-}CH_2{-}SO_2R + OH^-$$

This was titrated with standard acid.

1.4. Addition to the C=C—O—C=O group

Excess isoprenyl acetate, $CH_3{-}CO{-}O{-}C(CH_3){=}CH_2$, in carbon tetrachloride solution, was used by Volodina and Kon'kova (1968) to acylate thiols in a 10–15 min reaction at ambient temperature. Unused reagent was then estimated by adding excess bromine/carbon tetrachloride and evaluating the unconsumed bromine through addition of sodium iodide/acetone and spectrophotometry of the liberated iodine at 364 nm (at 0–5°C). Amino groups also reacted with the ester.

1.5. Addition to the O=C—C=C—C=O group

Two major groups of compounds fall under this heading, making it the most important and widely applied principle: derivatives of maleimide; and quinones and related compounds. The use of disodium maleate to prepare derivatives is mentioned also.

1.5.1. *Maleimide Derivatives*

Thiols add on to the carbon–carbon double bond in these compounds to form succinimide derivatives:

$$
RSH + \quad
\begin{array}{c}
| \\
C-CO \\
\| \qquad \rangle N-X \rightarrow \\
C-CO \\
|
\end{array}
\qquad
\begin{array}{c}
| \\
RS-C-CO \\
| \qquad \rangle N-X \\
H-C-CO \\
|
\end{array}
$$

The chief example is N-ethylmaleimide, where X is the ethyl group. Since its introduction in 1949 it has enjoyed considerable popularity, especially in quantitative determination of thiols.

(a) *N-Ethylmaleimide.* Friedmann *et al.* (1949) observed that free thiol groups of glutathione or mercaptoacetic acid were rapidly blocked in neutral solution at room temperature containing equimolecular amounts of certain maleimide derivatives, such as the N-ethyl derivative (NEM). Good yields of adducts of NEM (and also of N-phenylmaleimide) with mercaptoacetic and thiosalicyclic acids were reported by Marrian (1949). In 1952, Friedmann noted spectral changes in the solution during reaction of glutathione with NEM in phosphate buffer of pH 7·4. Reaction was evidently possible even at very low concentration. Gregory (1955) investigated the influence of pH on the rate of reaction between glutathione and NEM, and on the stability of the reagent itself. He found that in neutral solution the reaction rate was so much higher than the reagent decomposition rate that quantitative application was possible. The low molar absorptivity (only ca. 610) militates against sensitivity and is less than 10% of that of the *p*-mercuribenzoate reagents used by Boyer (see p. 69). Nevertheless, this principle has been the basis of much analytical work.

Most investigators have related thiol concentration to the diminution of light absorbance of the N-ethylmaleimide at 300 nm. Examples are: Roberts and Rouser (1958) for cysteine; Alexander (1958) for glutathione and thiol groups of rat liver extract; Cole *et al.* (1958) for thiols in haemoglobin, using Alexander's method among others; Benesch and Benesch (1961) for thiols in haemoglobin in a study of the "Bohr effect"; Leslie *et al.* (1962) for aliphatic and aromatic thiols; Leslie (1965) also for thiol groups in β-lactoglobulin (in 5 M urea); Hamm and Hofmann (1966) for thiols in meat, applying the Roberts and Rouser procedure; De Marco *et al.* (1966) for protein disulphides, treated with cyanide ion to yield thiols, then determined using Roberts and

Rouser's method; Narang *et al.* (1967) for milk thiols; Ambrosino *et al.* (1969) for protein thiols and disulphides after reduction. Gorin *et al.* (1966) studied 2-mercaptoethylamine derivatives, cysteine and glutathione at pH 4·78 to 6·98 observing that different thiols reacted at different rates and that the rates were highly sensitive to pH (2–3% change per 0·01 pH unit).

Other techniques are used to determine consumption of NEM in the reaction with thiols. Thus Weitzman and Tyler (1971) based their cysteine determination on the diminution of the polarographic reduction wave of NEM at −1·5 V versus the SCE; they stated that this was ten to a hundred times more sensitive than the spectrophotometric method.

Tsao and Bailey (1953) directly titrated the thiol groups of actin, myosin and glutathione in glycine citrate buffer of pH 9·7 in the presence of the protein denaturant guanidine hydrochloride. They employed nitroprusside as indicator, giving a colour change from rose pink to yellow as end-point. Some potassium cyanide stabilized the colour. Fraenkel-Conrat (1955) used this method in a study of the thiol groups of a tobacco mosaic virus. Hofmann (1971) recently published an indirect titration method for thiols, e.g. glutathione, cysteine, cysteine ethyl ester, mercaptopropionic acid. He treated 0·2 to 0·8 μmol with 1·0 μmol of NEM. After 2 h reaction, he added 1·0 μmol of glutathione and the unused glutathione was titrated amperometrically with 0·001 M silver nitrate.

Use has been made of radioactivity measurements on the reaction product. Tkachuk and Hlynka (1963) studied the reaction of thiol groups with NEM and ^{14}C-NEM and applied the latter to determine thiol groups in flour. After reaction for 3 h, dialysis and freeze drying to remove excess reagent, the radioactivity of the isolated (by hydrolysis) S-succinyl-L-cysteine was estimated. Lee and Samuels (1964) used ^{14}C-NEM in a closely similar procedure for determining protein cysteine groups. Lee and Lai (1967) estimated cysteine groups in flour proteins also with the help of the ^{14}C-reagent, isolating the product by hydrolysis using 6 N hydrochloric acid and then submitting to paper chromatography with butanol–pyridine–acetic acid–water (15 + 10 + 3 + 12). They located the spots with 0·1% ninhydrin in acetone, then cut out the S-succinyl-L-cysteine spots and evaluated their radioactivity. Klein and Robbins (1970) utilized the same principle for determining glutathione in HeLa cells. Flavin (1963) used ^{14}N-NEM for determining cysteine and homocysteine liberated enzymatically from cystathione. Lee and Samuels (1964) also determined cysteine containing ^{35}S through the radioactivity of the reaction product with NEM but the radioactivity is derived in this case from the sample and not the reagent.

Benesch *et al.* (1956) detected down to 1 μg of glutathione and 100 ng of other thiols on paper chromatograms by treating first with NEM in anhydrous isopropanol, drying the chromatogram and then treating with potassium hydroxide also in dry isopropanol. This yielded a red colour. Broekhuysen (1958) adapted this quantitatively, allowing 4–5 min for reaction in isopropanol, then rendering alkaline and evaluating colorimetrically at 515 nm, 10 min after having added the potassium hydroxide. He proposed a reaction scheme:

Like other thiol reagents, N-ethylmaleimide can be used to block thiol groups and permit separation from other compounds. Thus Hanes *et al.* (1950) treated glutathione with NEM and then performed paper chromatography using propanol–water as mobile phase. Gutcho and Laufer (1952) also blocked glutathione thiol groups with NEM before PC. Smith and Rodnight (1959) separated the NEM-adducts of non-protein thiols on columns of Amberlite CG-120 in the H^+-form; they examined the fractions with paper chromatography. Glutathione in blood and tissues was determined by Wernze and Koch (1965) by reacting with NEM and then deproteinizing with acetone. They evaporated down the acetone extract of the adduct in vacuo and finally submitted it to paper chromatography using propanol–water (81 + 19). The adduct was located with a ninhydrin–cadmium salt reagent, eluted and determined spectrophotometrically. Titov (1969) similarly concluded his determination of 2-mercaptoethylamine in urine and stomach contents. After reaction with NEM and PC using *n*-propanol-N-acetic acid (10 + 1), he visualized with a ninhydrin/cadmium acetate/water/acetic acid/acetone reagent. The spots, likewise those from standard amounts of thiol and reagent, were cut out, eluted with methanol and evaluated at 509 nm. States and Segal (1969) used thin-layer chromatography on silica gel or MN 300 cellulose, and electrophoresis for separating the NEM-adducts of non-protein thiols. Tietze (1969) and Wendell (1970) removed interfering glutathione with NEM in a biological determination of oxidized glutathione. Miyagawa *et al.* (1971) separated glutathione in rate epidermis by a PC procedure following reaction with NEM.

Differentiation between different thiols is sometimes possible quantitatively. Thus Roberts and Rouser (1958), as mentioned above, determined cysteine through the reduction in absorbance at ca. 300 nm of the N-ethylmaleimide reagent; glutathione in the same sample could be estimated through the increase in absorbance of the reaction mixture recorded at 248 nm. Smyth *et al.* (1960) studied the reaction of NEM with cysteine where, in alkaline solution, ring formation to a thiazane takes place:

$$\begin{array}{l} \text{CH}_2\!-\!\text{SH} \\ | \\ \text{HOOC}\!-\!\text{CH} \quad + \text{CH}\!=\!\text{CH}\!-\!\text{CO}\!-\!\text{N}\!-\!\text{C}_2\text{H}_5 \rightarrow \\ | \qquad\qquad | \\ \text{NH}_2 \qquad\quad \text{CO} \end{array}$$

$$\begin{array}{l} \qquad\qquad\qquad\qquad \text{CH}_2\!-\!\text{S} \\ \qquad\qquad\qquad\qquad\quad | \qquad | \\ \rightarrow \text{HOOC}\!-\!\text{CH} \quad \text{CH}\!-\!\text{CH}_2\!-\!\text{CO}\!-\!\text{NH}\!-\!\text{C}_2\text{H}_5 \\ \qquad\qquad\qquad | \qquad | \\ \qquad\qquad\quad \text{NH}\!-\!\text{CO} \end{array}$$

(b) *Other maleimide derivatives.* Many substituted maleimides have been prepared in recent years for various purposes, e.g. blocking, cross-linking or detecting thiol groups, mainly in proteins and similar materials. Not all of this work can be said to be genuinely analytical. Quoted below are some publications reporting the use of a maleimide reagent containing a chromophore or potential chromophore group to facilitate location and even quantitative determination of thiol groups.

Tsou *et al.* (1955) tested a number of compounds as histochemical reagents. They treated the tissue section at pH 8, removed the excess of reagent by extraction and reacted the thereby modified tissue with tetrazotized di-*o*-anisidine, obtaining intensely coloured coupling products. Among their maleimides may be mentioned:

N-(4-hydroxy-1-naphthyl)-maleimide

Corresponding isomaleimide

N-(1-naphthyl)maleimide

Price and Campbell (1957) used the isomaleimide derivative for visualization on paper chromatograms. They treated thiols with the reagent + sodium bicarbonate for 1 min in aqueous solution, then added hydrochloric acid to arrest the reaction. They then chromatographed on Whatman No. 20 paper, using seven different solvent systems. The dried chromatogram was exposed briefly to ammonia and then dipped into the diazonium salt solution, which yielded blue or red spots within a few seconds, fluorescing in UV light of 366 nm. They could detect down to ca. 0·1 μmol of cysteine, homocysteine, glutathione and thiocarboxylates, claiming a sensitivity of 100 times that with NEM. However, alkanethiols could not be detected in this way.

Directly chromophoric reagents were developed later, although not primarily for analytical work. Examples are:

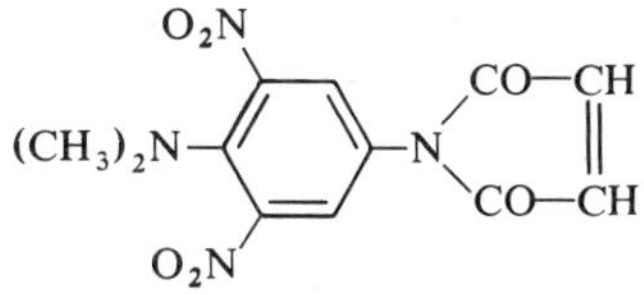

N-(4-dimethylamino-3,5-
dinitrophenyl)-maleimide
(Witter and Tuppy, 1960)

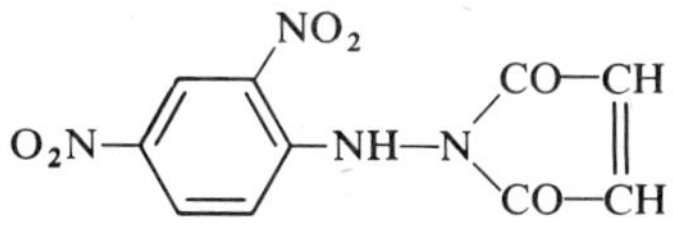

N-(2,4-dinitroanilino)-
maleimide
(Clark-Walker and Robinson, 1961)

Witter and Tuppy studied the reaction with cysteine and free thiol groups of human and bovine serum albumin and isolated the reaction products by adsorption on talc, paper ionophoresis and chromatography, their yellow colour proving useful in this. Clark-Walker and Robinson found that their substituted maleimide reacted readily with cysteine and glutathione at room temperature and over a pH range from 3 to 8. It was used to label protein thiol groups.

Fluorescent labels can be attached to protein thiol groups by reagents such as

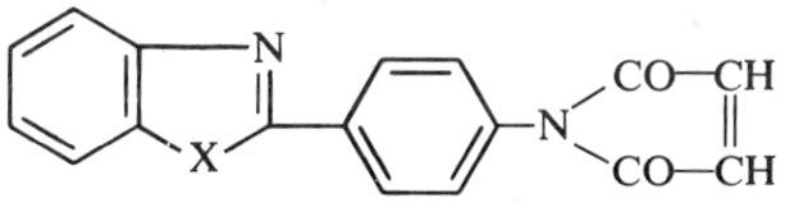

Kanaoka *et al.* (1967) introduced this reagent where X = S or O. Sekine *et al.* (1972) used the corresponding product where X = —NH, "BIPM", N-[*p*-(2-benzimidazolyl)phenyl]-maleimide, for quantitative work. This reagent does not fluoresce appreciably but the adduct with thiols shows intense fluorescence at 365 nm (excitation wavelength of 315 nm). A linear relation holds between concentration of adduct (and hence amount of thiol) and fluorescence intensity. Reaction is complete in 30 min at 0°C and pH 6·85. Micromole amounts can be estimated. Nara and Tuzimura (1973) found that N-(9-acridinyl)-maleimide gives strongly fluorescent products with thiols (at 426 nm, excitation at 362 nm). They used this to determine 0·4 to 16 μmol amounts of thiols. They studied the reaction with cysteine at pH 3 to 6 and found that other amino acids do not react.

Other interesting thiol reagents are a water-insoluble maleimide, prepared from poly(aminostyrene) and maleic anhydride, which reacted rapidly and irreversibly with thiols but not with nucleotides, soluble RNA or other non-thiols (Richards *et al.*, 1966); and those prepared to effect cross-linking of thiol groups on the same protein, *e.g.* azobenzenedimaleimide,

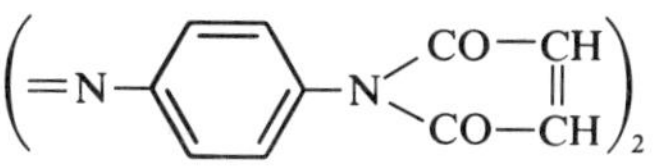

(Fasold *et al.*, 1963)

and

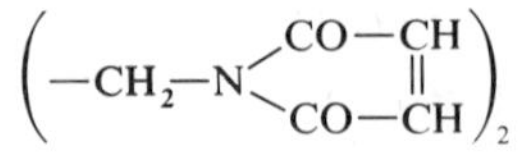

$$\left(-CH_2-N{\overset{CO-CH}{\underset{CO-CH}{}}}\right)_2$$

(Freedberg and Hardman, 1971). Recently Yamaguchi and Takechi (1972) studied N-(p-substituted-phenyl)maleimides, including the p-chloro derivatives; these were stated to give no coloured products and to react by chlorine-replacement rather than by addition.

1.5.2. *Quinones*

p-Benzoquinone is known to add on thiols to yield substituted hydroquinones and one may assume that the first step is addition to the C=C bond:

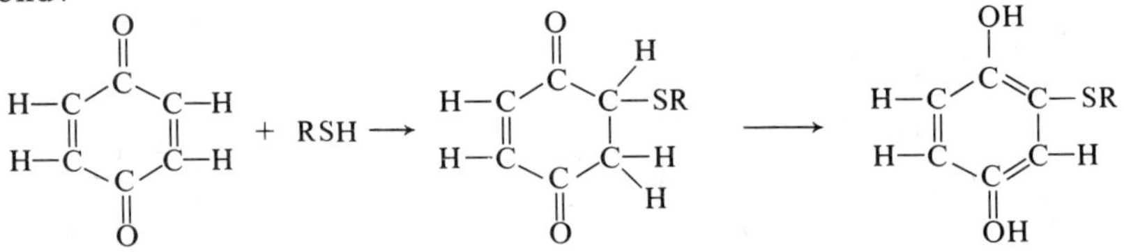

Subsequent reactions depend on the quinone, group R, and the conditions. For example, Kuhn and Beinert (1944) added p-benzoquinone to cysteine ester, stating that the substituted hydroquinone was oxidized by excess quinone to a substituted quinone, in which the amino group of the cysteine molecule reacts with the adjacent quinone carbonyl group:

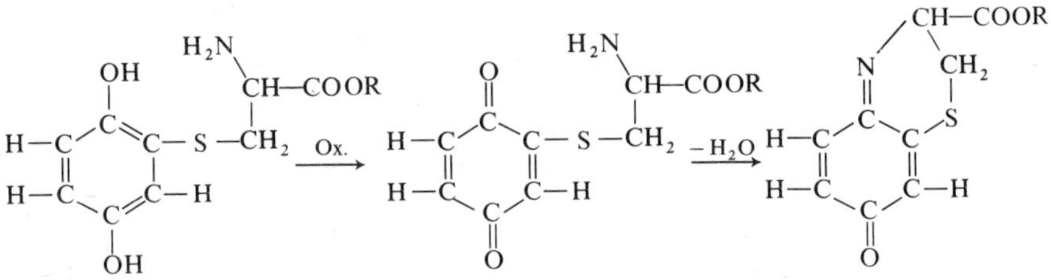

The product is a yellow precipitate.

An early colour test of this sort for cysteine was given by Dyer and Baudisch (1932; also Baudisch and Dyer, 1933). They found that the "o-benzoquinone" reagent, prepared by oxidation of pyrocatechol with silver oxide according to Willstätter's procedure, yields a product with cysteine hydrochloride which is soluble in chloroform to give a deep red solution (down to 10 parts/10^6 of cysteine). It permits cysteine to be detected in the presence of many other nitrogen- and sulphur-containing compounds. Hess and Sullivan (1932) criticized the test as being less specific than claimed, and they quoted a list of other compounds, mostly amines such as aniline, benzidine, indole and lower alkylamines, which also react; however, they found no other amino acid giving a positive response. Hazeloop (1934) adapted the reaction quantitatively for cysteine, comparing the colour intensity of the chloroform layer (after drying over sodium sulphate) with a standard cobalt/iron colour.

Fernandez and Henry (1965) determined cysteine through reaction with 2,6-dichlorobenzoquinone. The method was really for cystine in urine. This they first reduced to the thiol with copper(I) and then allowed to react for 30 min at 75°C in ethanol. After cooling the mixture, they acidified with sulphuric acid and left for 30 min before finally evaluating at 510 nm.

A TLC-visualization procedure for sulphur compounds, including L-cysteine and 2-thiolhistidine, has been worked out by Belliveau and Frei (1971). After exposing the dried layer for 15 sec to several min to bromine vapour and then removing excess of this reagent, they sprayed the plate with 0·1% 2,3-dichloro-4,5-dicyanobenzoquinone (DDQ) in benzene, absolute ethanol or aqueous ethanol. The sulphur compounds (from 0·1 to 1 µg) appear as blue fluorescent spots on a non-fluorescing background in u.v. light. Oxidation of the sulphur compounds probably leads to acids which protonate the DDQ reagent.

Naphthaquinones have found only little use in analytical work on thiols. Tu and Chou (1963) employed 1,4-naphthoquinone to detect certain thiols in the presence of mercaptoacetic acid. The adduct of the latter can be extracted quantitatively with ether. The adducts of the other thiols, e.g. ethanethiol, glutathione and in various proteins, show absorption maxima near 415 nm, with an approximately linear relationship of thiol concentration and absorbance. Hofmann (1965) observed that thiols gave stable coloured products with 2,3-dichloro-1,4-naphthoquinone. He shook a mixture of the reagent in chloroform with water, potassium carbonate solution and test solution for 1 h and then centrifuged. After drying the chloroform layer over sodium sulphate, he evaluated absorbance at a wavelength between 400 and 500 nm (435 for cysteine, 440 for glutathione, etc.). Devani et al (1973) found that 2,3-dichloro-1,4-naphthoquinone yielded coloured complexes with compounds possessing a thiol tautomeric form, such as thioureas or thiosemicarbazones. They used the reagent for detection by treating with an ethanolic solution of it, then adding ethanolic ammonia to yield red-violet colours, absorbing at 500–570 nm.

Among a number of compounds developed by Tsou et al. (1955) for histochemical work on protein thiols, was 1,4-naphthoquinone monobenzenesulphonimide,

$N—SO_2C_6H_5$

F*

After treating the tissue section with the reagent at pH 8, they removed excess reagent and then immersed the treated section in tetrazotised di-o-anisidine to yield intense colours.

Sawicki *et al.* (1967) used 7,7,8,8-tetracyanoquinodimethane

$$(CN)_2C{=}\!\!\left\langle\;\bigcirc\;\right\rangle\!\!{=}C(CN)_2,$$

to detect free radical precursors, including some thiols such as cysteine, 2-naphthalenethiol and thiosalicylic acid. They visualized in PC and TLC by spraying with 0·3% reagent in acetone or acetone/pyridine. They carried out quantitative estimation by treating the sample in pyridine–water (1 + 1) with 0·02% reagent in butanone. After 3 min at 60°C they read the absorption at 756 or 848 nm. The authors believe that the main chromogen is the free radical anion

$$(CN)_2\overset{(-)}{C}{=}\!\!\left\langle\;\bigcirc\;\right\rangle\!\!{=}\overset{\cdot}{C}(CN)_2.$$

Two other quinone reagents appear to react differently. Tillmans' reagent oxidizes and has been included under Chapter 1, "Oxidation" (Section 52) (p. 45).

$$O{=}\!\!\left\langle\;\overset{\text{Hal}}{\underset{\text{Hal}}{\bigcirc}}\;\right\rangle\!\!{=}N{-}\!\!\left\langle\;\bigcirc\;\right\rangle\!\!{-}OH + 2H^+ \rightarrow HO{-}\!\!\left\langle\;\overset{\text{Hal}}{\underset{\text{Hal}}{\bigcirc}}\;\right\rangle\!\!{-}NH{-}\!\!\left\langle\;\bigcirc\;\right\rangle\!\!{-}OH$$

Probably the best known quinone reagent is 1,2-naphthoquinone-4-sulphonate. This evidently reacts with thiols under expulsion of the $-SO_3^-$ group:

Although this does not involve addition to the C=C double bond, it is convenient to classify it here along with the other quinone reagents.

Reid *et al.* (1921) (also Hofmann and Reid, 1923, and Ellis and Reid, 1932) obtained such sulphides (thioethers) by heating sodium α-anthraquinone-1-sulphonate with thiols in alkaline solution. They serve as derivatives for identifying the thiols. The reaction is the basis of the Sullivan method for cystine determination *via* intermediate cysteine. It thus applies equally to cysteine, although in practice it has rarely been used for this determination. Sullivan (1924) added 0·1 N sodium hydroxide to a mixture of reagent and various thiols and amino acids, obtaining colours ranging from reddish orange to dark brown. Only that from cysteine was not discharged by reducing agents such as dithionite. Sullivan (1925) mentioned that cystine also gives the test but weakly, presumably as a result of slow conversion to the thiol by sulphite; this is accelerated by cyanide addition. Later (1926) he observed that the reaction with quinone, dithionite and cyanide was given neither by compounds containing only a thiol group, nor only an amino group, nor an amino group and sulphur, as in cystine, nor by mixes of amino acids and thiols, nor by compounds containing an amino and a thiol group further apart than in cysteine.

In the following years, much work was published discussing aspects and improvements and the quantitative adaptation. As stated, most of these methods are for cystine, which is converted to cysteine in a preliminary step, generally with cyanide ion, or a metallic reducing agent:

$$CySSCy + CN^- \rightarrow CyS^- + CySCN$$

$$CySSCy + 2e \rightarrow 2CyS^-$$

Examples of application in this way are: Sullivan (1929a) for cystine in casein hydrolysate, using cyanide and comparing with colour standards; Bushill *et al.* (1934), using the Pulfrich photometer; Sullivan and Hess (1936) for urinary cystine, using a higher quinone reagent concentration for full colour development; Rossouw (1940) for cystine in wool, precipitating as copper(I) cysteine derivative; Brand *et al.* (1940) for cystine in dog urine, also reducing and precipitating with Cu(I); Zittle and O'Dell (1941) for cystine in hydrolysates of biological materials, also first precipitating with copper(I); Hess *et al.* (1942) on hydrolysates of tobacco mosaic virus proteins; Sullivan *et al.* (1942), reducing with sodium amalgam giving two equivalents of cysteine (by carrying out the usual determination with cyanide furnishing only one equivalent of cysteine it was possible to analyse mixtures of these two amino acids); Csonka *et al.* (1944) on foods and proteins, evaluating the absorbance at 505 nm; Sanford and Humoller (1947) in altered human hair fibres; Matsuura and Szafarz (1965) for micro amounts in protein hydrolysates, evaluating at 504 nm; Kurtzman *et al.* (1965) using an AutoAnalyzer; MacDonald and Fellers (1967, 1968) on urine, measuring absorbance at 505 nm in the method of Csonka *et al.*; Lange *et al.* (1967) on urea, evaluating at 496 nm.

The reaction and influence of other compounds has been investigated extensively. For example, Sullivan (1929b) tested the influence of pyruvic acid, furfuraldehyde and cysteamine. Sullivan and Hess (1931) estimated cysteine in the presence of glutathione and isocysteine, observing that more reagent was then needed for good results. Rossouw and Wilken-Jorden

(1934) improved the method by increasing the cyanide concentration and modifying the interval between adding quinone and alkaline sulphite. A critical study was undertaken by Andrews and Andrews (1936) who found that reducing agents such as ascorbic acid, polyphenols, tannic acid, hydrogen sulphide and some aldehydes are powerful inhibitors. Neubeck and Smythe (1944) investigated the spectra of the products from several thiols, noting that those from cysteine and mercaptoacetic acid had similarly situated absorption maxima at 500–510 nm but that the latter yielded only about 2% of the colour intensity of the former. The absorption maxima of the products from glutathione and some other compounds were at lower wavelengths.

The colour reaction has been applied to determine cysteine and cysteinyl-glycine products of breakdown of glutathione. Thus Binkley and Nakamura (1948) developed colour according to Sullivan's 1926 procedure and then evaluated at 540 nm where the absorption curves of the two compounds intersect and they thus have the same absorptivity; this provided a value for the sum of the two. They measured also the ratio of absorbances at 500 and 580 nm to break down this total into the two individual amounts. Fodor *et al.* (1953) used the procedure of Sullivan *et al.* (1942) in this problem, also measuring absorbance at 540 nm but saying that the absorptivities of the cysteine and cysteinylglycine are the same only if the reaction mixture is at pH 12·7 before the quinone reagent is added; at lower pH values cysteine furnishes less colour, and its contribution becomes negligible at pH 9·9.

Two references to the use of the naphthoquinone reagent for visualization in chromatography may be quoted, although conditions do not correspond to those in the Sullivan cysteine determination. Giri and Najabhushanam (1952) dipped their paper chromatograms into 0·1% sodium 1,2-naphtho-quinone-4-sulphonate solution in acetone–water (30 + 1), then heating to 80–90°C for 3–5 min. They then immersed the strip in a reagent of 2 ml of 5 N sodium hydroxide made up to 100 ml with 95% ethanol and then mixed 10:3 with the first reagent solution. This intensified the colour obtained after 30–60 sec. Cysteine yields dark blue, turning brown; glutathione (and also cystine) yields dark blue. Down to 5 µg amounts could be detected. Freytag (1953) mixed the test solution with a few drops of 5·7% ethanolamine in water and adjusted the pH to 7·5. Some of this resulting solution was spotted on to Schleicher and Schüll 2190 paper, which he then dried and treated with a drop of 1% aqueous sodium salt of the quinone reagent. Numerous thiols (e.g. cysteine, mercaptoacetic acid, mercaptopropionic acid, sodium mer-captobutanesulphonate) and also ascorbic acid yielded products fluorescing light blue in u.v. light.

1.5.3. *Disodium Maleate*

Hendrickson and Hatch (1960) converted thiols (alkanethiols up to C_9

and some others) to *S*-substituted mercaptosuccinic acids by treating with disodium maleate in ethanol and subsequently acidifying.

$$\begin{array}{l} \text{CH—COONa} \\ \| \quad\quad\quad\quad\; + \text{RSH} \rightarrow \\ \text{CH—COONa} \end{array} \quad \begin{array}{l} \text{RS—CH—COONa} \\ \;\;\;\; | \\ \text{CH}_2\text{—COONa} \end{array} \xrightarrow{\text{H}^+} \begin{array}{l} \text{RS—CH—COOH} \\ \;\;\;\; | \\ \text{CH}_2\text{COOH} \end{array}$$

These acids can be used to identify the original thiols.

1.6. Addition to the C=C=O group

Some work, although not directly analytical, as been carried out on the blocking of thiol groups with ketene and carbon suboxide. For example, Desnuelle and Rovery (1947, 1949) tried the first to block reactive thiol groups in denatured egg albumin. They needed a large excess (ca. 50:1) in ether solution. The "inert" thiol groups did not react with ketene. Fraenkel-Conrat (1944) also tested both these reagents. Carbon suboxide reacted preferentially with the thiol group in crystalline egg albumin rather than with phenolic or amine groups. Ketene, however, acetylated amines more rapidly than thiols. The thio esters prepared with these reagents can be easily hydrolysed with alkali at room temperature to regenerate the thiols. Fraenkel-Conrat demonstrated this with cysteine and glutathione.

$$\text{RSH} + \text{CH}_2\text{=C=O} \rightarrow \begin{array}{l} \text{CH}_3\text{—C=O} \\ \quad\quad\;\; | \\ \quad\quad\; \text{SR} \end{array}$$

$$2\text{RSH} + \text{O=C=C=C=O} \rightarrow \begin{array}{l} \text{O=C—CH}_2\text{—C=O} \\ \quad\;\; | \quad\quad\quad\; | \\ \quad\; \text{RS} \quad\quad\quad \text{SR} \end{array}$$

2. ADDITION OF THIOLS TO —C=O BONDS

Like the addition to carbon–carbon double bonds, that to the carbonyl group has been known for many years. As long ago as 1885, Baumann prepared addition compounds of thiophenol with chloral, pyruvic acid, phenylglyoxalic acid and isatin.

$$\begin{array}{l} \text{R—C—R}' \\ \quad\; \| \quad\quad + \text{R}''\text{—SH} \rightarrow \\ \quad\; \text{O} \end{array} \quad \begin{array}{c} \text{R—C—R}' \\ \diagup \quad \diagdown \\ \text{HO} \quad\; \text{SR}'' \end{array}$$

This was later extended to other representatives of both classes and offers a method of preparing sulphides. In many cases an unstable reaction product is formed, easily decomposing into the starting materials.

Where certain other functional groups are present in the thiol, a further reaction of cyclization is possible, as, for instance, with cysteine:

$$
\begin{array}{ccccc}
\text{SH} & & \text{R} & \text{R} & \text{R}\\
| & & | & |\quad & |\quad\\
\text{CH}_2 & + & \text{O=C-R'} \longrightarrow & \text{S}\diagdown\text{C-R'} \longrightarrow & \text{S}\diagdown\text{C-R'}\\
| & & & \text{CH}_2\ \ \text{OH} & \text{CH}_2\\
\text{HOOC-CH-NH}_2 & & & \text{HOOC-CH-NH}_2 & \text{HOOC-CH-NH}
\end{array}
$$

Thiazolidine

Ratner and Clarke (1937) showed that the initial reaction of thiazolidine formation is indeed addition of the thiol group to the carbonyl double bond, and not condensation of amino and carbonyl groups. They observed that at pH 5·12, the optical activity of S-ethylcysteine (containing no thiol group) was not changed by addition of formaldehyde, but that that of N-acetylcysteine was modified.

Many thiazolidines were prepared from cysteine and various aliphatic and aromatic aldehydes and ketones, including even 3-ketosteroids; a prominent investigator was Schubert (1935, 1936, 1939). Ratner and Clarke also prepared the parent thiazolidine from formaldehyde and 2-mercaptoethylamine.

The kinetics of the reaction was also studied and Ratner and Clarke found that at pH values above 5 the rate of reaction of cysteine with formaldehyde became immeasurably fast.

There are relatively few analytical applications of the addition reaction to thiols. Birch and Harris (1930) noted that adding formaldehyde to cysteine led to the disappearance of the typical buffering action of the thiol group; Shinohara (1935) found that formaldehyde completely inhibited the reducing power of cysteine for tungstophosphate; Schubert (1939) added reducing sugars to cysteine in the presence of bicarbonate and observed that the solution no longer gave the nitroprusside test for the thiol group. These and many other findings indicate that carbonyl compounds could be used at least to eliminate or block thiol groups, especially those in compounds such as cysteine where the further reaction forces the addition reaction to completion.

Some examples of the use of this and of other analytical applications are given below. Some other carbonyl reagents are also conveniently quoted here, although it is not always certain if the initial reaction is addition of the thiol group to the carbon–oxygen double bond.

2.1. Formaldehyde

Lewin (1956, 1962) pointed out the contrasting results of addition of formaldehyde to amino acids (Sorensen method) and to thiols. With the former, the pH of the solution is lowered, with the latter, raised through removal of the acidic group:

$$RSH + HCHO \rightarrow R—S—CH_2OH$$

He titrated mercaptoethanol potentiometrically with sodium hydroxide in the presence and absence of formaldehyde, and observed the pH-gap between the two titration curves. A similar gap was found in the titration of calcium mercaptoacetate with hydrochloric acid. Lewin suggested the use of this for recognizing the thiol group.

Strickland et al. (1954) utilized thiazolidine formation to determine cysteine and cystine in proteins. They left the sample overnight with formaldehyde reagent and then precipitated protein with acetone. After centrifuging, complete conversion into the thiazolidine was ensured by heating 15 min at 100°C. They then hydrolysed the modified protein in the usual way with 6 N hydrochloric acid and precipitated cystine with a copper(I) reagent, ultimately determining it *via* its sulphur content (as barium sulphate). The thiazolidine was likewise oxidized and determined gravimetrically as barium sulphate to obtain a value for the cysteine. In an investigation of the McCarthy–Sullivan colorimetric methionine determination with nitroprusside, Strack et al. 1956b) found that certain condensation products of formaldehyde with cysteine and homocysteine also gave colour. This source of error was removed by a longer reaction time of 3 h at 30°C, probably through the cyclization reaction. They were able in fact to determine the thiols through the difference in light absorbance at 510 nm with and without this heating treatment. The same authors (1956a) made use of this reaction for facilitating paper chromatographic separation of cysteine, homocysteine, the two corresponding disulphides and methionine. They first converted the thiols to thiazolidines and then separated with butanol-formic acid–water (77 + 10 + 13).

In his series of studies of mercurimetric titration of sulphur-containing compounds, Wronski (1963, 1964a, 1965) inactivated cysteine with formaldehyde through this conversion to a thiazolidine. Thus in the first two mentioned publications, he refers to treatment of 25 ml of sample with 1 ml of 37% formaldehyde and 1 ml of N sodium hydroxide for 3–5 min at 22°C. He was then able to titrate mercaptoacetate in the mixture, with o-hydroxy-mercuribenzoate and using thiofluorescein as indicator (to disappearance of blue). The sum of both thiols he obtained from a similar titration without adding formaldehyde; cysteine was yielded by difference. In the third article, Wronski analysed a mixture of glutathione and cysteine. He obtained the total by the mercurimetric titration mentioned above. He then treated a 20 ml sample + 1 ml of N sodium hydroxide with 5 ml of 3% formalin for 1 min and titrated residual glutathione with the same reagent but to dithizone indicator (yellow → purple); a small correction was necessary for the glutathione which reacted with formaldehyde but this reaction is slow since it yields a 10-membered ring. Thibert and Ke (1971) also inactivated cysteine and other compounds containing a thiol and amino group by treating a 10 ml sample at pH 7·4 with 1 ml of 35% formaldehyde, heating for 2 min and then adding potassium hydroxide to bring the pH to 9. β-Mercapto-pyruvic acid from a cysteine transamination mixture was then titrated with o-hydroxymercuribenzoate to thiofluorescein.

In a recent procedure for analysing mixtures of cysteine, glutathione and ergothioneine, Kurzawa and Kurzawa (1974) blocked the thiol groups of the first two before applying the iodine/azide method. Zhantalai (1964) observed that cysteine and methionine yield polarographically active compounds on treatment with formaldehyde. The heights of the kinetic waves are pH-dependent. By measuring wave heights at pH 4·7 (where the cysteine product has a maximum value and the contribution from methionine is small) and at pH 8·2 (where the reverse obtains), each thiol could be determined. The small cysteine wave at pH 8·2 could be due to thiazolidine formation.

2.2. Glyoxalic acid

Gadal (1963) analysed cysteine/glutathione mixtures with the help of glyoxalic acid for blocking instead of formaldehyde. It yields thiazolidine-2,4-dicarboxylic acid in neutral or alkaline solution with cysteine. After reaction for 5 min he titrated the non-reacting glutathions with silver nitrate. The sum of both thiols was obtained from a similar titration without pre-treatment with glyoxalic acid.

Ball (1966) also used glyoxalic acid for inactivating cysteine in a mixture with glutathione (5 min at 60°C and pH 8). Cysteine was then estimated from the difference from assays, using Ellman's reagent II (p. 179) with and without this treatment.

2.3. Ninhydrin(1,2,3-triketohydrindane)

Ninhydrin yields a characteristic blue–violet product of absorption maximum at ca. 570 nm with α-amino acids and is a well known reagent for their detection and determination. Cysteine and other amino acids containing sulphur are often among the examples thus visualized or determined but the general reaction does not involve the thiol group so that it is not discussed further here.

However, Chinard (1952) noted that cysteine gives a pink colour on heating with ninhydrin/acetic acid/phosphoric acid. Gaitonde (1967) based a quantitative spectrophotometric determination of cysteine on this. Details may be given: From 0·05 to 0·5 μmol of cysteine (hydrochloride) in aqueous solution is treated with 0·5 ml of acetic acid and 0·5 ml of an acidic ninhydrin reagent. The reaction tubes are closed with glass marbles or aluminium caps and placed for 10 min in a bath of boiling water. The contents of the tubes are then rapidly cooled in tap water, diluted with 95% ethanol to 5 or 10 ml, and their absorbances measured at 560 nm. The molar absorptivity for the cysteine product is 28 000 and the Beer–Lambert law holds within the concentration range given. Neither glutathione nor homocysteine nor the other natural amino acids interfere. The acidic ninhydrin reagent is prepared by dissolving 250 mg of ninhydrin in a mixture of 6 ml of acetic acid and 4 ml of either 0·6 M phosphoric acid or concentrated hydrochloric acid; the former yields a solution stable for 2 weeks at 4°C, the latter must be prepared freshly before use.

Gaitonde made no proposal about the reaction but it is not unreasonable to suppose that the thiol group first adds to the central carbonyl group, followed by ring closure:

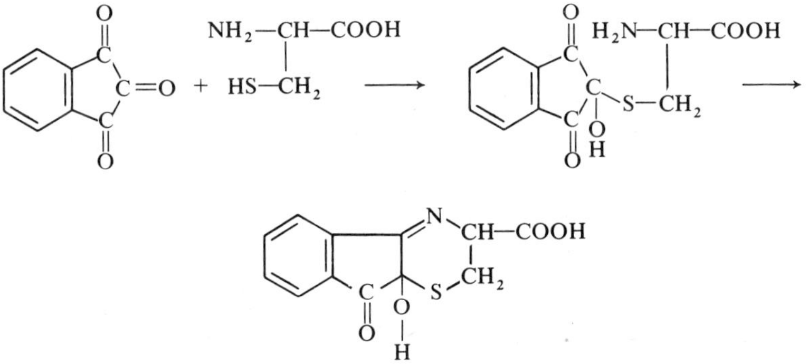

De Koning and Van Rooijen (1971) used Gaitonde's method to determine cysteine in hydrolysates from whey proteins; and States and Segal (1973) applied a modification to determine cysteine (and also cystine after treatment with dithiothreitol) in human fibroblasts.

Pasieka and Morgan (1955) detected homocysteine on PCs by spraying with ninhydrin, heating and then treating with mercuric nitrate solution. This yielded cherry-red spots on deep blue and appeared specific for homocysteine. Eighty other biological substances, including 28 sulphur-containing, gave no such response.

Walker et al. (1956) detected primary, secondary and tertiary thiols with ninhydrin in alkaline solution, in contrast to Gaitonde's method. The pinkish colour yielded at pH 4 gave way to a deep red at pH 9. An intense blue was yielded at high alkanity. Patil et al. (1972) also carried out their determination of cysteine in alkaline solution, a carbonate–bicarbonate buffer of pH 10. They heated for 3 min at 100°C and then evaluated at 470 nm, the absorption maximum of hydrintantin, evidently formed by reduction of the ninhydrin with the thiol:

Hydrintantin

2.4. Alloxan (2,4,5,6-tetraoxohexahydropyrimidine)

The chemistry of the reaction between alloxan and thiols has not yet been fully explained. Labes and Freisburger (1930) considered that alloxan oxidizes, being thereby converted into murexide, the ammonium salt of purpuric acid, a compound analogous to that derived from ninhydrin (see above, p. 164)

Alloxan   Murexide

Resnick and Wolff (1956) believed that a thiazine ring,

may be formed. They studied the reaction between alloxan and some thiols at pH 7·6 and found products of absorption maxima 305–317 nm, depending on the thiol (glutathione at the lower, mercaptoacetic acid at the higher wavelength). They considered that the usefulness of the reaction was marred by absorption at these wavelengths of other products in extracts.

Younathan and Rudel (1968) quote 312 nm for the absorption maximum of the product resulting from alloxan and reduced ribonuclease.

Patterson et al. (1949) based quantitative estimation of glutathione on the 305 nm maximum, as did also Kay and Murfitt (1960), for blood glutathione, reacting for 20–40 min in phosphate buffer of pH 7·6. Kuninori and Matsumoto (1964) applied Kay and Murfitt's procedure to determine glutathione in wheat and wheat flour after reduction with sodium borohydride.

2.5. Isatin (indole-2,3-dione)

Denigès (1889) noted the colour reaction of thiols with an isatin/concentrated sulphuric acid reagent. A 1 % solution yields green with the sample in alcoholic solution. The corresponding sulphides yield no colour but aldehydes and higher alcohols destroy it. Only little analytical use has been made of the reaction. Reith (1934) detected methanethiol in air with a reagent of 10 mg isatin in 100 ml of the concentrated acid, freshly made up each day. He was able to detect 50 µg of thiol in the presence of 100 mg of hydrogen sulphide, which he removed with lead acetate. He did not advise its use for semi-quantitative work. Tatiya (1941) based a test for S-methyl groups on de-methylation by alkali fusion to produce methanethiol which he then detected by the colour change from yellow to grass green of a solution of 10–20 mg of isatin in 100 ml of concentrated sulphuric acid. More recently, Morozova (1965) visualized amino acids on chromatograms with an isatin reagent, an acetone solution containing 0·2% of isatin, 4% of acetic acid and 0·06% of a metal salt; for sulphur-containing amino acids, cadmium, uranyl and stannous salts proved to be very good.

Little appears to be known about the chemistry of these tests. Isatin is reduced to a variety of bimolecular products by reduction. For example, hydrogen sulphide gives disulphisatide,

and thiols might form an analogous initial product with —SR instead of —SH.

2.6. Luminol(5-amino-2,3-dihydrophthalazine-1,4-dione

Luminol exhibits chemiluminescence under appropriate conditions. This is inhibited or quenched by certain compounds, e.g. cysteine, which can be utilized for their determination. Thus Ponomarenko and Amelina (1965) treated the test solution with 3 ml of 0·06 M hydrogen peroxide. After 1 min they added 12 ml of 0·002 M luminol in 0·1 M sodium hydroxide − 0·001 M cupric sulphate in 1 % ammonia (1 + 1). The quenching of chemiluminescence was related to the concentration of amino acid. Lukovskaya and Markova (1969) were also able to determine amounts of cysteine down to 10 ng in a similar way. Solutions of 10^{-4} M iodine and of cysteine (or sulphlde ion) in 0·1 M sodium hydroxide are treated with 5×10^{-5} M luminol in the same alkaline solution. They estimated the luminescence photographically and related amount of cysteine to the quenching.

2.7. Phthalaldehyde

Cohn and Lyle (1966) determined glutathione in deproteinized blood by adding phosphate buffer of pH 8 and a phthalaldehyde reagent in methanol. This yields a fluorescent product. After 15–20 min at room temperature they evaluated the fluorescence intensity at 420 nm (excitation at 350 nm).

Havu *et al.* (1967) based their determination of glutathione in mammalian islet tissue and McNeil and Beck (1968) based their method for liver and blood samples, on the same fluorescence-yielding reaction. Jocelyn and Kamminga (1970) used a tris buffer of pH 8·3 and found that many thiols, proteins and non-proteins gave fluorescence, e.g. 6,8-dimercaptooctanoic acid, cysteine and homocysteine. Maximum activity was found with compounds of the formula $XYHC—CH_2SH$ where X and Y were H; minimum activity when X or Y was a free thiol or amino group.

Roth (1971) studied the reaction of amino acids with phthalaldehyde. He treated ca. 100 μl amounts of 10^{-3} to 10^{-4} M solutions with phthalaldehyde in borate buffer of pH 9·5 for 5 to 25 min, then evaluating fluorescence intensities at 455 nm using excitation at 340 nm. Much higher intensities were obtained when the reagent contained a reducing agent such as potassium borohydride or 2-mercaptoethanol. Roth tabulated relative fluorescence intensities and it is noticeable that cysteine is one of the most productive with phthalaldehyde in the absence of a reducing agent but that little improvement is obtained in the presence of mercaptoethanol, where it is in fact the last in the series. Roth admits to having no idea of the nature of the reaction(s) involved, but suggests a possible similarity to that with ninhydrin, with which reducing agents are also sometimes used.

There appears to be no record of adapting the reaction to determine the reducing agent, in this case mercaptoethanol.

2.8. *o*-Diacetylbenzene

In the work just quoted, Roth (1971) also tested *o*-diacetylbenzene and found results similar to those with phthalaldehyde. He excited at 355 nm and measured fluorescence at 445 nm in this case. Here, too, relatively intense fluorescence was yielded by cysteine, but it was overtaken by most other amino acids when the reagent contained mercaptoethanol.

Later, Roth and Jeanneret (1972) reported a method for determining lysine, based on reaction with *o*-diacetylbenzene in the presence of 2-mercapto-ethanol at pH 10 and room temperature. The excitation and measuring wavelengths mentioned above (355, 445 nm, respectively) were used. The possibility of determining the mercaptoethanol through the fluorescence intensification might be worth considering here also.

2.9. 2-Nitroindane-1,3-dione

This acid reagent has been used for characterizing bases through melting points of the derivatives, e.g. by Wanag and co-workers (1936, 1937, 1942) and Christensen et al. (1949). No compound containing a thiol group was among the examples but L-cysteine was one of the amino acids studied by Larsen et al. (1948). They added the sample to a hot saturated solution of the reagent in water or alcohol and then boiled for 5 min. They quote melting point, crystal habit and other crystalline data which enable the amino acids to be identified.

2.10. Hexafluoroacetone

Leader (1970) treated some compounds containing active hydrogen (including alkanethiols) with hexafluoroacetone in aqueous acetic acid:

$$(CF_3)_2C{=}O + X{-}H \rightarrow F_3C{-}\overset{\displaystyle X}{\underset{\displaystyle OH}{C}}{-}CF_3$$

This causes a chemical shift in the ^{19}F-NMR spectrum, relative to the hexafluoroacetone and ranging from 0·87 to 1·5 parts/10^6 (only 0·04 for benzenethiol). This shift permitted the original compounds to be identified.

2.11. Phenothiazine-3-one

The method of thiol determination of Raileanu and Dobre (1970) may also depend on addition of the thiol to a carbonyl group. They allowed the sample (e.g. cysteine, mercaptoacetic acid) in 80% acetic acid to react with phenothiazine-3-one in acetic acid. The mixture was heated to 90–100°C and the excess of phenothiazine-3-one was back-titrated with thiosulphate to a colour change from red to green.

2.12. Furfuraldehyde,5-hydroxymethylfurfuraldehyde

The first stage in the colour reactions between certain thiols (e.g. cysteine) and sugars and related compounds (pentoses, methylpentoses, hexoses, heptoses, glucuronic acid) in the presence of concentrated acids is probably addition of the thiol group to the carbonyl group in a degradation product from the action of the acid on the carbohydrate (furfuraldehyde, 2-hydroxyfurfuraldehyde, glutaconaldehyde). Colorimetric methods based on evaluation of the coloured final product were developed for the carbohydrate, especially in the late forties and fifties by Dische and co-workers. However, this does not appear to have been applied in the reverse direction to determine thiols, although this must be possible.

Recent work by Kunovits (1971) indicates that the coloured products from cysteine and mercaptoacetic acid are pentamethine dyes,

$$RS-CH\!\!=\!\!CH-CH\!\!=\!\!CH-CH\!\!=\!\!\overset{(+)}{SR} \leftrightarrow$$

$$\leftrightarrow \overset{(+)}{RS}\!\!=\!\!CH-CH\!\!=\!\!CH-CH\!\!=\!\!CH-SR$$

also with a hydroxyl group in the 2-position of the chain.

2.13. Salicylaldehyde

Schlütz (1953) tested for down to 50 µg of cysteine by heating the sample to 100°C (water bath) with 5% salicylaldehyde/15% sodium hydroxide. A red, orange or wine (Burgundy?) colour within 1 min was the positive outcome. The test was applicable to cysteine in proteins, and also to glutathione.

3. ADDITION OF THIOLS TO —C=N— BONDS

Phenyl isocyanate is among compounds tested in earlier years for inactivating thiol groups:

$$C_6H_5-N\!\!=\!\!C\!\!=\!\!O + RSH \rightarrow C_6H_5-\underset{H}{\overset{}{N}}-\underset{SR}{\overset{}{C}}\!\!=\!\!O$$

Thus Fraenkel-Conrat (1944) treated a crystalline egg albumin with it, stirring for 16–24 h. He found that it reacted at pH 5–6 preferentially with thiol groups rather than with phenolic or amine groups. The products could be easily hydrolysed with alkali at room temperature to regenerate the thiols. He demonstrated this in model experiments with cysteine and glutathione. Desnuelle and Rovery (1947, 1949) also blocked "reactive" thiol groups in denatured egg albumin using phenyl isocyanate in acetone. The "inert" thiol groups reacted only slowly.

Stark (1964) used potassium cyanate to block cysteine and other thiol groups, quoting the equation:

$$RS^- + HN\!\!=\!\!C\!\!=\!\!O + H_2O \rightarrow H_2N-\underset{SR}{\overset{}{C}}\!\!=\!\!O + OH^-$$

The products were stable at pH values below 6 but the reverse reaction was favoured in alkaline medium.

Trichloromethyl isocyanate was used by Butler and Mueller (1966) to characterize aliphatic and aromatic mono- and dithiols. The monothiolo-carbamates, CCl_3—NH—COSR, yielded with drops of reagent, were dissolved in deuterochloroform or hexadeuterodimethyl sulphoxide and their NMR-spectra were recorded.

Scola *et al.* (1969) treated compounds containing active hydrogen (including thiols) with *trans*-vinylidene diisocyanate, obtaining bis(alkyl)- or bis(aryl)-*trans*-vinylidene-bis(thiolocarbamates). They identified the original thiols through the melting points and infrared data of these derivatives.

Methyl isothiocyanate was used by Toniolo and Jori (1970) to determine thiols. They carried out the reaction with cysteine at pH 4 and with glutathione at pH 5·5, for example, leaving the mixture overnight or for 2 h, respectively:

$$CH_3—N{=}C{=}S + RSH \rightarrow CH_3—\underset{\underset{H}{|}}{N}—\underset{\underset{SR}{|}}{C}{=}S$$

The circular dichroism spectra of the reaction mixtures show a band near 320 nm (323 for cysteine, 318 for glutathione). This Cotton effect is related to the $n \rightarrow \pi^*$ transition within the C=S chromophore. This transition of low absorbance is in addition in a spectral region where all amino acids are transparent.

Another compound containing a —C=N— group which reacts with thiols is N-propionyldiphenylketimine, $(C_6H_5)_2C{=}N—COC_2H_5$. Gunsalus *et al.* (1956) found it useful for distinguishing primary from secondary thiol groups, e.g. in dihydrolipoic acid, $HOOC—(CH_2)_4—\underset{\underset{SH}{|}}{CH}—CH_2—CH_2SH$. It reacts only with primary thiol (and alcohol) groups, and they were able to demonstrate the presence of a primary group in the acid. There appears to be no analytical application of the reagent.

REFERENCES

Alexander, N. M. (1958). *Anal. Chem.* **30**, 1292.
Ambrosino, C., Vancheri, L., Lausarot, P. M. and Papa, G. (1969). *Ric. Sci.* **39**, 924.
Andrews, J. C. and Andrews, K. C. (1936). *Am. J. Med. Sci.* **191**, 594.
Axberg, G. and Holmberg, B. (1933). *Chem. Ber.* **66**, 1193.
Ball, C. R. (1966). *Biochem. Pharm.* **15**, 809.
Baudisch, O. and Dyer, E. (1933). *J. Biol. Chem.* **99**, 485.
Baumann, E. (1885). *Chem. Ber.* **18**, 258, 883.
Belliveau, P. E. and Frei, R. W. (1971). *Chromatographia*, **4**, 189.
Benesch, R. and Benesch, R. E. (1961). *J. Biol. Chem.* **236**, 405.
Benesch, R., Benesch, R. E., Gutcho, M. and Laufer, L. (1956). *Science*, **123**, 981.

Binkley, F. and Nakamura, K. (1948). *J. Biol. Chem.* **173**, 411.
Birch, T. W. and Harris, L. J. (1930). *Biochem. J.* **24**, 1080.
Brand, E., Cahill, G. F. and Kassell, E. (1940). *J. Biol. Chem.* **133**, 431.
Broekhuysen, J. (1958). *Anal. Chim. Acta,* **19**, 542.
Bushill, J. H., Lampitt, L. H. and Baker, L. C. (1934). *Biochem. J.* **28**, 1293.
Butler, P. E. and Mueller, W. H. (1966). *Anal. Chem.* **38**, 1407.
Cavins, J. F., Krull, L. H., Friedman, M., Gibbs, D. E. and Inglett, G. R. (1972). *J. Agr. Food Chem.* **20**, 1124.
Chinard, F. P. (1952). *J. Biol. Chem.* **199**, 91.
Christensen, B. E., Wang, C. H., Davies, I. W. and Harris, D. (1949). *Anal. Chem.* **21**, 1573.
Clark-Walker, G. D. and Robinson, H. C. (1961). *J. Chem. Soc. (London),* 2810.
Cohn, V. H. and Lyle, J. (1966). *Anal. Biochem.* **14**, 434.
Cole, R. D., Stein, W. H. and Moore, S. (1958). *J. Biol. Chem.* **233**, 1359.
Csonka, F. A., Lichtenstein, H. and Denten, C. A. (1944). *J. Biol. Chem.* **156**, 571.
De Koning, P. J. and Van Rooijen, P. J. (1971). *Milchwissenschaft,* **26**, 1.
De Marco, C., Graziani, M. T. and Mosti, R. (1966). *Anal. Biochem.* **15**, 40.
Denigès, G. (1889). *J. Pharm. Chim.* [5], **19**, 276.
Desnuelle, P. and Rovery, M. (1947). *Biochim. Biophys. Acta,* **1**, 497; also (1949). *ibid.* **3**, 62.
Devani, M. B., Shishoo, C. J. and Shah, M. G. (1973). *Analyst (London),* **98**, 759.
Dyer, E. and Baudisch, O. (1932). *J. Biol. Chem.* **95**, 483.
Earle, T. E. (1953). *Anal. Chem.* **25**, 769.
Ellis, L. M. and Reid, E. E. (1932). *J. Amer. Chem. Soc.* **54**, 1674.
Eshelman, L. R., Manzo, E. Y., Marcus, S. J., Decoteau, A. E. and Hammond, E. G. (1960). *Anal. Chem.* **32**, 844.
Fasold, H., Groeschel-Stewart, U. and Turba, F. (1963). *Biochem. Z.* **337**, 425.
Fernandez, A. A. and Henry, R. J. (1965). *Anal. Biochem.* **11**, 190.
Flavin, M. (1963). *Anal. Biochem.* **5**, 60.
Fodor, P. J., Miller, A. and Waelsch, H. (1953). *J. Biol. Chem.* **202**, 551.
Fraenkel-Conrat, H. (1944). *J. Biol. Chem.* **152**, 385.
Fraenkle-Conrat, H. (1955). *J. Biol. Chem.* **217**, 373.
Freedberg, W. B. and Hardman, J. K. (1971). *J. Biol. Chem.* **246**, 1439.
Freytag, H. (1953). *Z. Anal. Chem.* **138**, 259.
Friedman, M. and Krull, L. H. (1969). *Biochem. Biophys. Res. Commun.* **37**, 630.
Friedman, M. and Noma, A. T. (1970). *Text. Res. J.* **40**, 1073.
Friedman, M., Krull, L. H. and Cavins, J. F. (1970). *J. Biol. Chem.* **245**, 3868.
Friedmann, E. (1952). *Biochim. Biophys. Acta,* **9**, 65.
Friedmann, E., Marrian, D. H. and Simon-Ruess, I. (1949). *Brit. J. Pharmacol.* **4**, 105.
Gadal, P. (1963). *Compt. Rend.* **256**, 4311.
Gaitonde, M. K. (1967). *Biochem. J.* **104**, 627.
Giri, K. V. and Nagabhushanam, A. (1953). *Naturwissensch.* **39**, 548.
Gorin, G., Martic, P. A. and Doughty, G. (1966). *Arch. Biochem. Biophys.* **115**, 593.
Greenhow, E. J. and Loo, L. H. (1974). *Analyst (London),* **99**, 360.
Gregory, J. D. (1955). *J. Amer. Chem. Soc.* **77**, 3922.
Gunsalus, I. C., Barton, L. S. and Gruber, Q. (1956). *J. Amer. Chem. Soc.* **78**, 1763.
Gutcho, M. and Laufer, L. (1953). *Glutathione, Proc. Symp. Ridgefield, Conn., U.S.A.* 79; *Chem. Abs.* **49**, 5566.
Hamm, R. and Hofmann, K. (1966). *Z. Lebensm.-Untersuch. Forsch.* **130**, 85.
Hanes, C. S., Hird, F. R. and Isherwood, F. A. (1950). *Nature (London),* **166**, 288.

Havu, N., Lindberg, B. and Falkner, S. (1967). *Diabetes, Proc. Congr. Int. Diabetes Feb.* 6, 203; *Chem. Abs.* 76, 123761.

Hazeloop, E. (1934). *Ann. P. van der Weilen* 159; *Chem. Abs.* 29, 3364.

Hendrickson, J. G. and Hatch, L. F. (1960). *J. Org. Chem.* 25, 1747.

Hess, W. C. and Sullivan, M. X. (1932). *J. Biol. Chem.* 99, 485.

Hess, W. C., Sullivan, M. X. and Palmes, E. D. (1942). *Proc. Soc. Exp. Biol. Med.* 48, 353.

Hofmann, K. (1965). *Naturwissensch.* 52, 428.

Hofmann, K. (1971). *Z. Anal. Chem.* 256, 187.

Hofmann, W. S. and Reid, E. E. (1923). *J. Amer. Chem. Soc.* 45, 1831.

Hoog, H. and Eichwald, E. (1939). *Rec. Trav. Chim.* 58, 481.

Jacot-Guillarmod, A. and Ceschini, P. (1959). *Helv. Chim. Acta,* 42, 713.

Jocelyn, P. C. and Kamminga, A. (1970). *Anal. Biochem.* 37, 417.

Kalan, E. B., Neistadt, A., Weil, L. and Gordon, W. G. (1965). *Anal. Biochem.* 12, 488.

Kanaoka, Y., Machida, M., Ban, Y. and Sekine, T. (1967). *Chem. Pharm. Bull.* 15, 1738.

Kay, W. W. and Murfitt, K. C. (1960). *Biochem. J.* 74, 203.

Klein, P. and Robbins, E. (1970). *J. Cell. Biol.* 46, 165.

Koenig, N. H. and Swern, D. (1957). *J. Amer. Chem. Soc.* 79, 362.

Krull, L. H., Gibbs, D. E. and Friedman, M. (1971). *Anal. Biochem.* 40, 80.

Kuhn, R. and Beinert, H. (1944). *Chem. Ber.* 77, 604.

Kuninori, T. and Matsumoto, H. (1964). *Cereal Chem.* 41, 252.

Kunovits, G. (1971). *Anal. Chim. Acta,* 55, 221.

Kurtzman, C. H., Smith, jr., D. and Snyder, D. G. (1965). *Anal. Biochem.* 12, 282.

Kurzawa, Z. and Kurzawa, J. (1974). *Chem. Anal. (Warsaw),* 19, 755.

Labes, R. and Freisburger, H. (1930). *Arch. Exptl. Pathol. Pharmakol.* 156, 226.

Lange, J., Freund, K. and Buechl, H. (1967). *Arzneimittelforsch.* 17, 856.

Larsen, J., Witt, N. F. and Poe, C. F. (1948). *Mikrochim. Acta,* 34, 1.

Leader, G. R. (1970). *Anal. Chem.* 42, 16.

Lee, G. C. and Lai, T-S. (1967). *Cereal Chem.* 44, 620.

Lee, G. C. and Samuels, E. R. (1964). *Can. J. Chem.* 42, 164.

Leslie, J. (1965). *Anal. Biochem.* 10, 162.

Leslie, J., Williams, D. L. and Gorin, G. (1962). *Anal. Biochem.* 3, 257.

Lewin, S. (1956). *Biochem. J.* 64, 30P, 31P.

Lewin, S. (1962). *J. Chem. Soc. (London),* 1462.

Lukovskaya, N. M. and Markova, L. V. (1969). *Zh. Anal. Khim.* 24, 1862.

Macdonald, W. B. and Fellers, F. X. (1967). *Tech. Bull. Regist. Med. Technol.* 37, 299; *Chem. Abs.* 68, 112070.

Macdonald, W. B. and Fellers, F. X. (1968). *Am. J. Clin. Path.* 49, 123.

McNeil, T. L. and Beck, L. V. (1968). *Anal. Biochem.* 22, 431.

Marrian, D. N. (1949). *J. Chem. Soc* (London), 1515.

Matsuura, T. and Szafarz, D. (1965). *Experientia,* 21, 737.

Miyagawa, T., Kaji, H., Sakata, Y. and Minakawa, S. (1971). *Kurume Med. J.* 18, 177. *Chem. Abs.* 76, 137660.

Morozova, R. P. (1965). *Ukr. Biokhim. Zh.* 37, 290; *Chem. Abs.* 63, 3294.

Nara, Y. and Tuzimura, K. (1973). *Bunseki Kagaku,* 22, 451.

Narang, A. S., Singh, J., Rao, R. V. and Bhalerao, V. R. (1967). *Milchwissensch.* 22, 682.

Neubeck, C. E. and Smythe, C. V. (1944). *Arch. Biochem.* 4, 435.

Obtemperanskaya, S. I. and Nguyen Dyk Hoc (1970). *Vestn. Mosk. Univ., Khim.* 11, 369; *Chem. Abs.* 74, 82828.

Obtemperanskaya, S. I., Terent'ev, A. P. and Buzlanova, N. M. (1957). *Vest. Mosk. Univ.* 12, *Ser. Mat. Mekh. Astron. Fiz. Khim.* No. 3, 145; *Chem. Abs.* **52**, 4414.

Obtemperanskaya, S. I., Kareva, G. P. and Zhukovskaya, O. S. (1973). *Zh. Anal. Khim.* **28**, 2066.

Obtemperanskaya, S. I., Kareva, G. P. and Zhukovskaya, O. S. (1974). *Zh. Anal. Khim.* **29**, 175.

Obtemperanskaya, S. I., Kalinina, N. N., Speranskaya, T. N. and Solov'eva, N. M. (1974). *Zh. Anal. Khim.* **29**, 949.

Pasieka, A. E. and Morgan, J. F. (1955). *Biochim. Biophys. Acta,* **18**, 236.

Patil, M. B., Jeyakumar, S. G., Roberts, S. and Kalyankar, G. D. (1972). *Indian J. Biochem. Biophys.* **9**, 217.

Patterson, J. W., Lazarow, A., Lemm, F. J. and Levey, S. (1949). *J. Biol. Chem.* **177**, 197.

Ponomarenko, A. A. and Amelina, L. M. (1965). *Zh. Obshch. Khim.* **35**, 2252.

Posner, T. (1905). *Chem. Ber.* **38**, 646.

Price, C. A. and Campbell, C. W. (1957). *Biochem. J.* **65**, 512.

Raileanu, M. and Dobre, E. (1970). *Revue Roum. Chim.* **15**, 823; *Anal. Abs.* **22**, 196; *Chem. Abs.* **75**, 71135.

Ratner, S. and Clarke, H. T. (1937). *J. Amer. Chem Soc.* **59**, 203.

Reid, E. E., Mackail, C. M. and Miller, G. E. (1921). *J. Amer. Chem. Soc.* **43**, 2104.

Reith, J. F. (1934). *Rec. Trav. Chim* **53**, 18.

Resnick, R. A. and Wolff, A. R. (1956). *Arch. Biochem. Biophys.* **64**, 33.

Richards, E. G., Snow, D. L. and McClare, C. W. F. (1966). *Biochem.* **5**, 485.

Roberts, E. and Rouser, G. (1958). *Anal. Chem.* **30**, 1291.

Rossouw, S. D. (1940). *Onderstepoort, J. Vet. Sci. Animal Ind.* **14**, 461; *Chem. Abs.* **35**, 2443.

Roussouw, S. D. and Wilken-Jorden, T. J. (1934). *Onderstepoort, J. Vet. Sci.* **2**, 361; *Chem. Abs.* **28**, 4240.

Roth, M. (1971). *Anal. Chem.* **43**, 880.

Roth, M. and Jeanneret, L. (1972). *Z. Physiol. Chem.* **353**, 1607.

Sanford, D. and Humoller, F. L. (1947). *Ind. Eng. Chem., Anal. Ed.* **19**, 404.

Sawicki, E., Engel, C. R. and Elbert, W. C. (1967). *Talanta,* **14**, 1169.

Schlütz, G. O. (1953). *Z. Physiol. Chem.* **293**, 48.

Schubert, M. P. (1935). *J. Biol. Chem.* **111**, 671; Also (1936). *ibid.* **114**, 347; and (1939). *ibid.* **130**, 601.

Scola, D. A., Adams, jr., J. S. and Lopiekes, D. V. (1969). *J. Chem. Eng. Data,* **14**, 490.

Sekine, T., Ando, K., Machida, M. and Kanaoka, Y. (1972). *Anal. Biochem.* **48**, 557.

Shinohara, K. (1935). *J. Biol. Chem.* **110**, 263.

Smith, M. J. and Rodnight, R. (1959). *Biochem. J.* **72**, 1P.

Smyth, D. G., Nagamatsu, A. and Fruton, J. S. (1960). *J. Amer. Chem. Soc.* **82**, 4600.

Stark, G. R. (1964). *J. Biol. Chem.* **239**, 1411.

States, B. and Segal, S. (1969). *Anal. Biochem.* **27**, 323.

States, B. and Segal, S. (1973). *Clin. Chim. Acta,* **43**, 49.

Strack, E., Friedel, W. and Hambsch, A. (1956a). *Z. Physiol. Chem.* **305**, 166.

Strack, E., Friedel, W. and Hambsch, A. (1956b). *Z. Physiol. Chem.* **305**, 237.

Strickland, R. D., Martin, E. L. and Reibsomer, J. L. (1954). *J. Biol. Chem.* **207**, 903.

Sullivan, M. X. (1924). *J. Biol. Chem.* **59**, 1 (letter).

Sullivan, M. X. (1925). *Abs. Biochem.* **9**, 37; *Chem. Abs.* **19**, 2834.

Sullivan, M. X. (1926). *U.S. Publ. Health Repts.* **41**, 1030.

Sullivan, M. X. (1929a). *U.S. Publ. Health Repts.* Suppl. No. 78.

Sullivan, M. X. (1929b). *U.S. Publ. Health Repts.* **44**, 1421.

Sullivan, M. X. and Hess, W. C. (1931). *U.S. Publ. Health Repts.* **46**, 390.

Sullivan, M. X. and Hess, W. C. (1936). *J. Biol. Chem.* **116**, 221.

Sullivan, M. X., Hess, W. C. and Howard, H. W. (1942). *J. Biol. Chem.* **145**, 621.

Tatiya, Y. (1941). *J. Agr. Chem. Soc. Japan* **17**, 465; also *Bull. Agr. Chem. Soc. Japan* **17**, 48; *Chem. Abs.* **36**, 3752.

Thibert, R. J. and Ke, P. J. (1971). *Mikrochim. Acta*, 531.

Tietze, F. (1969). *Anal. Biochem.* **27**, 502.

Titov, A. V. (1969). *Vop. Med. Khim.* **15**, 295; *Chem. Abs.* **71**, 67814.

Tkachuk, R. and Hlynka, I. (1963). *Cereal. Chem.* **40**, 704.

Toniolo, C. and Jori, G. (1970). *Biochim. Biophys. Acta*, **214**, 368.

Tsao, T. C. and Bailey, K. (1953). *Biochim. Biophys. Acta*, **11**, 102.

Tsou, K-Ch., Barrnett, R. J. and Seligman, A. M. (1955). *J. Amer. Chem. Soc.* **77**, 4613.

Tu, Y-T. and Chou, C-L. (1963). *Sheng Wu Hua Hsueh Yu Sheng Wu Wu Li Hsueh Pao*, **3**, 89; *Chem. Abs.* **59**, 3074.

Volodina, M. A. and Kon'kova, I. V. (1968). *Vest. Mosk. Univ. Ser. II* **23**, 153.

Walker, G. T., Freeman, F. M. and Hirsch, F. (1956). *Seifen-Öle-Fette-Wachse*, **82**, 161.

Wanag, G. and Dombrowski, A. (1942). *Chem. Ber.* **75**, 82.

Wanag, G. and Lode, A. (1937). *Chem. Ber.* **70**, 547.

Wanag, G. and Walbe, U. (1936). *Chem. Ber.* **69**, 1054.

Weil, L. and Seibles, T. S. (1961). *Arch. Biochem. Biophys.* **95**, 470.

Weitzman, P. D. J. and Tyler, H. J. (1971). *Anal. Biochem.* **43**, 321.

Wendell, P. L. (1970). *Biochem. J.* **117**, 661.

Wernze, H. and Koch, W. (1965). *Klin. Wochenschr.* **43**, 454.

Witter, A. and Tuppy, H. (1960). *Biochim. Biophys. Acta*, **45**, 429.

Wronski, M. (1960). *Analyst* (*London*), **85**, 527.

Wronski, M. (1961). *Acta Chim. Acad. Sci. Hung.* **28**, 87.

Wronski, M. (1963). *Analyst* (*London*), **88**, 562.

Wronski, M. (1964a). *Analyst* (*London*), **89**, 800.

Wronski, M. (1964b). *Z. Anal. Chem.* **206**, 352.

Wronski, M. (1965). *Analyst* (*London*), **90**, 697.

Yamaguchi, R. and Takechi, M. (1972). *Hoshi Yakka Daigaku Kiyo*, **14**, 42; *Chem. Abs.* **78**, 156182.

Younathan, E. S. and Rudel, L. L. (1968). *Biochim. Biophys. Acta*, **168**, 11.

Zhantalai, B. P. (1964). *Biokhimiya*, **29**, 1009; *Anal. Abs.* **13**, 1426.

Zittle, C. A. and O'Dell, R. A. (1941). *J. Biol. Chem.* **139**, 753.

5. EQUILIBRIUM REACTIONS WITH DISULPHIDES

Thiols and disulphides evidently exist in equilibrium:

$$RSH + R'SSR' \rightleftharpoons R'SSR + R'SH$$

The first intimation of the equilibrium seems to have been made by Lecher (1920) who reacted o-nitrobenzenesulphenyl chloride with the sodium salt of thiophenol. He obtained both the symmetrical and unsymmetrical disulphides. In subsequent years the exchange reaction was studied and became interesting in connection with the denaturation of proteins. Among such studies may be mentioned that of Bersin and Steudel (1938) who followed polarimetrically the reaction between L-cystine and mercaptoacetic acid and between L-cysteine and dithiodiacetic acid, at 30°C in solutions of pH 4·5 and 6; of Kolthoff *et al.* (1955) who determined the equilibrium constants, using solubility and polarographic data, of exchanges between cysteine, gluthathione and mercaptoacetic acid and their disulphide forms; of Fava *et al.* (1957), who used disulphides containing ^{35}S, initiating the reaction with alkali and stopping it with acid, precipitating mercaptides as silver salts and ultimately converting them to barium sulphate containing the ^{35}S, measured radiometrically; and of Eldjarn and Pihl (1957) who also determined equilibrium constants of thiol/disulphide systems (including cysteine-oxidized glutathione) with the help of ^{35}S, separating the three disulphide species electrophoretically.

Mixed disulphides were reported not only with cysteine or glutathione groups (e.g. Wikberg, 1953 or Livermore and Mueck, 1954) but even with alkane groups. Thus Gorin *et al.* (1949) heated propyl disulphide in a sealed tube with n-decanethiol and showed that the weight of the precipitate with silver ion decreased progressively with reaction time and approached the value expected from the equilibrium mixture of propane- and decanethiols.

By judicious choice of reagent and conditions the equilibrium can be displaced to the right and the new thiol or disulphide detected or determined in an analytical procedure. Thus Hopkins (1925) oxidized proteins with gluta-

176

thione in neutral or faintly alkaline solution, showing the disappearance of protein thiol groups by absence of response to nitroprusside. Goddard and Michaelis (1934) split the disulphide groups in wool keratin with sodium sulphide or cyanide, or with mercaptoacetic acid through reaction at pH 12 for 3 h at 30°C. Probably the first genuinely analytical application was by Mirsky and Anson (1935) who treated protein thiol groups with cystine and estimated the liberated cysteine colorimetrically with tungstophosphoric acid.

Mostly, however, the disulphide reagent contains a chromophore or potential chromophore group; this facilitates detection or determination of the new thiol.

1. THIAMINE DISULPHIDE

The first analytical application of this reagent appears to have been by Zima et al. (1941). They determined blood cysteine by reaction for 2 h with this disulphide reagent. The aneurine thiol yielded was oxidized with ferricyanide in alkaline solution, the thiochrome product from this was extracted with butanol and estimated fluorometrically in the usual way. Sahashi and Shibasaki (1951–2) used the method of Zima et al. for determining thiols in denatured proteins, reacting for 90 min at 37°C. Kiermeier and Hamed (1962) used a closely similar procedure for thiol groups in sweet and acid milk products. They treated the aqueous sample with reagent and dilute hydrochloric acid for 2 h at 21°C before ultimate oxidation with ferricyanide, extraction with isobutanol and fluorometric estimation. Spanyar et al. (1964) compared three methods for microthiol estimation in plant extracts and considered Kiermeier and Harned's procedure to be very sensitive although less accurate than argentometric amperometric titration. Kono (1966) also used thiamine disulphide to determine low molecular weight thiols and protein thiol groups; he carried out the reaction in a phosphate buffer of pH 7·8 to 8, containing EDTA, for 10–30 min at 30°C.

2. 2,2′-DIHYDROXY-6,6′-DINAPHTHYL DISULPHIDE ("DDD") (2,2′-DITHIOBIS (6-HYDROXYNAPHTHALENE))

DDD was introduced by Barrnett and Seligman (1952, 1953, 1954). They treated the sample at pH 8·5 with a 0·16% solution of reagent in absolute ethanol. After 10 min at room temperature, they acidified with dilute acetic acid to convert unused reagent and the thiol reaction product to free naphthols. These were extracted with alcohol and, after washing with ether and water, the naphthol residue introduced into the sample was coupled for 2 min with tetrazotized di-o-anisidine in a phosphate buffer of pH 7·4. The monocoupling product is red, the dicoupling product blue, both enabling the original thiol group to be identified.

$$X{-}SH \; + \; \text{(DDD disulphide naphthol structure)} \longrightarrow \text{(6-mercapto-2-naphthol)} + \text{(mixed disulphide X structure)}$$

This method was used or recommended by numerous authors in the following years for histochemical localization of protein-bound thiol groups, e.g. Gomori (1956); Gabe and Martoja-Pierson (1956); Bahr and Moberger (1958); Hyde and Palival (1959) and Roberts (1960) for plant meristems; Ackerman and Sneeringer (1960); Pomeranz and Shellenberger (1961) for wheat kernel thiol groups.

Other coupling reagents have been used. Thus Pomeranz and Shellenberger (1962) proposed Fast Blue RR (C.I. Azoic Diazo Component 24) in phosphate buffer of pH 7·4. The green fluorescence of the product served for a spot test of a few micrograms of thiols such as cysteine, homocysteine, glutathione and mercaptoacetic acid. The DDD reagent alone, in the absence of thiols, gave a violet colour. Zwaan (1966) used Fast Blue Salt B or Fast Black Salt K for the visualizing coupling in detection of protein-bound thiols with DDD.

In quantitative work, the thiol reaction product, 6-mercapto-2-naphthol, is estimated. Flesch *et al.* (1954), for instance, determined protein-bound thiols by treating a 5–15 mg sample with 2 ml of a solution of 1 mg DDD in 4 ml of absolute ethanol and diluting to 10 ml with a 0·1 M barbiturate buffer of pH 8·5. After 1 h reaction at 50°C, they centrifuged to remove protein, and mixed 0.2 ml of supernatant with 0·5 ml of sodium laurylsulphate as colloid stabilizer and 0·5 ml of reagent made by dissolving 1 mg of the tetrazotized di-*o*-anisidine in 2 ml of water. This solution was diluted with 10 ml of water before evaluating the absorbance at 560 nm. Bahr and Moberger (1958) preferred diazotized 4-amino-2,5-dimethoxy-4′-nitroazobenzene as coupling component. It reacted best at pH 7·0 and gives only one product; quantitative adaptation was possible. Linko (1962) determined protein-bound thiols in wheat flour using a procedure closely resembling that of Flesch *et al.* but he coupled with diazotized 4-benzamido-2,4-dimethoxyaniline and then evaluated at 555 nm. In more recent years further quantitative applications and studies have been reported, e.g. by Bahr (1966), Sillevis-Smitt *et al.* (1969) and Esterbauer (1972).

3. DITHIO-BIS(4-NITROBENZENE)

This is the first of Ellman's reagents, developed for quantitative work and yielding for this purpose a thiol product which can be directly determined colorimetrically without coupling. Ellman (1958, 1959a) used it at pH 8 (phosphate buffer) in acetone. The reaction product, 4-nitrobenzenethiol, has its absorption maximum at 412 nm, where the disulphide reagent absorbs negligibly. Ellman found that simple thiols reacted quantitatively within ca. 10 s whereas cysteine and related compounds needed 60–90 min.

$$O_2N\!-\!\langle\ \rangle\!-\!S\!-\!S\!-\!\langle\ \rangle\!-\!NO_2 \ + \ RSH \longrightarrow O_2N\!-\!\langle\ \rangle\!-\!SH \ + \ O_2N\!-\!\langle\ \rangle\!-\!SR$$

Stevenson *et al.* (1960) used this principle to determine glutathione in erythrocytes. Maier (1969) also applied it to estimate volatile thiols (and hydrogen sulphide) in foods, finally evaluating colorimetrically at 435 nm. Saville (1959) gives the reaction as a test for thiols of pK_a equal to or greater than 7 (e.g. 2-mercaptoethanol and mercaptoacetic acid); they yield the yellow 4-nitrobenzenethiolate ion in basic solvents such as morpholine, piperidine or triethylamine.

The reagent suffers, however, from the disadvantage of poor solubility in water and has now been largely replaced by Ellman's second reagent, 5,5'-dithio-bis(2-nitrobenzoic acid).

4. 5,5'-DITHIO-BIS(2-NITROBENZOIC ACID)

Ellman (1959b) introduced this reagent later. The corresponding thiolate anion also has an absorption maximum at 412 nm, with molar absorptivity of ca. 13 000. Slow colour fading due to autoxidation can be hindered by adding EDTA to complex metallic catalysts.

Jocelyn (1962) distinguished non-protein from protein thiols. Only the former react at pH 6·8; both react at pH 7·6. Distinction through variation of conditions was achieved also by Sedlak and Lindsay (1968), Sedlak (1970), and van Caneghem and De Bruyn (1971).

Some references to the application of the reagent with colorimetric estimation of the thiol anion at 412 nm are:

Protein thiols in hydrolysed wool by du Toit *et al.* (1965) and Meichelbeck *et al.* (1968); in skim milk by Koka *et al.* (1968) and Kalab (1970), who studied factors influencing the method. Beveridge *et al.* (1974) used the reagent in tris-glycine-EDTA (pH 8) to determine thiols in proteins such as flour, egg white, and milk.

Glutathione in blood was determined thus by Beutler *et al.* (1963); Kaplan and Dreyfus (1964); Owens and Belcher (1965); Weller (1965), who compared the method with the nitroprusside and iodate methods; Güntherberg and Rapoport (1968); Tietze (1969); Wendell (1970); Roberts and Agar (1971), who evaluated at 420 nm; Ladenson and Purdy (1971), who compared with a coulometric silver titration. Glutathione in dialysates of wheat flour was determined by Kuninori et al. (1968). Owens and Belcher, Güntherberg and Rapoport, Kuninori et al., Tietze and Wendell all carried out the reaction in the presence of glutathione reductase and nicotinamide–adenine-dinucleotide phosphate sodium salt (reduced form, $NADPH_2$), so that oxidized glutathione present was reduced to the thiol form:

$$GSSG + NADPH_2 \rightarrow 2GSH + NADP$$

Thiols in other biological material were determined by Rootwelt (1967) in urine; by Ellman and Lysko (1967) in human blood and plasma; by Gabay et al. (1968) in brains; by Sobrinko et al. (1968) in cell homogenates and fragments; by Sedlak (1970) in thyroid and adrenals of rats; and by Robyt et al. (1971) in a case study of an α-amylase, papain and lysozyme.

Butterworth et al. (1967) modified Ellman's procedure for determining protein thiols by separating the thiophenylated protein from the reaction mixture with the help of a molecular filter, by gel filtration or by precipitation with 60% perchloric acid. They then estimated the amount of nitrothiobenzoic acid bound to the protein by treatment with, e.g. dithiothreitol or alkali, and evaluating absorbance at 412 nm.

Maurice (1973) determined traces (down to 0·02 parts/10^6) of thiols in light hydrocarbons by treating with the reagent in phosphate buffer, pH 7, diluted with a citrate/phosphate buffer, pH 8. He used a special device for shaking (30 min) together under pressure to accelerate the otherwise slow reaction. The final absorbance measurements were made at 412 nm.

Vermeij (1972) determined penicillinic acid, containing a thiol group, in the presence of various penicillins by reaction for 2 min in phosphate buffer of pH 7·2 before evaluating at 412 nm.

The use of Ellman's reagent has not always been conducted by spectrophotometric estimation of thiolate ion. Thus Snyder and Moehl (1970) quenched with the thiolate ion the fluorescence of a plastic or liquid scintillator mixed with a β-emitter (^{14}C). They found for glutathione a linear quenching activity for amounts ranging from 50 to 200 ng. Steinert et al. (1974) reacted protein thiols with ^{35}S-containing Ellman reagent for 3 h at room temperature in a phosphate/tris hydrochloride mixture of near neutral pH, then separating added protein from coloured anion and excess reagent by chromatography on Sephadex G-50 equilibrated with the tris buffer. Fractions were collected and evaluated by a liquid scintillation technique.

Thiols such as cysteine were visualized as yellow spots appearing within a few seconds and persisting for at least 15 h by Glaser *et al.* (1970) on paper and thin-layer chromatograms by spraying with a 0·1% disulphide reagent solution in ethanol–0·45 M tris buffer, pH 8·2 (1 + 1).

The method has of course been employed to determine thiols resulting from reduction of disulphides in their estimation. Examples of this come from the work of Rootwelt (1967), who treated disulphides with thiolated Sephadex; of Zahler and Cleland (1968) who reduced them with dithioerythritol or dithiothreitol and complexed unused reducing agent with arsenic(III); of Beveridge *et al.* (1974) who reduced with 2-mercaptoethanol: and of Cavallini *et al.* (1966), Ellman and Lysko (1967), Kalab (1970) and Habeeb (1973), all of whom reduced with sodium borohydride.

5. DITHIO-BIS(2- AND 4-PYRIDINE)

These reagents were introduced by Grassetti and Murray (1967a). The pyridinethiol products exist virtually only in the thione form:

$$\text{(pyridine)}-S-S-\text{(pyridine)} + X-SH \longrightarrow$$

$$\text{(pyridine)}-S-X + \text{(pyridine)}-SH \rightleftharpoons \text{(pyridinethione)}$$

The thiones display absorption maxima at 343 (2-pyridine) and 324 nm (4-pyridine)· with molar absorptivities of 7060 and 19 800, respectively. They carried out the reaction in Krebs–Ringer phosphate buffer of pH 7·2 to determine thiols in tissue homogenates or urine. After reaction times of up to 15 min they measured absorbances at the wavelengths indicated. Down to 1·5 μg (2-pyridine) or 0·5 μg (4-pyridine) thiol could be determined.

The authors applied the method to determine glutathione, for example, using the dithio-bis (2-pyridine) reagent (1967b). Their results include oxidized glutathione since they worked in the presence of glutathione reductase and nicotinamide-adenine dinucleotide phosphate (reduced) sodium salt (NADPH$_2$) which, as mentioned above, functions according to:

$$\text{GSSG} + \text{NADPH}_2 \rightarrow 2\text{GSH} + \text{NADP}$$

Ampulski *et al.* (1969) used dithio-bis (4-pyridine) to estimate reactive thiols in haeme proteins, recording absorbances at 324 nm.

Morell and Taketa (1969) recommended these reagents if substances such as haemoglobin, which absorb at ca. 400 nm, are present.

G

Brocklehurst and Little (1970) used dithio-bis (2-pyridine) as a specific reagent for papain (thiol groups), finding that the rate of reaction was a maximum at pH 3·75. In acidic medium the reaction was far faster than with L-cysteine or mercaptoethanol.

6. 2,2'-DITHIO-BIS(5-NITROPYRIDINE)

This reagent also was proposed by Grassetti and Murray (1969a) for selective detection of thiol groups in chromatography or electrophoresis. Its thione reaction product has absorption maxima at 386 and 470 nm, so that it is visibly coloured. They could detect as little as 0·2 μg of cysteine, glutathione, cysteamine, etc., through the yellow colour which persists for 24 h. The sole interference was from strong reducing agents. Grassetti (1971) patented the reagent and indicated the possibilities of quantitative determination through reaction in pH 7·4 buffer. The 5-nitropyridine-2-thione product has a molar absorptivity of 12 900 at 386 nm.

Swatditat and Tsen (1972) used the reagent to determine low molecular weight thiols and contents of biological samples. They compared it favourably with the second Ellman reagent [5,5'-dithio-bis-(2-nitrobenzoic acid)] in that it yields colour also in acid solution (pH range 5·5 to 8·9) and is more sensitive. Their reaction time was 20 min at room temperature, then evaluating at 386 nm.

7. OTHER HETEROCYCLIC DISULPHIDES

Grassetti and Murray (1969b) gave a table of 10 disulphide reagents similar to those just mentioned. They are pyridine, quinoline, thiazole and pyrimidine compounds, all containing the —S—C group, and are used under

$$-S-C \overset{\diagdown}{\underset{\diagup}{}} N$$

much the same reaction conditions. The absorption maxima range from 240 to 430 nm. Some of them were quoted in a patent by the same authors (1971), in which glutathione reductase and NADPH$_2$ were also used to include oxidized glutathione in the result; these were 6,6'-dithiodinicotinic acid; 6,6'-dithio-bis (isonicotinic acid); 2,2'-dithio-bis (dipyrimidine). In a later patent publication (1972) Grassetti and Murray used the 6,6'-dithiodinicotinic acid for spectrophotometric determination of thiol groups in serum albumin through reaction at pH 7·5. This reagent has the advantage of being more soluble in buffers of near neutral pH than the original dithio-bis (pyridines). Further heterocyclic disulphides were also proposed, e.g. 2,2'-dithio-bis (4-hydroxypyrimidine).

8. OTHER DISULPHIDES

In the earlier periods of the thirties and forties, cystine, oxidized glutathione and dithiodiacetic acid were the reagents commonly used for reaction with protein thiols. There was generally greater interest in the reverse reaction of fission of protein disulphide groups with thiol reagents.

Most work was non-analytical, e.g. Hopkins (1925) investigated the reaction of protein thiol groups with oxidized glutathione observing the disappearance of thiol in neutral or alkaline solution (nitroprusside test no longer positive). An analytical example from this period is the work of Mirsky and Anson (1935) who treated protein thiols with cystine and estimated the cysteine formed by titration with ferricyanide.

Klotz et al. (1958) titrated photometrically protein thiols (also cysteine and glutathione) with 2,2'-(2-hydroxy-6-sulphonaphthyl-1-azo)-diphenyl disulphide, in acetate buffer of pH 6·1. They recorded the absorbance at 530 nm,

where the reaction product absorbs less. A break in the curve was found, after which the absorbance increased more rapidly because the excess reagent has an absorption maximum at 515 nm.

Bitny-Szlachto et al. (1963) used 2-hydroxyethyl-2,4-dinitrophenyl disulphide,

to determine thiols in albumins through reaction in phosphate buffer of pH 6·85 for 15 min. The 2,4-dinitrobenzenethiolate anion was colorimetrically estimated at 408 nm.

Neims *et al.* (1966) measured radioactivity to determine the thiol reaction product from tetraethylthiuram disulphide, labelled with ^{35}S,

$$(C_2H_5)_2N—CS—S—S—CS—N(C_2H_5)_2$$

On acidification to pH 4, the diethyldithiocarbamate product yields the unstable free acid which decomposes into diethylamine and carbon disulphide. The latter, containing the ^{35}S, was diffused into alkaline piperidine to give another dithiocarbamate, the ^{35}S activity of which they evaluated to give a measure of the original protein thiol amount.

$$(C_2H_5)_2N—CS—S^- + H^+ \rightarrow (C_2H_5)N—CSSH \rightarrow (C_2H_5)_2NH + CS_2$$

The interchange examples above are all reactions carried out in fairly near-neutral solution. Disulphide fission is probably due to nucleophilic attack of RS$^-$. Examples have been found, however, of disulphide interchange in strongly acid solution, e.g. by Sanger (1953). Benesch and Benesch (1958) proposed a mechanism involving a sulphenium ion, RS$^+$. The initiation reaction is then,

$$RSSR + H^+ \rightarrow RS^+ + RSH$$

and is followed by the interchange reaction with the other disulphide,

$$RS^+ + R'SSR' \rightarrow RSSR' + R'S^+$$

Inhibition is caused by thiols which capture the sulphenium ions to yield further disulphide. If the thiol is R'SH, the same mixed disulphide is formed as that from the disulphide interchange.

Glazer and Smith (1961) developed a method for determining the cysteine and half the cystine content of proteins. They treated the protein preparation in 9·6 N hydrochloric acid at 39° for periods of up to several days with bis(2,4-dinitrophenyl)-L-cystine. Unused reagent was then extracted with ether and the mixed disulphide, mono(2,4-dinitrophenyl)-L-cystine, in the aqueous phase, was estimated spectrophotometrically at 357 nm.

REFERENCES

Ackerman, G. A. and Sneeringer, S. C. (1960). *Lab. Invest.* **9**, 356.
Ampulski, R. S., Ayers, J. E. and Morell, S. A. (1969). *Anal. Biochem.* **32**, 163.
Bahr, G. F. (1966). *Introd. Quant. Cytochem.* 469; *Chem. Abst.* **68**, 719.
Bahr, G. F. and Moberger, G. (1958). *Acta. Pathol. Microbiol. Scand.* **42**, 109.
Barrnett, R. J. and Seligman, A. M. (1952). *Science,* **116**, 323.
Barrnett, R. J. and Seligman, A. M. (1953). *Glutathione, Proc. Symp. Ridgefield Conn. U.S.A.* 89; *Chem. Abst.* **49**, 5551.
Barrnett, R. J. and Seligman, A. M. (1954). *J. Natl. Cancer Inst.* **14**, 769.
Benesch, R. E. and Benesch, R. (1958). *J. Amer. Chem. Soc.* **80**, 1666.
Bersin, T. and Steudel, J. (1938). *Chem. Ber.* **71**, 1015.

Beutler, E., Duron, O. S. and Kelly, B. K. (1963). *J. Lab. Clin. Med.* **61**, 882.
Beveridge, T., Toma, S. J. and Nakai, S. (1974). *J. Food Sci.* **39**, 49.
Bitny-Szlachto, S., Kosinski, J. and Niedzielska, M. (1963). *Acta Polon. Pharm.* **20**, 365.
Brocklehurst, K. and Little, G. (1970). *FEBS Lett.* **9**, 113.
Butterworth, P. H. W., Baum, H. and Porter, J. W. (1967). *Arch. Biochem. Biophys.* **118**, 716.
van Caneghem, P. and De Bruyn, M. (1971). *Arch. Int. Physiol. Biochim.* 79.
Cavallini, D., Graziani, M. T. and Dupre, S. (1966). *Nature (London)* **212**, 294.
Eldjarn, L. and Pihl, A. (1957). *J. Amer. Chem. Soc.* **79**, 4589.
Ellman, G. L. (1958). *Arch. Biochem. Biophys.* **74**, 443.
Ellman, G. L. (1959a). *U.S. Patent No.* 3,119,668 of Jan. 26.
Ellman, G. L. (1959b). *Arch. Biochem. Biophys.* **82**, 70.
Ellman, G. L. and Lysko, H. (1967). *J. Lab. Clin. Med.* **70**, 518.
Esterbauer, H. (1972). *Acta Histochem.* **42**, 351.
Fava, A., Iliceto, A. and Camera, E. (1957). *J. Amer. Chem. Soc.* **79**, 833.
Flesch, P., Golomb, S. and Satanove, A. (1954). *J. Lab. Clin. Med.* **43**, 957.
Gabay, S., Cabral, A. M. and Healy, R. (1968). *Proc. Soc. Exp. Biol. Med.* **127**, 1081.
Gabe, M. and Martoja-Pierson, M. (1956). *Ann. Histochem.* **1**, 230.
Glaser, C. B., Maeda, H. and Meienhofer, J. (1970). *J. Chromatogr.* **50**, 151.
Glazer, A. N. and Smith, E. L. (1961). *J. Biol. Chem.* **236**, 416.
Goddard, D. R. and Michaelis, L. (1934). *J. Biol. Chem.* **106**, 605.
Gomori, G. (1956). *Quart. J. Microscop. Sci.* **97**, 1.
Gorin, G., Dougherty, G. and Tobolsky, A. V. (1949). *J. Amer. Chem. Soc.* **71**, 3551.
Grassetti, D. A. (1971). *U.S. Patent No.* 3,597,160 of Aug. 3.
Grassetti, D. R. and Murray, J. F.(1967a). *Arch. Biochem. Biophys.* **119**, 41.
Grassetti, D. R. and Murray, J. F. (1967b). *Anal. Biochem.* **21**, 427.
Grassetti, D. R. and Murray, J. F. (1969a). *J. Chromatogr.* **41**, 121.
Grassetti, D. R. and Murray, J. F. (1969b). *Anal. Chim. Acta,* **46**, 139.
Grassetti, D. R. and Murray, J. F. (1971). *U.S. Patent No.* 3, 627, 643 of Dec. 14.
Grassetti, D. R. and Murray, J. F. (1972). *U.S. Patent No.* 3, 698, 866 of Oct. 17.
Güntherberg, H. and Rapoport, S. (1968). *Acta. Biol. Med. Germ.* **20**, 559.
Habeeb, A. F. S. A. (1973). *Anal. Biochem.* **56**, 60.
Hopkins, F. G. (1925). *Biochem. J.* **19**, 787.
Hyde, B. B. and Paliwal, R. L. (1959). *Stain Technol.* **34**, 175.
Jocelyn, P. C. (1962). *Biochem. J.* **85**, 480.
Kalab, M. (1970). *J. Dairy Sci.* **53**, 711.
Kaplan, J. C. and Dreyfus, J. C. (1964). *Bull. Soc. Chim. Biol.* **46**, 775.
Kiermeier, F. and Hamed M. G. E. (1962). *Nahrung,* **6**, 639.
Klotz, I. M., Ayers, J., Ho, J. Y. C., Horowitz, M. G. and Heiney, R. F. (1958). *J. Amer. Chem. Soc.* **80**, 2132.
Koka, M., Mikolajcik, E. M. and Gould, J. A. (1968). *J. Dairy Sci.* **51**, 217.
Kolthoff, I. M., Stricks, W. and Kapoor, R. C. (1955). *J. Amer. Chem. Soc.* **77**, 4733.
Kono, K. (1966). *J. Vitaminol. (Kyoto),* **12**, 137.
Kuninori, T., Yagi, M., Yoshino, D. and Matsumoto, H. (1968). *Cereal Chem.* **46**, 480.
Ladenson, J. H. and Purdy, W. C. (1971). *Clin. Chem.* **17**, 908.
Lecher, H. (1920). *Chem. Ber.* **53**, 591.
Linko, P. (1962). *Suomen Kem.* **35**, 77.
Livermore, A. H. and Mueck, E. C. (1954). *Nature (London),* **173**, 265.
Maier, H. G. (1969). *Z. Anal. Chem.* **247**, 46.
Maurice, J. (1973). *Erdöl Kohle Erdgas Petrochem. Brennst.-Chem.* **26**, 140.

Meichelbeck, H., Hack, A. G. and Sentler, C. (1968). *Z. Gesamte Text.-Ind.* **70**, 242.
Mirsky, A. E. and Anson, M. L. (1935). *J. Gen. Physiol.* **18**, 307.
Morell, S. A. and Taketa, F. (1969). *Anal. Biochem.* **32**, 169.
Neims, A. H., Coffey, D. S. and Hellerman, L. (1966). *J. Biol. Chem.* **241**, 3036.
Owens, C. W. I. and Belcher, R. V. (1965). *Biochem. J.* **94**, 705.
Pomeranz, Y. and Shellenberger, J. A. (1961). *Cereal Chem.* **38**, 133.
Pomeranz, Y. and Shellenberger, J. A. (1962). *Anal. Chim. Acta* **26**, 301.
Roberts, L. (1960). *Am. J. Botany* **47**, 110.
Roberts, J. and Agar, N. S. (1971). *Clin. Chim. Acta* **34**, 475.
Robyt, J. F., Ackerman, R. J. and Chittenden, C. G. (1971). *Arch. Biochem. Biophys.* **147**, 262.
Rootwelt, K. (1967). *Scand. J. Clin. Lab. Invest.* **19**, 325.
Sahashi, Y. and Shibasaki, H. (1951–2). *J. Agr. Chem. Soc. Japan,* **25**, 57.
Sanger, F. (1953). *Nature (London)*, **171**, 1025.
Saville, B. (1959). *Proc. Chem. Soc.* (*London*), 160.
Sedlak, J. (1970). *Endocrinol. Exp.* **4**, 3; *Chem. Abs.* **73**, 128472.
Sedlak, J. and Lindsay, R. H. (1968). *Anal. Biochem.* **25**, 192.
Sillevis-Smitt, P. A. E., James, J. and Wisse, J. H. (1969). *Acta Histochem.* **33**, 53.
Snyder, F. and Moehl, A. (1970). *Org. Scintill. Liquid Scintill. Counting, Proc. Int. Conf.* 419; *Chem. Abs.* **76**, 41769.
Sobrinko, S. M., Pinto de Barras, J. and Hipolito-Reis, C. (1968). *Arq. Port. Bioquim.* **10**, 475; *Chem. Abs.* **73**, 52911.
Spanyar, P., Kevei, J. and Blazovich, M. (1964). *Kozp. Elelmiszeripari Kutatointez. Kozlem.* 21; *Chem. Abs.* **64**, 656.
Steinert, P. M., Nalega, B. S. and Munro, H. N. (1974). *Anal. Biochem.* **59**, 416.
Stevenson, T. D., McDonald, B. L. and Roston, S. (1960). *J. Lab. Clin. Med.* **56**, 157.
Swatditat, A. and Tsen, G. C. (1972). *Anal. Biochem.* **45**, 349.
Tietze, F. (1969). *Anal. Biochem.* **27**, 502.
du Toit, E. H., van Rensburg, N. J. and Swanepoel, O. A. (1965). *J. S. African Chem. Inst.* **18**, 52.
Vermeij, P. (1972). *Pharm. Weekblad,* **107**, 249.
Weller, G. (1965). *Rev. Franc. Etudes Clin. Biol,* **10**, 547.
Wendell, P. L. (1970). *Biochem. J.* **117**, 661.
Wikberg, E. (1953). *Nature (London)*, **172**, 398.
Zahler, W. L. and Cleland, W. W. (1968). *J. Biol. Chem.* **243**, 716.
Zima, O., Ritsert, K. and Moll, T. (1941). *Z. Physiol. Chem.* **267**, 210.
Zwaan, J. (1966). *Anal. Biochem.* **15**, 369.

6. REACTION WITH BASES

Analytical benefit has been obtained from the acidic properties of the thiol group in two ways: separation from other, non-acidic compounds; and determination through direct titration.

1. SEPARATION OF THIOLS WITH ALKALIES

This extractive separation with alkalies has been a standard procedure for many years but it assumed its present importance with the development of the petroleum industry. It is used there to free petroleum and similar products from thiols. This is a widely practised procedure, nowadays so self-evident that publication is no longer justified. Some concrete, published examples are given below to illustrate the types of procedure and follow-up.

An early publication is that of Sabrou and Renaudie (1937) who removed thiols from internal combustion engine fuels with 10% aqueous sodium hydroxide. Koons (1941) scrubbed refinery products with caustic soda solution and then regenerated the thiols by acidification. He extracted them into naphtha for subsequent analysis using conventional methods. Arnold *et al.* (1952) identified cyclohexanethiol in virgin naphtha after extracting it with a potassium hydroxide/methanol mixture, then hydrolysing the extract by steam distillation at ca. 160°C and removing the methanol by washing with water. Hopkins and Smith (1954) diluted petroleum distillates with isopentane and then extracted their thiols with sodium aminoethoxide in ethylenediamine. They recovered the thiols by acidification and steam distillation. Liberti and Cartoni (1957) extracted thiols, e.g. in petroleum distillates, with ethanolic potassium hydroxide. They regenerated them by hydrolysis, then extracted with isopentane and separated them by gas chromatography (see p. 240). The individual thiols in the issuing gases were titrated coulometrically with silver ion (see p. 101). A potassium hydroxide–potassium isobutryrate solution in water was used by Hoogendonk and Porsche (1960) to extract thiols (up to C_7) from naphthas by shaking for 5 min at $-5°C$. They then acidified the aqueous solution with concen-

Table VIII. Direct titration of thiols in non-aqueous solution

Sample	Reagent	Solvent for sample	End-point indication	References
E.g. thiophenol	KOH/isopropanol	Benzene–isopropanol (1 + 1)	Potentiometric	Lykken et al. (1944)
E.g. thiophenol, mercaptobenzothiazole (MBT)	$NaOCH_3$/benzene–methanol	Butylamine	Thymol blue	Fritz and Lisicki (1951)
Thiouracils	$NaOCH_3$/benzene–methanol	Pyridine or dimethylformamide	Thymol blue; azo violet	Backe-Hansen (1955)
α-Toluenethiol; MBT; p-chlorothiophenol	$[N(C_4H_9)_4]OH$ in benzene/methanol	Pyridine	Potentiometric; thymol blue; azo violet	Cundiff and Markunas (1956)
Thiophenol; octanethiol (study)	$[CH_3-N(C_4H_9)_3]OH$	Pyridine	Potentiometric	Dahmen (1958)
(No examples specified)	$[CH_3-N(C_4H_9)_3]OH$ in benzene/isopropanol (4 + 1)	90% Acetone	Potentiometric; thymol blue; azo violet	Malmstadt and Vassallo (1959)
Methylthiouracil	$NaOCH_3$	Dimethylformamide	Thymol blue	Uchida (1959)
Methylthiouracil in tablets	$NaOCH_3$	Dimethylformamide	Thymol blue	Ionescu et al. (1962)
Methyl- and other thiouracils	NaOH/benzene–methanol	Dimethylformamide also plus benzene or acetone	Potentiometric or thymol blue	Rapaport and Verzina (1966)
Thiophenols (determination of pK_a values)	NaH/dimethyl sulphoxide, giving Na derivative of DMSO	Dimethyl sulphoxide	Potentiometric	Courtot-Coupez and Le Démézet (1969)
Thiols in presence of other acids	1. KOH/isopropanol 2. $N(C_4H_9)_4OH$	1. Acetone 2. Dimethylformamide + acrylonitrile (reacts with the thiols)	Differential thermometric, based on chem. reaction with acetone (1) or the nitrile (2)	Greenhow and Loo (1974)

trated hydrochloric acid, extracted the liberated thiols into isooctane and analysed them by mass spectrometry (see p. 225). LeRosen (1961) extracted thiols from refinery gases using aqueous alkali. He then just acidified the solution (to methyl orange) with sulphuric acid and collected the gaseous thiols under reduced pressure for gas chromatography (see p. 240); he stated that this was better than extracting the thiols, e.g. with *n*-heptane, from the acidified solution, since they are evolved too quickly. Bloemberger and Vermaak (1965) employed a potassium hydroxide–diethlene glycol– water (49 + 10 + 41 w/w) reagent to remove thiols from light hydrocarbons at 0°C. After acidifying the solution they extracted it with toluene and ulti-mately determined the thiols by gas chromatography (see p. 242) Staszeweski and Zygmunt (1973) separated thiols from gases by passing through sodium hydroxide solution, finally determining them mercurimetrically (see p. 80).

Thiols have been separated also from effluents, especially of the paper industry, using alkalies, e.g. by Felicetta *et al.* (1952) and Thoen *et al.* (1968) for kraft gases. Oehme (1960) extracted thiols (and hydrogen sulphide) from technical gases with sodium hydroxide, removing excess alkali by passing the solution through an ion exchanger in the H^+-form and finally titrating with lead acetate (see p. 64). Methanethiol and hydrogen sulphide were stripped with nitrogen from an industrial effluent by Bethge *et al.* (1968); they separated the methanethiol by absorption in aqueous sodium hydroxide/ zinc acetate, ultimately determining it *via* nitrosation (see p. 35).

Hopkins and Smith (1955) separated primary and secondary thiols from tertiary thiols by treating an ethereal solution with liquid ammonia to yield insoluble ammonium salts. They filtered these after 30 min in a dry ice bath. The precipitates gave ammonia and the original thiols on warming.

2. DIRECT TITRATION OF THIOLS WITH BASES

Thiols are only weakly acidic so that they usually have to be titrated in non-aqueous basic solvents with strong bases, such as alkoxides or quater-nary ammonium hydroxides. Thymol blue or azo violet are suitable colour indicators, or instrumental end-point indication, mostly potentiometric, is used. Table VIII contains some examples.

A few earlier examples of titration with sodium hydroxide in aqueous solution can also be given. Smith (1950) reported a collective study of the titration of propylthiouracil in tablets, using standard sodium hydroxide and phenolphthalein or thymolphthalein as indicator. Anastasi *et al.* (1951) also titrated thiouracil and alkyl derivatives potentiometrically. Scheele and Gensch (1953) titrated mercaptobenzothiazole conductimetrically; and

Pavlova and Faerman (1955) titrated phenylmercaptotetrazole potentio-metrically or to methyl red.

3. DETERMINATIONS BY UV MEASUREMENTS OF ALKALINE SOLUTIONS

This heading is not entirely relevant here but short reference is made to the examples of mercaptobenzothiazole (at 309 nm) (Gurvich and Kostikova, 1962); methylthiouracil (261 nm) (Nekrasov, 1966); methyl- and propyl-thiouracils (260 nm) and 2-thioorotic acid (263 nm) (Brueggeman and Schole, 1967).

REFERENCES

Anastasi, A., Mecarelli, E. and Novacic, L. (1951). *Mikrochem. ver. Mikrochim. Acta* **38**, 160.

Arnold, R. C., Launer, P. J. and Lien, A. P. (1952). *Anal. Chem.* **24**, 1741.

Backe-Hansen, K. (1955). *Medd. Norsk Farm. Selskap.* **17**, 63.

Bethge, P. O., Carlson, M. and Rodestrom, R. (1968). *Svensk Papperstidn.* **71**, 864.

Bloemberger, R. H. and Vermaak, C. (1965). *Erdöl u. Kohle,* **18**, 185.

Brueggeman, J. and Schole, J. (1967). *Landwirt. Forsch.* No. 21,134; *Chem. Abs.* **68**, 2099.

Courtot-Coupez, J. and Le Démézet, M. (1969). *Bull. Soc. Chim. France,* 1033.

Cundiff, R. H. and Markunas, P. C. (1956). *Anal. Chem.* **28**, 792.

Dahmen, E. A. M. F. (1958). *Chim. Anal. (Paris),* **40**, 378.

Felicetta, V. E., Peniston, Q. P. and McCarthy, J. L. (1952). *Can. Pulp Paper Ind.* **5**, Nos. 12, 16, 18, 20, 22, 24, 26–7, 30, 41; also (1953). *Tappi,* **36**, 425; *Chem. Abs.* **47**, 5115 and 12810.

Fritz, J. S. and Lisicki, N. M. (1951). *Anal. Chem.* **23**, 589.

Greenhow, E. J. and Loo, L. H. (1974). *Analyst (London),* **99**, 360.

Gurvich, Ya. A. and Kostikova, V. P. (1962). *USSR Patent No.* 144,636 of Feb. 15; *Anal. Abs.* **11**, 1807.

Hoogendonk, W. P. and Porsche, F. W. (1960). *Anal. Chem.* **32**, 941.

Hopkins, R. L. and Smith, H. M. (1954). *Anal. Chem.* **26**, 206.

Hopkins, R. L. and Smith, H. M. (1955). *Anal. Chem.* **27**, 1832.

Ionescu, I. S., Popescu, D. and Constantinescu, T. (1962). *Farmacia (Bucharest)* **10**, 491.

Koons, R. D. (1941). *Refiner Natural Gasoline Mfr.* **20**, 393; *Chem. Abs.* **36**, 1474.

LeRosen, H. D. (1961). *Anal. Chem.* **33**, 973.

Liberti, A. and Cartoni, G. P. (1957). *Chim. Ind. (Milan)* **39**, 821.

Lykken, L., Porter, P., Ruliffson, H. D. and Tuemmler, F. D. (1944). *Ind. Eng. Chem., Anal. Ed.* **16**, 219.

Malmstadt, H. V. and Vassallo, D. A. (1959). *Anal. Chem.* **31**, 862.

Nekrasov, V. I. (1966). Aptech. Delo **15**, 37; *Chem. Abs.* **65**, 8676.

Oehme, F. (1960). *Erdöl u. Kohle* **13**, 394.

Pavlova, V. A. and Faerman, G. P. (1955). *Tr. Leningr. Inst. Kinoinzh. No.* 3. 175. *Chem. Abs.* **53**, 6888.

Rapaport, L. I. and Verzina, G. V. (1966). *Farmatsevt. Zh. (Kiev)*, **21**, 33.
Sabrou, L. and Renaudie, M. (1937). *Compt. Rend. 17me Congr. Chim. Ind. Paris*, 98 ; *Chem. Abs.* **32**, 6437.
Scheele, W. and Gensch, C. (1953). *Kautschuk u. Gummi*, **6**, WT 147.
Smith, G. (1950). *J. Assoc. Offic. Agr. Chemists*, **33**, 196.
Staszewski, S. and Zygmunt, B. (1973). *Chem. Anal. (Warsaw)*, **18**, 85.
Thoen, G. N., De Haas, G. G. and Austin, R. R. (1968). *Tappi*, **51**, 246.
Uchida, I. (1959). *Takamine Kenkyusho Nempo*, **11**, 220; *Chem. Abs.* **55**, 5853.

7. REDUCTION

The analytically useful reduction product of thiols is hydrogen sulphide. Feigl (1966a) detected alkanethiols by heating with fused benzoin at 150–200°C:

$$RSH + C_6H_5CO—CHOH—C_6H_5 \rightarrow C_6H_5CO—CO—C_6H_5 + RH$$
$$+ H_2S$$

He used lead acetate to demonstrate the formation of the sulphide. Neither arenethiols nor disulphides react. Another of Feigl's tests (1966b) can be mentioned here although he considers that the hydrogen sulphide is yielded by hydrolysis. He heated the sample to 100°C with a drop of concentrated ammonium hydroxide, whereby primary and secondary thiols give hydrogen sulphide likewise detected with lead acetate:

$$R—CH_2SH + H_2O \rightarrow R—CH_2OH + H_2S$$

Up to 100 µg of several thiols can be detected in this way.

Surprisingly, in view of the improbable uniformity of the reaction, there are several quantitative methods based on the formation of sulphide. Oldach and Field (1946) determined sulphur compounds, including methane- and ethanethiols, in gas mixtures by passing with hydrogen gas over aluminium oxide at 900°C. They estimated the hydrogen sulphide colorimetrically as a suspension of bismuth sulphide, Bi_2S_3.

Hydrazine hydrate has been used as reducing agent. Kuratomi *et al.* (1957) carried out a micro-determination of cysteine and cystine in protein by heating for 5–7 h at 115–120°C with this reagent. They swept the hydrogen sulphide into a mixed reagent of 4-dimethylaminoaniline, iron(III) and sulphuric acid and evaluated the coloured dye formed at 630 nm. Goa (1961) claimed to have increased the sensitivity of this method one hundred-fold by using a different method for estimating the hydrogen sulphide; he passed it into a mixed reagent of bismuth(III) nitrate, mannitol, glycerol, gum acacia and acetate buffer of pH 5·2, and read the absorbance of the bismuth sulphide suspension at 400 nm.

Granatelli (1959) reduced sulphur compounds, including thiols such as butanethiol, with Raney nickel to yield nickel sulphide, NiS. He liberated hydrogen sulphide from this with hydrochloric acid, absorbed it in acetone–N sodium hydroxide (1 + 1) and titrated the sulphide ion with mercuric acetate to dithizone indicator.

An unusual method was published by Falgout and Harding (1968) for determining hydrogen sulphide and methanethiol in air. They drew the air through a silver metal membrane filter which became tarnished through the formation of silver sulphide. They found the accompanying decrease in reflectance to be proportional to the sulphur content. Neither ozone nor nitrogen oxides interfered.

Loo and Michael (1958) reduced substituted purines, e.g. 6-mercapto-purine, with zinc amalgam but the subsequent procedure depended on reaction of the amine formed; they diazotized the amine and coupled to give an azo compound of absorption maximum at 505 nm, estimated spectrophotometrically.

Classifiable here is the method of Ando et al. (1963) for separating benzenethiol from other sulphur(II) compounds formed in the reaction of benzene with sulphur. They passed the benzene solution through a column containing spongy active copper powder, which evidently reacted with the thiol to yield cuprous phenyl mercaptide. Since the sulphur(II) compounds possess absorption maxima at 280 nm, they were able to estimate the amount of thiol from the diminution in absorbance after passage through the column.

REFERENCES

Ando, W., Sugimoto, K. and Oae, S. (1963). *Bull. Chem. Soc. Japan,* **36**, 477.
Feigl, F. (1966a). Unpublished work with Cohen, R. *"Spot Tests in Organic Analysis"*, p. 224; Elsevier London.
Feigl, F. (1966b). Unpublished work with Jungreis, E. *"Spot Tests in Organic Analysis"*, p. 223, Elsevier, London
Falgout, D. A. and Harding, C. I. (1968). *J. Air Pollution Contr. Ass.* **18**, 15; *Chem. Abs.* **68**, 89694.
Goa, J. (1961). *Acta Chem. Scand.* **15**, 853.
Granatelli, L. (1959) *Anal. Chem.* **31**, 434.
Kuratomi, K., Ohno, K. and Akabori, S. (1957). *J. Biochem. Tokyo* **44**, 183; *Chem. Abs.* **51**, 10626.
Loo, T. L. and Michael, M. E. (1958). *J. Biol. Chem.* **232**, 99.
Oldach, C. S. and Field, E. (1946). *Ind. Eng. Chem., Anal. Ed.* **18**, 669.

8. CATALYSED REACTIONS

The catalysis of certain reactions by thiols may be utilized in their detection and quantitative determination. Three reactions can be quoted:

1. AZIDE–IODINE REACTION

This reaction is immeasurably slow in the absence of catalysts,

$$2N_3^- + I_2 \rightarrow 3N_2 + 2I^-$$

Several types of sulphur(II) compounds function catalytically in the reaction and of these, thiols are among the most effective. The occurrence and progress of catalysis is demonstrated by nitrogen evolution and disappearance of the iodine colour. Analytical procedures are based on both.

Feigl (1934) appears to have been the first to apply the catalysis for analytical purposes. He proposed the mechanism:

$$RS^- + I_2 \rightarrow RSI + I^-$$
$$RSI + 2N_3^- \rightarrow RS^- + I^- + 3N_2$$

Evolution of nitrogen was taken by him as the positive outcome of the test and he was able to detect nanogram amounts of thiols such as mercaptoacetic acid. The amount of evolved nitrogen has been estimated in quantitative adaptations. Thus Ishii (1952–3) determined thiols such as cysteine, mercaptoacetic acid and 2,3-dimercapto-1-propanol by mixing with azide–iodine reagent in phosphate buffer, pH7, at 30°C and recording the nitrogen volume after 1 h. Akiba and Ishii (1952) applied this to bacteria. Whitman and Whitney (1953) studied the catalytic activity of cysteine and related compounds in an acetate (pH 4·63) buffer and established that the volume of nitrogen was proportional to the thiol amount. They reported the influence of other factors and considered that further investigation was needed before the procedure could be accepted analytically. Recently, Atkinson and Natoli (1974) determined thiols and sulphide in waters, industrial effluents and sediments through reaction with an azide/iodine reagent and gas chromatographic estimation of the nitrogen. Any normal gas chromatograph will serve, provided access of nitrogen to the column is prevented, e.g. with a molecular sieve. The reagent is injected after the air (oxygen, nitrogen) peaks have passed.

The more usual analytical route is *via* iodine disappearance or consumption. The reagent is quoted in standard text books of chromatography for the visualization of many organic sulphur compounds. Thus Chargaff *et al.* (1948) visualized such compounds, including cysteine, by spraying with an azide/iodine reagent. The compounds showed as white spots on a light brown background, the contrast being improved in u.v. light. It is especially useful for the faster reacting thiols; cysteine, for example, can be identified almost immediately. Other authors who used this visualization were: Fischer and Otterbeck (1958) in PC of pharmaceuticals, including thiouracils; Kumar and Sen (1969) in TLC of sulphur compounds, including cysteine.

Kurzawa and Suszka (1960) based a quantitative determination of cysteine on participation in the azide/iodine reaction at pH 6·9. After 20–30 s they titrated residual iodine with standard arsenite and drew a calibration curve relating back-titration and cysteine amount from which they estimated sample amounts. Later Kurzawa (1961a) applied the principle to the estimation of cysteine in protein hydrolysates, and then Kurzawa and Suszka also applied it (1962) to glutathione determinations. The different rates of catalysis of cysteine, cystine and methionine enabled Kurzawa (1961b) to determine these in the presence of one another. He obtained the cysteine value by back-titration after 20 s, during which the other two exerted negligible catalytic influence. Dahl and Purdue (1965) studied the kinetics of catalysis of the reaction and remarked incidentally that some thiols were highly active. Markova and Glasivtsova (1970) carried out the reaction in acetate buffers of pH 4·1 or 5·71 and recorded the absorbance at 530 nm for 3 min; they could then read the thol (e.g. cysteine) content from a calibration curve. Down to ca. 10 ng ml^{-1} of cysteine could be estimated. Gershkovich (1971) determined microgram amounts of ethanethiol in air by back-titrating unused iodine with arsenite after 1 h. A linear relation between iodine consumption at pH 5 and methylthiouracil concentration was found by Kossakowski *et al.* (1972a); they back-titrated after 10 min and were able to estimate 5–60 µg amounts. Kossakowski *et al.* (1972b) likewise determined 10–60 µg amounts of 1-methyl-2-thioimidazole in substances and tablets; they reacted for 10 min. at pH 5·6 with azide and excess 0·02 M iodine, back-titrating with 0·02 M arsenite.

2. SILVER(I)–IRON(II) REACTION

The reduction of silver ions by iron(II) is catalysed by cysteine and this was utilized by Babko *et al.* (1968) to determine the amino acid. They first mixed silver nitrate, sample and acetate buffer of pH 5·7 and left the mixture for 5 min. A ferrous ammonium sulphate-ferric alum-sulphuric acid reagent was then added, plus glycerol to stabilize the silver colloid formed. After 5 min, the absorbance was measured at 530 nm, giving a measure of the silver formed and hence of the original thiol.

3. DECARBOXYLATION OF MESOXALIC ACID

Brunel-Capelle (1951) observed that some biological decarboxylation reactions are catalysed by proteins containing thiol groups; the optimum pH was 2·2. Glutathione and cysteine were also active, but not cystine and non-sulphur-containing amino acids. Creach (1961) applied this to the decarboxylation of mesoxalic acid in acid solution to yield glyoxalic acid which he estimated colorimetrically

$$HOOC-CO-COOH \rightarrow OCH-COOH + CO_2$$

The work was a kinetic study but it is surprising that no analytical utilisation appears to be known.

REFERENCES

Akiba, T. and Ishii, K. (1952). *Japan J. Exptl. Med.* **22**, 229; *Chem. Abs.* **47**, 5983.

Atkinson, L. P. and Natoli, J. G. (1974). *Anal. Chem.* **46**, 1316.

Babko, A. K., Markova, L. V. and Maksimenko, T. S. (1968). *Zh. Anal. Khim.* **23**, 1268.

Brunel-Capelle, G. (1951). *Compt. Rend.* **233**, 1658.

Chargaff, E., Levine, C. and Green, C. (1948). *J. Biol. Chem.* **175**, 67.

Creach, Y. (1961). *Compt. Rend. Soc. Biol.* **155**, 1575.

Dahl, W. E. and Purdue, H. L. (1965). *Anal. Chem.* **37**, 1382.

Feigl, F. (1966). "*Spot Tests in Organic Analysis*" p. 219. Elsevier, London from (1934). *Mikrochemie* 15, 1.

Fischer, R. and Otterbeck, N. (1958). *Sci. Pharm.* **26**, 184.

Gershkovich, E. E. (1971). *Gig. Tr. Prof. Zabol.* **15**, 58; *Chem. Abs.* **76**, 17449.

Ishii, K. (1952–3). *J. Japan Biochem. Soc.* **24**, 118.

Kossakowski, J., Klopocki, T. and Zbikowski, B. (1972a). *Farm. Pol.* **28**, 707.

Kossakowski, J., Klopocki, T., Kuryl, T. and Zbikowski, B. (1972b) *Acta Pol. Pharm.* **29**, 469.

Kumar, R. and Sen, A. K. (1969). *Indian J. Biochem.* **6**, 82.

Kurzawa, Z. (1961a). *Cheml Anal.* (*Warsaw*), **6**, 813.

Kurzawa, Z. (1961b). *Chem. Anal.* (*Warsaw*), **6**, 1025.

Kurzawa, Z. and Suszka, A. (1960). *Chem. Anal.* (*Warsaw*), **5**, 327.
Kurzawa, Z. and Suszka, A. (1962). *Chem. Anal.* (*Warsaw*), **7**, 645.
Markova, L. V. and Glasivtsova, N. Z. (1970). *Anal. Khim. Ekstr. Protsessy*, 96; *Chem. Abs.* **74**, 150941.
Whitman, D. W. and Whitney, R. McL. (1953). *Anal. Chem.* **25**, 1523.

9. MISCELLANEOUS CHEMICAL PROCEDURES

A rather heterogeneous collection of reagents appears under this heading. These are used in:

(a) qualitative and quantitative methods, the chemistry of which is uncertainly known or even unknown, or is complex and not clearly classifiable under any preceding heading.
(b) methods based on groups in the molecule other than the thiol group.

Group (b) is practically boundless and only some especially interesting or more widely used reagents are mentioned. Probably the best example is ninhydrin for amino acids; it usually reacts independently of the thiol group. These reagents are listed below in alphabetical order.

1. ACIDS

This includes carboxylic acids and the principal concentrated mineral acids as sub-headings:

1.1 Carboxylic acids

Yoshimura and Tamura (1971) recently studied the direct conductimetric titration of various sulphur compounds with carboxylic acids, e.g. benzoic and maleic acids and also EDTA. They used dimethylformamide and sulpholan as solvents. Among their compounds were ethanethiol, thiophenol, 2-mercaptoethanol, 2-mercaptobenzothiazole and 8-mercaptoquinoline. Complexes were evidently formed, e.g. thiophenol–benzoic acid of 2:1 and 1:1. An analytical application might be possible.

1.2 Hydrochloric acid

Racker (1955b) determined homocysteine *via* cyclization to the thiolactone with hydrochloric acid, presumably according to:

$$\begin{array}{ccc}
\text{CH}_2\!-\!\text{CH}_2 & & \text{CH}_2\!-\!\text{CH}_2 \\
| \quad\quad | & \xrightarrow{-\text{H}_2\text{O}} & | \quad\quad | \\
\text{H}_2\text{N}\!-\!\text{CH} \quad \text{SH} & & \text{H}_2\text{N}\!-\!\text{CH} \quad \text{S} \\
| & & \diagdown\diagup \\
\text{COOH} & & \text{CO}
\end{array}$$

He heated a 0·5 ml sample with 2·5 ml concentrated acid for 5 min at 100°C and then read the absorbance at 240 nm after cooling.

1.3 Sulphuric acid

Jocelyn (1967) determined reduced glutathione by heating at 100°C in 10·5 M sulphuric acid for 30 min and then comparing the absorbance at 265 nm with that of a similarly treated but unheated solution. Eriksson and Tabova (1968) likewise heated 0·1 to 0·6 μmol samples of glutathione for 30 min at 100°C with 2 ml of concentrated sulphuric acid and evaluated the absorbance at 265 nm.

The determination of cysteine (and glycine) by Kjeldahl digestion (30 min reaction) of Irion (1951) can be included here but it is not related to the thiol group.

1.4 Nitric acid

Rulfs and Mackela (1953) included cysteine hydrochloride among their examples of sulphur determination in organic compounds by Carius digestion with concentrated nitric acid. They followed this by precipitation of the sulphate with lead nitrate, dissolving the lead sulphate in acetate buffer of pH 4·2 and titrating the lead(II) amperometrically with bichromate.

No doubt many such examples from elemental analysis could be found where a thiol was among the compounds tested.

1.5 Nitric–Sulphuric acid mixture

Zafir (1951) reported a test for thiophenols in liquid paraffin through the violet colour yielded with sulphuric acid containing nitric acid.

Tsareva and Kuleshova's work (1969), mentioned in Section 1.32, may be interpolated here. Their new tests for 6-methyl-2-thiouracil included treatment with sodium nitrate and then concentrated sulphuric acid, which yielded a yellow colour quickly turning red. Oxidation and/or nitrosation may be involved.

2. ALUMINIUM CHLORIDE

Talsky (1962) tested aluminium chloride as a qualitative reagent for many compound classes, including the thiol benzyl mercaptan; in suspension or solution in chloroform, this gave a yellow colour with the dry reagent. The test is too unspecific to be of special value.

3. BENZOIC ACID

See "Acids" (Section 1).

4. BENZOYL PEROXIDE

This reagent is used in Feigl's spot test for the thiol group (1957). It is specific for mercaptoacetic acid, for which he quoted the equation:

$$HS-CH_2-COOH + (C_6H_5CH_2)_2O_2 \rightarrow$$
$$(C_6H_5CH_2)_2O + OCH-COOH + H_2S$$

One drop of test solution is heated at 100°C with a few mg of reagent and the hydrogen sulphide evolved is detected through the blackening of lead acetate paper.

5. CARBINOL,4,4'-BIS(DIMETHYLAMINO)DIPHENYL-

In acid aqueous buffers, this reagent dissociates to yield a resonance-stabilized carbenium-immonium ion with absorption maximum at 606 nm and molar absorptivity of ca. 70 000.

The reagent reacts with thiol groups to give derivatives carrying the RS— residue on the central carbon atom, which is accompanied by reduced light absorbance. Rohrbach *et al.* (1973) applied this to quantitative determination, using acetate buffer medium of pH 5·5 in the presence of 4 M guanidine hydrochloride. They related the decrease in absorbance after 25 min reaction to the amount of thiol.

6. DEOXYRIBONUCLEIC ACID

See "Nucleic Acids" (Section 15).

7. DIAZONIUM SALTS

Ershov (1934) determined thiosalicylic acid by titrating in sodium carbonate solution with a diazonium salt reagent.

8. *p*-N-DIMETHYLAMINOANILINE

The reaction of certain sulphur compounds with dimethylaminoaniline in the presence of oxidation agents, especially iron(III), has been known for about 100 years, often under the name of "Lauth" or "methylene blue" reaction. A phenothiazine ring is evidently formed, e.g. with hydrogen sulphide:

Methylene blue

Fleming (1930a) appears to have been the first to try this reagent on organic sulphur compounds and he observed a stable deep blue colour on heating cysteine hydrochloride with the dimethylaminoaniline (hydrochloride) in the presence of ferric chloride. He claimed to be able to detect 50 µg or 50 parts/10^6 of cysteine. Neither mercaptoacetic acid, thiolactic acid nor glutathione yielded the colour under these conditions. Toyoda (1934) also noted a blue colour under more strongly acidic conditions with sulphuric acid.

Freytag (1954) undertook a detailed study of numerous tests for thiols and included the Fleming and Toyoda versions just mentioned. He was able to confirm the claims of the authors regarding cysteine detection but observed that the colour tint and intensity depended greatly on the concentrations of the two reagent components and of the sample, their mutual proportions, the nature of the thiol, the temperature and the time of reaction. The complexity of the reaction is reflected in the experimental details and information from the many quantitative procedures subsequently based on the colour formation.

Fujita and Numata (1939b) appear to be the first to work out a quantitative method. They treated deproteinized filtrate from tissue with the amine hydrochloride in 4 N sulphuric acid and ferric alum in N sulphuric acid, warming for 40 min at 100°C. Cysteine was determined from the intensity of the dark red–violet colour evaluated with the help of a Pulfrich photometer. They found that glutathione yielded a colour only at much higher concentrations. Vassel (1941) determined cysteine and also cystine which he reduced with zinc dust. His reagents were also in sulphuric acid solution, but rather more concentrated than those of Fujita and Numata. He mixed the reagents and sample in a definite sequence and used the zinc dust even in the cysteine determination, without which no colour was produced. Final evaluation was at 575–580 nm but Vassel stated that the absorption maximum of the solution was displaced to higher wavelengths at larger cysteine concentrations, ultimately reaching 655 nm. Since neither homocysteine nor glutathione yielded colour, Vassel believed that postively responding thiols must contain an amino group separated from the thiol group by two carbon atoms, and suggested the possible formation of a benzothiazine.

Mecham (1943) modified Vassel's procedure, claiming increased precision, and Robinson (1959) improved these methods for cysteine by extracing the coloured product, formed after reaction for ca 50 min at 100°C with isoamyl alcohol. He re-extracted into dilute ammonium hydroxide to enhance the colour, and evaluated also at 570 nm. Snell and Snell (1953) also give a procedure for cysteine, in which 2 ml of deproteinized filtrate are mixed with 2 ml of 2% amine hydrochloride in 1 : 4 sulphuric acid (ca. 4 N) and 0·2 ml of 10% ferric alum in 1 : 35 sulphuric acid (these amounts are evidently the same as in the method of Fujita and Numata). After 40 min in boiling water, they evaluate at 610 nm.

Many applications of the colour reaction have been to determine thiols in materials such as beer and foodstuffs, and also in hydrocarbons and air. These differ from the earlier work in that absorbances have been measured at or near 500 nm, on red coloured solutions. Thus Brenner *et al.* (1954) determined thiols in brewing materials, collecting them in aqueous zinc acetate, to which a sulphuric acid-containing amine reagent, ferric chloride and further zinc acetate were added to give the red colour. Ikeya (1964) aspirated volatile thiols from beer into mercuric acetate solution to which he subsequently added the amine hydrochloride in hydrochloric acid solution and a ferric chloride/nitric acid mixture ("Reissner's reagent") to yield the final colour. Sainsbury and Maw (1967) modified Ikeya's original procedure for separating thiols and hydrogen sulphide; after 30 min reaction, they determined the absorbance at 500 nm also. Steffen (1968) employed a closely similar procedure for beer thiols but Sinclair *et al.* (1969) considered that Ikeda's method lacked precision and they determined thiols in beer with a sulphuric acid-containing amine reagent, extracting the coloured product into nitrobenzene before evaluation.

Other determinations of thiols in foodstuffs based on this reaction have been by the following: Sliwinski and Doty (1958), for methanethiol yielded by γ-irradiated meat, drove the thiols with nitrogen into mercuric acetate solution at 0–4°C, subsequently adding the amine hydrochloride and ferric chloride/nitric reagents and evaluating the red colour at 500 nm; Ocker and Rotsch (1959), for thiols in bread and baked goods, who used Sliwinski and Doty's procedure; Segal and Proctor (1959), for thiols in coffee extract. Maier and Diemair (1967) investigated the behaviour of various sulphur compounds, including several lower alkanethiols and mercaptofurfuraldehyde, using a 0·7% amine hydrochloride reagent in concentrated nitric acid and a N ferric chloride–10% nitric acid–water $(1 + 2 + 1)$ reagent. They found that, in the presence of zinc acetate, the reaction product with hydrogen sulphide has several absorption maxima (500, 670, 750 nm); in contrast that with thiols has only a single maximum, near 500 nm. They estimated thiols in foodstuffs, such as coffee beans, radishes, onions and cabbage by distilling 38 ml from the sample + pH 7 phosphate buffer into 10 ml of 3% zinc or mercuric acetate and then adding 1 ml of the amine reagent and 1 ml of the ferric reagent. After 1 h at room temperature, the absorbance was read at 500 nm, with a correction for hydrogen sulphide based on absorbance at 670 nm. The absorbances were essentially the same with zinc and mercuric acetates, except for the mercaptofurfuraldehyde, where no colour was yielded in the presence of the latter salt (addition to the double bonds?).

Brychta and Rudolf (1956) based their determination of thiols in natural gas on absorption in neutral mercuric cyanide or chloride solution and subsequent treatment of this solution with amine hydrochloride and ferric chloride reagents. They removed hydrogen sulphide with cadmium chloride.

They formulated the red product as:

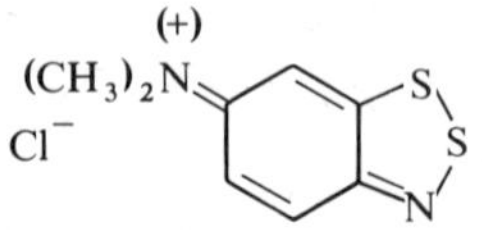

The colour reached its maximum intensity after 2·5 to 3·5 h. Romováček and Bednař (1958) compared the method using p-dimethylaminoaniline–iron(III) with direct and indirect silver ion titration for determining thiols in hydrocarbon mixtures.

Thiols in air were determined by Moore *et al.* (1960) by absorption in 5 % mercuric acetate, followed by dilution and treatment with amine hydrochloride and ferric chloride–nitric acid reagents. After 30 min, they evaluated at 500 nm. Lipina (1969) also determined methanethiol in air with a reaction mixture of mercuric acetate, amine and ferric chloride; and Uvarova and Siridchenko (1969) published a patent on the determination of thiol, e.g. ethanethiol, in gas; the method was based similarly on absorption in mercuric acetate and subsequent treatment with the dual reagent.

9. *p*-N-DIMETHYLAMINOBENZALDEHYDE

Menzie (1956) treated amino acids with several mg of *p*-N-dimethylamino-benzaldehyde + a mixture of 0·3 ml of toluene and 0·02 ml of concentrated sulphuric acid. After 1 min he added ethanol and obtained characteristic colours, e.g. intense red from cysteine and bright orange from glutathione.

Aminobenzenethiols, including cysteine, homocysteine and glutathione, were visualized by Papke and Pohloudek-Fabini (1967) on paper and thin-layer chromatograms through spraying with 1% solutions of *p*-dimethyl-aminobenzaldehyde in ethanol/hydrochloric acid (19 + 1). Yellow or orange-yellow spots were yielded. These tests are, however, almost certainly based on reaction of the reagent's aldehyde group with the amino groups in the samples.

10. DINITROBENZENES

Many compound classes are known to yield a colour with an alkaline *m*-dinitrobenzene reagent, or other reagent in which two nitro groups are meta to one another. For example, Péronnet and Truhaut (1933) report that cysteine gives a green–yellow, turning brown. This reaction has been studied and utilized analytically for compounds containing an active methylene group, but little work has been done on others. As a test for cysteine it suffers from lack of specificity.

Lukáts (1968) determined thiol groups, e.g. in L-cysteine or its N-acetyl derivative, by treating a 1 ml sample containing 150–200 µg with 5 ml of 0·15% o-dinitrobenzene in 96% ethanol and 1 ml of 4% potassium carbonate. After 2 h, he measured absorbance at 386 nm. Disulphides do not interfere but easily oxidizable substances do so, which suggests that the reaction may depend on the reducing properties of the thiol group.

11. ETHYLENEDIAMINETETRAACETIC ACID (EDTA)

(See "Acids" (Section 1).

12. MALEIC ACID

See "Acids" (Section 1).

13. NITROPRUSSIDE $[Fe(CN)_5NO]^{2-}$

Sodium nitroprusside has been known for many years as a colour-forming reagent with thiols and some other sulphur-containing compounds or ions such as sulphide ion. Probably the first mention of the use of this reagent for thiols was by Morner (1899) in a colorimetric procedure for cysteine. In the present work only the last fifty years are covered, during which the principal development of the methods took place.

The nature of the reaction does not appear to have been satisfactorily explained as yet. The suggestion has been made that the —NO group is converted into an —NOS group. This is not discussed here but mention may be made of a recent article by Toropova and Rybkina (1973) who studied photometrically the reaction between nitroprusside and mercaptoacetate and found evidence for a 1:1 complex, given as $[Fe(CN)_5NOS—CH_2COO—]^{4-}$.

There are three chief uses of nitroprusside: for detection; for quantitative colorimetric determination; as an indicator for the disappearance of thiol in titration. A composite reagent containing nitroprusside was introduced by Grote (1931) and is dealt with separately below.

The colour with thiols is given only in basic solution, for which alkali hydroxide or carbonate, ammonium hydroxide, borax and organic bases such as pyridine and piperidine have been used. The pH-dependence has been studied. The colour tends to fade quickly and various stabilizing additives have been proposed in both qualitative and quantitative work. The best known is cyanide ion. This converts disulphide to thiol. Any disulphide present in the sample will thus be included in the result which may or may not be welcome. Some compounds have been added to the reaction mixture with the

Table IX. Detection of thiols with nitroprusside

Sample	Comments	References
Cysteine	+ $ZnCl_2$ + reagent, then + NaOH or NH_4OH until $Zn(OH)_2$ just precipitates; gives ruby red colour	Okuda and Nishijima (1928)
Tissue sections (staining)	+ Zn salts → more stable colour	Giroud and Bulliard (1932)
Petroleum fractions and solvents	gives red-violet; more sensitive than "Doctor" test, detecting down to 100 parts/10^6 thiol sulphur	Herrera and Bermejo (1933)
E.g. cysteine, glutathione	+ $NiCl_2$ to stabilize colour	Zimmet and Perrenoud (1936)
Cysteine; also cystine + sulphite	at pH 7·5 to 9, + CN^-/NaOH, ultimately + reagent	Jonnard and Thompson (1945)
Petroleum products	+ reagent + 2 N NaOH; found more sensitive than "Doctor" test	Mapstone (1946)
Mercaptoacetic acid on hair	+ 10% reagent + alkali → reddish or purple	Semco (1950)
In paper chromatography	Dipped in ammoniacal methanolic reagent → red	Toennies and Kolb (1951)
Comparison of several thiol tests	1% reagent solution, + NH_4OH or NH_3 vapour in spot test → red-violet; also reagent saturated with NaCl; detected down to 3–50 µg	Freytag (1953)
Cysteine from cystine in cystinuria test	Reagent + $(NH_4)_2SO_4$ + Na_2CO_3 + NaCN, + urine sample → cherry red	Fischl et al. (1961)
Also for —S—S— after reduction	+ reagent + $ZnCl_2$ + pyridine/ethanol at pH 5–9 → pink colour or ppt; detects down to 2 µg	Pohloudek-Fabini and Papke (1964)
Cysteine, cysteamine; also cystine + sulphite	+ conc NH_4OH + reagent + Na pyruvate → purple, fading in 1–3 min	Konrad and Thamm (1965)

Sample	Procedure	References
In thin-layer chromatography	Reagent/alkali → red spots on white background	Prinzler et al. (1965, 1966)
Cysteine from cystine, homocysteine from homocystine ($NaBH_4$) in tests for the disulphides	Solution from reduction, + alkali, saturated with NH_4Cl, then + reagent	Rosenthal and Yaseen (1969)
Cysteine from cystine in cystinuria test (reduced with $NaBH_4$)	+ NaOH + NH_4Cl, + reagent → red persisting for 1 min	Kelly et al. (1972)

Table X. Thiol determination through colour with nitroprusside

Sample	Procedure	References
E.g. cysteine, glutathione, thiolactic acid	+1% reagent + NaCl + NH_4OH at room temperature; compared with neutral red standards after 30 sec reaction time	Fleming (1930b, 1931)
Protein cystine, via conversion to cysteine	In presence of KCN + NH_4OH	Blakenstein (1930)
Glutathione	Interference of acetone eliminated at pH 9, where it gives no colour with the reagent	Zimmet (1933)
Glutathione in tissue	Marked fading mentioned	Bierich and Rosenbohm (1933)
Cysteine	+ stabilizing Zn acetate (also prevents H_2S colour) + NH_4OH + reagent; compared with standards	Shinohara and Kilpatrick (1934)
Glutathione	Essentially method of Bierich and Rosenbohm, but Na carbonate instead of NH_4OH; photometric evaluation	Uhlenbroock (1935)

Table X (*cont.*)

Sample	Procedure	References
Tissue glutathione	Nitroprusside procedure preferred to iodometric	Fujita and Iwatake (1935)
Glutathione	In presence of alkali	Royston Maloeuf (1936)
	Sample + conc NH_4OH + reagent + $(NH_4)_2SO_4$; quickly matched with standards from cysteine	Hammett and Chapman (1938)
Glutathione; cysteine	+ NH_4OH; saturated with Na_2SO_4 or $MgSO_4$ for cysteine	Mentzer (1938)
Cysteine, in study of stability	+ NH_4OH at ca. 0°C, + reagent	Micheel (1939)
Glutathione in deproteinized extracts	+ saturated NaCl + reagent + NH_4OH; photometric evaluation	Fujita and Numata (1939a)
Biological thiols	+ cyanide	Anson (1941)
Cysteine from cystine estimation	+ HCl + NaCN + Na_2SO_3, then + NH_4OH and $ZnSO_4$ after 1 min; finally + 5% reagent; compared with standards after 5 min	Krishnaswamy (1942)
Glutathione in eye tissues	Essentially method of Fujita and Numata, + cyanide or oxine to prevent colour fading; + NH_4OH and evaluated after 1 min	Herrmann and Moses (1945)
Blood also tissue glutathione	Ultimately + satd. NaCl + reagent in $(NH_4)_2SO_4/Na_2CO_3$; evaluated at 500–530 nm after 1 min	Brückmann and Wertheimer (1947)
Blood glutathione	Modification of Brückmann and Wertheimer's procedure	Grunert and Phillips (1949)
Thiols from alkaline hydrolysis of acyl mercaptans in their determination	Reagent stabilized with $(NH_4)_2SO_4$; evaluated at 546 nm	Lynen (1951)

Glutathione	Sample + satd. NaCl + reagent, then + Na_2CO_3 + NaCN to prevent fading; evaluated at 520 nm within 1 min	Grunert and Phillips (1951)
	Stabilized colour by adding cyanide or ammonium salts or satd. NaCl	Thompson and Watson (1952)
Mercaptoacetic acid in cosmetic products	+ NaOH + reagent and colour evaluated with Pulfrich colorimeter within 2 min	Provvedi and Camozzo (1952)
Glutathione in presence of nitrite	Nitrite removed with sulphamic acid, then Grunert and Phillip's method (1951)	Mortensen (1953)
Lower thiols	+ reagent + Na_2CO_3 + NaCl + stabilizing KCN in modification of Gruner and Phillip's (1951) method	Rausch and Hamer (1955)
	Compared use of Tillmans' reagent favourably with nitroprusside method as having more stable reagent, less easily destroyed colour and higher sensitivity	Basford and Huennekens (1955)
Glutathione	Slightly modified Grunert and Phillips' (1951) procedure	Racker (1955a)
	+ reagent, with 5 min/40° reaction, then evaluated at 530 nm	Sterescu and Simionovici (1956)
Methanethiol from methionine estimation in pills through alkali degradation	+ NaOH + reagent; after 10 min, + acetic acid–methanol (4 + 1) and water; estimated green after 10 min	Vacek (1960)
6-Mercaptopurine	Modification of Grunert and Phillips' (1951) procedure	Barbieri and Brauckmann (1961)
Glutathione in extracted tissue	Sample at 10°C + K_2CO_3/NaCN + reagent and evaluated at 525 nm at from 2–10 min after last addition	Mortensen (1964)
Blood glutathione	Method of Mortensen (1964), confirming that lower temp. increases colour and stability	Watson (1964)
Blood glutathione		

Sample	Procedure	References
Blood glutathione	Compared own method (1964) with Thompson and Watson's (1952), saying that former yields more intense and stable colour	Mortensen (1965)
Blood glutathione	+ reagent in $(NH_4)_2SO_4$ solution, + NH_4OH; evaluated at 510 nm within 30 s	Garcia Canturri and Sanchez de Rivera (1966)
Urinary amino acids (cysteine)	+ reagent + $(NH_4)_2CO_3$ + NaCN; pink or red evaluated after 10–15 min	Hyanek (1967)
Cysteine	Studied pH-dependence, recommending pH 7–8 for urine	Georges and Politzer (1970)
Biological thiols, e.g. cysteine	Colour stabilized with pyrocatechol	Titaev and Balabolkin (1971)

aim of increasing sensitivity, stability and even selectivity. These include zinc and nickel salts, neutral sodium and magnesium salts and pyrocatechol, probably to reduce the influence of atmospheric oxygen; sodium sulphite is used also for this last purpose, and like cyanide ion, converts disulphide to thiol.

Quantitative evaluation is generally performed within 10 min at the most, and often within 1 min of the time of mixing reagents. The pink colour is usually measured at 510–540 nm.

Detection and determination of cysteine and reduced glutathione have been the principal goals, that of the former sometimes as the final stage of a cystine method. Other applications have been to lower thiols in petroleum products and some other individual compounds such as mercaptoacetate and cysteamine; or to thiols in general, where one may mention visualization in PC and TLC as a special field (see Chapter 12. Section 1).

Tables IX and X summarize information from a selection of references.

13.1 Grote's reagent

Grote (1931) introduced a reagent for various sulphur-containing compound classes. He prepared it by mixing sodium nitroprusside and hydroxyl-ammonium chloride, then adding sodium bicarbonate. After the evolution of carbon dioxide ceased, he added some drops of bromine. Excess bromine was removed by aeration and the reagent solution was found to yield intense green to blue with the thiocarbonyl group and purple-red with the thiol group. He was able to detect 1 part/10^6 of thiourea. Thiols tested by him included cysteine, glutathione, mercaptoacetic and thiosalicylic acids and butanethiol. Grote observed that a similarly reacting product was obtained by exposure of a sodium nitroprusside solution to sunlight or daylight. The product probably contains sodium pentacyanoaquoiron(III),

$$[Fe(CN)_5NO]^{2-} \xrightarrow[NaHCO_3]{NH_2OH} [Fe(CN)_5H_2O]^{3-} \xrightarrow{Br_2} [Fe(CN)_5H_2O]^{2-}$$

The nature of the coloured products with sulphur compounds has not yet been fully elucidated.

The reagent has undergone some later modifications, for example excess bromine has been removed with phenol. There appears to be few examples of its use with thiols. Di Capua (1935) reported detecting cysteine with it but otherwise it probably offers no special advantage over the usual nitroprusside reagent.

Danowski (1944), using a 560–610 nm filter, and Chesney (1944), who evaluated at 580 nm, based quantitative colorimetric determinations of thiourea on use of the reagent. Its chief use has been for determining thiouracils, however, generally at 660 nm after up to 15 min reaction in near neutral solution. Numerous publications of determinations on tissue and body fluids were made in the years following 1944, of which some are given here: Anderson (1944), for thiouracil in urine, after 1 h reaction; Williams *et al.* (1944), for thiouracil in tissue and body fluids; Christensen (1945), for thiouracil in human serum, evaluating at 660 nm after 5–10 min reaction in a barbital–glycine–hydrochloric acid buffer of pH 8; Mørch (1945), for urinary 4-methylthiouracil, by the method of Williams *et al.*; Moscovici (1946), for 6-methyl-2-thiouracil; Olson *et al.* (1947), for thiouracil in serum, plasma and urine, using Chesney's method of measuring at 600 nm after 5–15 min reaction at 50°C in a phosphate buffer of pH 6·8; van Genderen *et al.* (1948), for 4-methyl-2-thiouracil in animal tissue and blood; von Asperen (1950), for 6-methyl-2-thiouracil, using Mørch's method; Doden and Kopf (1951), for 2-thiouracil and its 4-methyl- and 4-propyl derivatives in urine using the Olson/Chesney procedure; Tiecco and Pugliese (1963), for 4-methyl-2-thiouracil using von Asperen's method. Determinations of the antithyroid thiouracils in feeds have been carried out by Bucci and Cusmano (1962), for 2-thiouracil and its 6-methyl and 6-propyl derivatives, using Christensen's reagent and measuring absorbance at 690 nm after 10–25 min reaction; and by Brueggeman and Schole (1967) after extraction with hydrochloric acid and evaluating at 644 nm.

Christensen (1946), in a study of thyroid-inhibiting agents, investigated the use of his modified Grote reagent of 1945. For 4-thiouracil, he found a reaction medium of pH 7·2 and evaluation at 700 nm better, and determined also 2-mercapto-5-amino-1,3,4-thiadiazole in this way. In addition he developed a procedure for 2-thiouracil and alkyl derivatives in the presence of further glycine, in which the final evaluation was made at 500 nm. Recently, Kossakowski *et al.* (1973) determined 1-methyl-2-mercaptoimidazole (methimazole) by treating with diluted (1:20) Grote reagent and phosphate buffer, pH 6·2, for 15 min at 50° and measuring absorbance at 560 nm.

As an indicator of thiol compounds in their direct titration with various reagents, nitroprusside has found some limited use. Table V on pp. 68–80 contains about ten examples of its use in titration with mercury(II) reagents. Other reagents with which it has served as an indicator are: iodine; N-ethylmaleimide; porphyrindine; ferricyanide; tetrathionate; silver(I). Examples can be found under these headings. Additives such as saturated salt or cyanide have been used here also to stabilize the indicator.

14. β-NITROSO-α-NAPHTHOL (ALSO 6-NITROSO-*m*-CRESOL)

Malowan (1963) claimed that β-nitroso-α-naphthol is a specific reagent for amino acids containing thiol groups, e.g. glutathione or cysteine. He mixed 1 drop of reagent solution in sodium carbonate with a crystal of the sample and examined the product in a microscope within a few minutes. The formation of large, fine, needle-like dark violet crystals is the positive response. 6-Nitroso-*m*-cresol can also be used.

15. NUCLEIC ACIDS

Dische (1944) studied the colour reactions between thymonucleic acid and thiols, such as cysteine, at 40°C in the presence of sulphuric acid. He was able to detect both cysteine and glutathione in parts/10^6 amounts. Stumpf (1947) determined deoxyribonucleic acid through the pink colour yielded with 5% aqueous cysteine and 70% sulphuric acid giving a final concentration of 60% acid) for 10 min at room temperature. He evaluated the colour at 490 nm. The reaction should be applicable to the determination of cysteine, although there seems to be no reference to this.

16. OXINE–VANADIUM(V)

Blair and Pantony (1955) developed a test for several types of "active hydrogen", namely, that joined to oxygen, nitrogen and sulphur. The reagent is a mixture of oxine (8-hydroxyquinoline) and vanadium(V), and is accorded the formula

$(C_9H_5ON)_2V$—OH   possibly

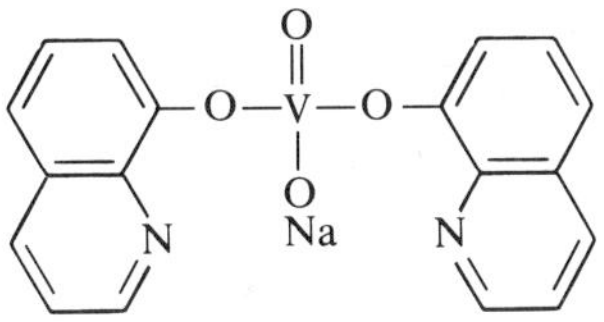

(Bielig and Bayer, 1953)

It yields a deep blue-black solution in certain solvents such as tetrachloroethane or *o*-dichlorobenzene. With it, thiols give a colour change *via* green to lemon yellow. 1–2 ml of a ca 2×10^{-4} M reagent in the organic solvent are mixed with 2–3 drops or 0·1 g of sample and observed in the cold. The authors tested alkane- and arenethiols.

17. PEROXIDE, BENZOYL

See "Benzoyl Peroxide" (Section 4).

18. PHOTOGRAPHY

Methods of detection of thiols based on the use of photographic paper

H

might be expected since thiols are adsorbed on silver salts. Heating reduces the contrast of photographs but Schwarz (1940) found that several compound classes prevented this brightening. Prominent among his examples are thiols e.g. mercaptobenzothiazole, cysteine, mercaptopropionic and -acetic acids. A small drop of the solution to be tested is placed on the exposed, developed and fixed photographic paper and is best allowed to evaporate. It is then dipped into water at 70–90°C. The test is positive if the spot where the drop was placed remains black against a grey background. He was able to detect down to a few parts/10^6 of most thiols.

Board (1951) referred to ''histochemography'', in which thiols yield images on Eastman NTB photographic emulsion. This was inhibited by iodoacetic acid which reacts with and blocks thiol groups. Compounds such as cysteine, glutathione, dimercaptopropylurea, etc., could be detected in this way; other sulphur compounds such as methionine or cystine were ineffective.

Schwarzenbach (1959) placed drops of a cysteine-containing solution on a photographic film, leaving about 3 min. He then exposed it for 8 s and ultimately developed to obtain a black film with pale spots where the solution had been placed. Glutathione showed the same effect but neither cystine nor methione, indicating that a thiol group must evidently be present. As expected, oxidizing agents interfered. He was able to detect down to 24 μg of cysteine in solution or on chromatograms.

19. PHOTOPRODUCT FROM 2-BENZYLPYRIDINE

On exposure to daylight or ultraviolet light, 2-benzylpyridine yields a green, reactive product which in turn gives a red colour with mercaptoacetic acid (Freytag and Müller, 1933). Cysteine behaves less convincingly. The test was not recommended by Freytag (1954) in his critical study of thiol detection.

20. PYROLYSIS

Lobanov (1966) detected down to 0·6 mg of methylthiouracil by heating to 300–400°C and demonstrating the presence of thiocyanate ion through the red or red–orange colour yielded on adding 1 to 2 ml of 1% ferric chloride.

21. RUTHENIUM CHLORIDE

Reinhardt (1953) determined 6-propyl-2-thiouracil by heating the solution in concentrated hydrochloric acid with 0·1% ruthenium chloride for 90 min at 100°C, then diluting and evaluating the colour at 520 nm. Later (1954) he utilized the colour reaction to visualize thiouracil derivatives (6-methyl, 6-propyl-, 6-amino-) on paper chromatograms after separation from thiourea and mercaptobenzoimidazole. The dried paper strips were sprayed with a reagent solution in dilute trichloroacetic acid and warmed at 80°C; after 5–10 min, the compounds appeared as pale green to blue zones on a red background.

22. TETRANITROMETHANE

A spot test for sulphur(II) compounds was given by Kawanami (1964). It is based on the intense yellow yielded with tetranitromethane in chloroform or n-hexane solution. The colour is even more intense than that from the reaction with double bonds and is especially marked with thiols. Probably

the product $(NO_2)_3C-\overset{\displaystyle\underset{\displaystyle NO_2}{|}}{\overset{\displaystyle R}{\diagup}}S-H$ is formed. It can be used for visualization in

PC or TLC.

23. TETRODOTOXIN

Cysteine/glutathione mixtures were analysed by Fujii *et al.* (1966), making use of the reaction of the former with tetrodotoxin at pH 7. After reaction for 5 min at 40°C, they determined residual glutathione by iodate titration (see p. 20). A similar titration of an untreated sample gave the sum of both thiols.

24. THIOCYANATE

Inukai *et al.* (1943, 1944) detected sulphur-containing amino acids by heating 0·5 to 1 mg of the acid slowly with 0·15 to 0·2 g of ammonium or potassium thiocyanate. Cysteine, also cystine, methionine and lanthionine, all yield a red colour if the mixture is kept some time in the molten and bubbling condition.

25. THYMONUCLEIC ACID

See "Nucleic Acids" (Section 15).

26. *o*-TOLUIDINE-CUPRIC CHLORIDE

Nilsson (1957) proposed a new reagent for detecting thiourea or thiouracils.

It is a mixture of 0·2 g *o*-toluidine in 5% acetic acid, 1% aqueous cupric chloride and 5% aqueous sodium acetate and yields an intense blue with these compounds. There seems to be no quantitative application.

REFERENCES

Anderson, A. B. (1944). *Lancet II*, 242.
Anson, M. L. (1941). *J. Gen. Physiol.* **24**, 399.
van Asperen, K. (1950). *Biochim. Biophys. Acta*, **6**, 187.
Barbieri, F. P. and Brauckmann, E. S. (1961). *Arch. Bioquím.. Quim. Pharm., Tucumán*, **9**, 85; *Chem. Abs.*, **57**, 7542.
Basford, R. E. and Huennekens, F. M. (1955). *J. Amer. Chem. Soc.*, **77**, 3873.
Bielig, H.-J. and Bayer, E. (1953). *Ann.* **584**, 96.
Bierich, R. and Rosenbohm, A. (1933). *Z. Physiol. Chem.* **215**, 151.
Blair, A. J. and Pantony, D. A. (1955). *Anal. Chim. Acta*, **13**, 1.
Blankenstein, A. (1930). *Biochem. Z.* **218**, 321.
Board, F. A. (1951). *J. Cellulose Comp. Physiol.* **38**, 377; *Chem. Abs.* **46**, 8176.
Brenner, M. W., Owades, J. L., Gutcho, M. and Golzyniak, R. (1954). *Amer. Soc. Brew. Chem., Proc. Ann. Mtg.* 88; *Chem. Abs.* **49**, 16328.
Brückmann, G. and Wertheimer, E. (1947). *J. Biol. Chem.* **168**, 241.
Brueggeman, J. and Schole, J. (1967). *Landwirt Forsch.* No. 21, 134.
Brychta, M. and Rudolf, J. (1956). *Paliva*, **36**, 307.
Bucci, F. and Cusmano, A. M. (1962). *Rend. Ist. Super Sanitá*, **25**, 518.
Chesney, L. C. (1944). *J. Biol. Chem.* **152**, 571.
Christensen, H. N. (1945). *J. Biol. Chem.* **160**, 425.
Christensen, H. N. (1946). *J. Biol. Chem.* **162**, 271.
Danowski, T. S. (1944). *J. Biol. Chem.* **152**, 201.
Di Capua, C. B. (1935). *Boll. Soc. Ital. Biol. Sper.* **10**, 428.
Dische, S. (1944). *Proc. Soc. Exptl. Biol. Med.* **55**, 217.
Doden, W. and Kopf, R. (1951). *Arch. Exptl. Path. Pharmakol.* **213**, 51.
Eriksson, B. and Tabova, R. (1968). *Anal. Biochem.* **24**, 350.
Ershov, A. P. (1934). *Anilino-Krasochnaya Prom.* **4**, 303; *Chem. Abs.* **28**, 7535.
Feigl, F. (1957). *J. Chem. Educ.* **34**, 457.
Fischl, J., Sason, I. and Segal, S. (1961). *Clin. Chem.* **7**, 674.
Fleming, R. (1930a) *Biochem. J.* **24**, 965.
Fleming, R. (1930b). *Compt. Rend. Soc. Biol.* **104**, 831.
Fleming, R. (1931). *Compt. Rend. Soc. Biol.* **106**, 259.
Freytag, H. (1953). *Z. Anal. Chem.* **138**, 259.
Freytag, H. (1954). *Z. Anal. Chem.* **143**, 401.
Freytag, H. and Müller, A. (1933). *Naturwissensch.* **21**, 720.
Fujii, M., Harada, K. and Matsuda, M. (1966). *Saga Daigaku Nagaku Iho No.* 23, 27; *Chem. Abs.* **68**, 802.
Fujita, A. and Iwatake, D. (1935). *Biochem. Z.* **277**, 284.
Fujita, A. and Numata, I. (1939a). *Biochem. Z.* **300**, 246.
Fujita, A. and Numata, I. (1939b). *Biochem. Z.* **300**, 264.
Garcia Canturri, F. L. and Sanchez de Rivera, M. P. (1966). *Medna Segur. Trab.* **14**, 50; *Anal. Abs.* **14**, 5630.
van Genderen, H., van Lier, K. L. and de Beus, J. (1948). *Biochim. Biophys. Acta* **2**, 482.
Georges, R. J. and Politzer, W. M. (1970). *Clin. Chim. Acta*, **30**, 737.

Giroud, A. and Bulliard, H. (1932). *Bull. Soc. Chim. Biol.* **14**, 278.
Grote, I. W. (1931). *J. Biol. Chem.* **93**, 25.
Grunert, R. R. and Phillips, P. H. (1949). *J. Biol. Chem.* **181**, 820.
Grunert, R. R. and Phillips, P. H. (1951). *Arch. Biochem. Biophys.* **30**, 217.
Hammett, F. S. and Chapman, S. S. (1938). *J. Lab. Clin. Med.* **24**, 293.
Herrera, J. J. and Bermejo, L. (1933). *Anales Soc. Españ. Fis. Quím.* **31**, 267.
Herrmann, H. and Moses, S. G. (1945). *J. Biol. Chem.* **158**, 33.
Hyanek, J. (1967). *Z. Med. Labortech.* **8**, 189.
Ikeya, T. (1964). *Jozo Kagaku Kenkyu Hokoku* **10**, 23; *Chem. Abs.* **64**, 8895.
Inukai, F., Tsurumi, M. and Sakai, S. (1943). *Bull. Inst. Phys. Chem. Res. (Tokyo)* **22**, 919; *Chem. Abs.* **41**, 5916.
Inukai, F., Tsurumi, M. and Sakai, S. (1944). *Proc. Imp. Acad. (Tokyo)*, **20**, 310; *Chem. Abs.* **48**, 2523.
Irion, W. (1951). *Z. Pflanzenernähr. Düngung Bodenk.* **52**, 193.
Jocelyn, P. C. (1967). *Anal. Biochem.* **18**, 493.
Jonnard, R. and Thompson, M. R. (1945). *J. Amer. Pharm. Assoc.* **34**, 293.
Kawanami, J. (1964). *Mikrochim. Acta*, 106.
Kelly, S., Leikhim, E. and Desjardins, L. (1972). *Clin. Chim. Acta*, **39**, 469.
Konrad, E. and Thamm, E. (1965). *Parfüm Kosmetik*, **46**, 64.
Kossakowski, J., Klopocki, T., Zbikowski, B. and Kuryl, T. (1973). *Chem. Anal. (Warsaw)*, **18**, 207.
Krishnaswamy, T. K. (1942). *Proc. Indian Acad. Sci.* **15A**, 135.
Lipina, T. G. (1969). *Gig. Tr. Prof. Zabol.* **13**, 50; *Chem. Abs.* **72**, 96472.
Lobanov, V. I. (1966). *Zh. Anal. Khim.* **21**, 1506.
Lukáts, B. (1968). *Acta Pharm. Hung.* **38**, 213.
Lynen, F. (1951). *Ann.* **574**, 33.
Maier, H. G. and Diemair, W. (1967). *Z. Anal. Chem.* **227**, 187.
Malowan, L. (1963). *Chemist-Analyst*, **52**, 16.
Mapstone, G. E. (1946). *Austr. Chem. Inst., J. & Proc.* **13**, 269.
Mecham, D. K. (1943). *J. Biol. Chem.* **151**, 643.
Mentzer, C. (1938). *J. Pharm. Chim.* **27**, 145.
Menzie, C. (1956). *Anal. Chem.* **28**, 1321.
Micheel, F. (1939). *Chem. Ber.* **72**, 68.
Moore, H., Helweg, H. L. and Graul, R. J. (1960). *Am. Ind. Hyg. Assoc.* **21**, 466.
Mørch, P. (1945). *Acta Pharmacol. Toxicol.* **1**, 106.
Morner, K. A. H. (1899). *Z. Physiol. Chem.* **28**, 595.
Mortensen, R. A. (1953). *J. Biol. Chem.* **203**, 855.
Mortensen, E. (1964). *Scand. J. Clin. Lab. Invest.* **16**, 87.
Mortensen, E. (1965). *Scand. J. Clin. Lab. Invest.* **17**, 93.
Moscovici, R. (1946). *Arquiv. Biol. São Paulo*, **30**, 66; *Chem. Abs.* **41**, 1578.
Nilsson, G. (1957). *Sci. Rev. (Holland)*, **89**, 86.
Ocker, H. D. and Rotsch, A. (1959). *Brot u. Gebäck*, **13**, 165.
Okuda, Y. and Nishijima, Y. (1928). *Bull. Sci. Fukultato Terkultra, Kjusu Imp. Univ.* **2**, 209; *Chem. Abs.* **22**, 2901.
Olson, K. J., Ely, R. E. and Reineke, E. P. (1947). *J. Biol. Chem.* **169**, 681.
Papke, K. and Pohloudek-Fabini, R. (1967). *Pharmazie*, **22**, 485.
Péronnet, M. and Truhaut, R. (1933). *Bull Soc. Chim. France*, **53**, 1464.
Pohloudek-Fabini, R. and Papke, K. (1964). *Mikrochim. Acta*, 876; also *Z. Anal. Chem.* **206**, 28.
Prinzler, H. W., Pape, D. and Teppke. M. (1965). *J. Chromatogr.* **19**, 375.

Prinzler, H. W., Pape, D., Tauchmann, H., Teppke, M. and Tzscharnke, C. (1966). *Ropa Uhlie*, **8**, 13.
Provvedi, F. and Camozzo, S. (1952), *Chim. e Ind.* (*Milano*), **34**, 517.
Racker, E. (1955a). *J. Biol. Chem.* **217**, 855.
Racker, E. (1955b) *J. Biol. Chem.* **217**, 867.
Rausch, L. and Hamer, K. (1955). *Klin. Wochschr.* **33**, 899.
Reinhardt, F. (1953). *Z. Physiol. Chem.* **293**, 268.
Reinhardt, F. (1954). *Mikrochim. Acta* 219.
Robinson, E. A. (1959). *Anal. Chim. Acta*, **21**, 190.
Rohrbach, M. S., Humphries, B. A., Yost jr., F. J., Rhodes, W. G., Boatman, S., Hiskey, R. G. and Harrison, J. H. (1973). *Anal. Biochem.* **52**, 127.
Romováček, J. and Bednář, J. (1958). *Paliva*, **38**, 9.
Rosenthal, A. F. and Yaseen, A. (1969). *Clin. Chim. Acta*, **26**, 363.
Royston Maloeuf, N. S. (1936). *Nature* (*London*), **138**, 75.
Rulfs, C. L., and Mackela, A. A. (1953). *Anal. Chem.* **25**, 660.
Sainsbury, D. M. and Maw, G. A. (1967). *J. Inst. Brew.* **73**, 293.
Schwarz, G. (1940). *Ind. Eng. Chem., Anal. Ed.* **12**, 369.
Schwarzenbach, F. H. (1959). *Helv. Chim. Acta*, **42**, 1133.
Segal, S. and Proctor, B. E. (1959). *Food Technol.* **13**, 679.
Semco, M. B. (1950). *U.S. Patent* No. 2,529,886 of Nov. 14.
Shinohara, K. and Kilpatrick, M. (1934). *J. Biol. Chem.* **105**, 241.
Sinclair, A., Hall, R. D. and Burns, D. T. (1969). *Eur. Brew. Conv. Proc. Congr.* **12**, 427; *Chem. Abs.* **74**, 139464.
Sliwinski, R. A. and Doty, D. M. (1958). *J. Agr. Food. Chem.* **6**, 41.
Snell, F. D. and Snell, C. T. (1953). "Colorimetric Methods of Analysis", Vol. III, Organic—I, p. 483. Von Nostrand.
Steffen, P. (1968). *Nahrung*, **12**, 701.
Sterescu, M. and Simionovici, R. (1956). *Rev. Chim.* (*Bucharest*), **7**, 299.
Stumpf, P. K. (1947). *J. Biol. Chem.* **169**, 367.
Talsky, G. (1962). *Z. Anal. Chem.* **188**, 417.
Thompson, R. H. S. and Watson, D. (1952). *J. Clin. Path.* **5**, 25.
Tiecco, G. and Pugliese, A. (1963). *Atti. Soc. Ital. Sci. Vet.* **16**, Pt. II, 313.
Titaev, A. A. and Balabolkin, I. I. (1971). *Lab. Delo*, 618.
Toennies, G. and Kolb, J. J. (1951). *Anal. Chem.* **23**, 823, (1095, correction).
Toropova, V. F. and Rybkina, A. A. (1973). *Izv. Vyssh. Ucheb. Zaved., Khim. Khim. Tekhnol.* **16**, 1800; *Chem. Abs.* **80**, 103546.
Toyoda, H. (1934). *Bull. Soc. Chem. Japan*, **9**, 263.
Tsareva, V. A. and Kuleshova, M. I. (1969). *Farmatsiya* (*Moscow*) **18**, 76.
Uhlenbroock, K. (1935). *Z. Physiol. Chem.* **236**, 192.
Uvarova, E. I. and Siridchenko, A. K. (1969). *USSR Patent No.* 246,152 of June 11.
Vacek, J. (1960). *Česk. Farm.* **9**, 126.
Vassel, B. (1941). *J. Biol. Chem.* **140**, 323.
Watson, D. (1964). *Scand. J. Clin. Lab. Invest.* **6**, 587.
Williams, R. H., Jandorf, B. J. and Lay, G. A. (1944). *J. Lab. Clin. Med.* **29**, 329.
Yoshimura, C. and Tamura, K. (1971). *Bunseki Kagaku*, **20**, 957.
Zafir, M. (1951). *Folia Pharm.* **2**, No. 1, 21.
Zimmet, D. (1933). *Compt. Rend. Soc. Biol.* **113**, 984.
Zimmet, D. and Perrenoud, J. P. (1936). *Bull. Soc. Chim. Biol.* **18**, 1704

10. POLAROGRAPHY AND RELATED METHODS

Thiols can be detected or determined through catalytic or anodic polarographic waves. The principal field of application has been biological, namely of compounds such as mercaptopurine, penicillamine, homocysteine and especially glutathione and cysteine; and also of thiols in protein hydrolysates for example. Other individual compounds, for instance mercaptobenzothiazole, unithiol, BAL, mercaptoacetic and -propionic acids, mercaptoethanol and lower alkanethiols also have been objects of polarographic and related analytical work. Further there are publications of thiol determinations on petroleum products, effluents, fruit juices, radioprotective substances, etc.

The first polarographic determination of a thiol dates back to Brdička (1933). During investigations of proteins he observed a catalytic polarographic wave, subsequently attributed to thiol and disulphide groups, in buffered solutions of cobalt salts. The best results are obtained with cobalt(II) chloride, ammonium hydroxide and ammonium chloride of respective concentrations 0·001, 0·1 and 0·1 N. Brdička utilized this to determine cysteine and cystine in hydrolysates of certain proteins (1934). The principle has been applied by others, *e.g.* de Lange and Hintzer (1955) to determine the apparent thiol content of wheat proteins which they dispersed in the ammoniacal cobalt medium; Zhantalai (1964) for determining cysteine in the presence of methionine; Soběslavský and Kůtova (1967) for thiols in serum in the diagnosis of rheumatic diseases; Blazsek and Bukaresti (1967) to estimate thiols in histone fractions.

In 1940 Kolthoff and Barnum reported a polarographic cysteine wave in 0·1 M perchloric acid; the diffusion current was proportional to thiol concentration, thus permitting quantitative determination. Above pH 2 irregular waves were obtained. The authors stated that the wave did not correspond to oxidation to the disulphide but to a one-electron change, evidently,

$$RSH + Hg \rightarrow RSHg + H^+ + e$$

Among others who have studied or determined thiols polarographically in such fairly strong acid solution may be mentioned: Gerber (1947), for gasoline thiols using 85–90% alcohol in the presence of 0·025 N sulphuric acid after removing hydrogen sulphide with cadmium(II); Konupčík et al. (1960), for mercaptothiamine, in dilute sulphuric acid; Beníšek (1960, 1961), for cysteine in wool hydrolysates, in 1% sulphuric and medium; Ratovskaya et al. (1961, 1964), for thiols in straight-run gasoline, using benzene–methanol–acetone (55 + 30 + 15), 0·025 N in sulphuric acid; Matsuura et al. (1964), for cysteine, in a sulphuric acid-containing medium; Obolentsev et al. (1964), for thiols in crude oils, using benzene–methanol (3 + 2) solvent, 0·015 N in sulphuric acid; Zhustareva et al (1972), for thiols (and sulphide ion) in waste water from the pulp and paper industry, using 80% alcohol containing hydrochloric acid.

Wenck et al (1972) compared five methods for thiol determination and found polarographic estimation in 0·6 M perchloric acid very reliable; a linear relation was found between diffusion current and concentration, the slope depending on the particular thiol.

Coult (1958) determined thiols by cathode-ray polarography, using 0·05 N hydrochloric acid as medium. His examples included cysteine, homo-cysteine, glutathione and penicillamine from penicillin in its estimation. Concentration was related to the peak height from a plot of current against sweep voltage.

Others have carried out polarographic work at acidic pH values but nearer to neutral. Thus Zuman (1951, 1952) observed that the anodic wave of thiols in acetate buffer of pH 407 was 0·3 V more negative than that of ascorbic acid, enabling the thiols in fruit juices to be estimated in the presence of the acid. He used a glutathione standard. Zumanová and Zuman (1954) deter-mined BAL(2,3-dimercapto-1-propanol) also in acetate buffer of pH 4·7, plus ethanol/chloroform. Kashiki and Ishida (1965, 1967) determined thiols in petroleum naphtha by square-wave polarography using aqueous metha-nol containing 2% acetic acid (by volume) and 1·4 M in sodium acetate as medium. Ueno (1968) also used an acetate buffer (pH 5·2, 0·5 N) for thiol groups in some radioprotective substances. Charles and Knevel (1968) determined non-protein thiol groups in biological systems by means of alternating-current polarography in a medium of potassium nitrate and borate/succinate buffer of pH 5. Tiwori et al (1973) recently determined this lactic acid through its diffusion-controlled anodic wave in aqueous or aqueous-alcoholic medium containing acetate buffer of pH 4·35.

Sensibly neutral solutions were employed by Vacek (1960, 1965) for 6-mercaptopurine (McIlvaine buffer of pH 7·1, measuring at −0·26 V); Čerňák et al. (1961) determined cysteine in the presence of sulphite and thio-sulphate by polarography in a neutral 0·1 M potassium nitrate medium, in which it yielded two waves, and Hird *et al.* (1968) for low molecular weight thiols in flour (tris buffer + hydrochloric acid, measuring at -0.5 V).

Distinctly alkaline solution was favoured by some, e.g. Stricks *et al.* (1954), who estimated cystine by conversion with sulphite ion to cysteine which they then evaluated through the anodic wave at -0.35 V in borax solution. Kaláb (1956) studied the oscillographic polarography of some amino acids, including cysteine, and developed a fast quantitative method. He found that the magnitude of the cut-ins on the elliptical curve of dv/dt against V, measured in a definite way to the long axis of the ellipse, was inversely proportional to the concentration of amino acid; he used potassium hydroxide solution as medium. Čerňák *et al.* (1961) in the work already cited above, observed a clear separation of the cysteine polarographic wave from those of sulphide ion and cystine in Britton–Robinson buffer, pH 11.56; they evaluated at -0.5 V against N calomel. Holzapfel and Stottmeister (1967) determined methane- and ethanethiols polarographically in dilute sodium hydroxide solution; as stable standard they used S-methylisothiouronium sulphate which, under these conditions, is hydrolysed rapidly and quantitatively to methanethiol. Zakharchenko and Antropov (1971) found wave heights proportional to concentration of unithiol in ammoniacal buffer. In their determination of thiolactic acid, mentioned above, Tiwari *et al.* (1973) also used borate buffer of pH 9.4.

Some studies of the influence of pH have been made, e.g. Stricks and Kolthoff (1952) for glutathione in the pH range 1 to 10.82. One of the two waves observed evidently resulted from reaction giving GSHg and enabled quantitative determination to be performed. Zumanová *et al.* (1954, 1955) tested BAL at various pH-values, quoting the electron process as

$$HS—X—SH + Hg \rightarrow [S—X—S]Hg + 2H^+ + 2e$$

Srivastave and Prakash (1973) state that mercaptopropionic acid can be determined polarographically in the pH range from 4.5 to 9.1. Mairesse-Ducarmois *et al.* (1974) studied the d.c., a.c. and differential pulse polarography of cysteine and cystine at pHs ranging from 1.82 to 11.60. They found that the last-named technique, using a medium of borax, potassium nitrate of pH 9.2, was the best for determining the acids alongside each other. Concentrations of cysteine down to 10^{-6} M could be estimated. Cysteine and glutathione (also cystine) were detected by Paleček (1958) also by oscillopolarography by plotting dV/dt as a function of V and using a stationary mercury, mercury coated platinum or copper electrodes. Dušinský and Faith (1967) also identified drugs, including thiouracils, by oscillopolarography in dilute sodium hydroxide, potassium chloride and sulphuric acid solutions. They published tables giving the Q-quotient in these three media.

H*

Some other electrochemical methods for thiols may be mentioned here. Voorhies and Parsons (1959) determined oxidizable substances, e.g. mercaptobenzothiazole, by anodic chronopotentiometry with a straight wire platinum electrode. They plotted E against t at constant current, working in a borate buffer of pH 10·4. Under these conditions, quantitative determination was based on the relation $c \propto t_x$ is the time to reach 0·84 V.

Engel et al. (1965) observed that some ions and organic nitrogen and sulphur compounds catalyse polarographic reduction of $In(H_2O)_6^{3+}$ ions. They carried out polarography in M perchloric acid and found that the catalytic current was proportional to the concentration of the ion or compound. Thiol examples were mercaptoacetic acid and 2-mercaptoethanol. This is given also under "Catalysed Reactions", Chapter 8.

Berge and Jeroschewski (1965) collected various compounds (including cysteine,2-mercaptobenzothiazole,2,5-dimercapto-1,3,4-thiadiazole   and thionalide) on the hanging mercury drop and then determined them voltametrically.

Phillips' method (1967) for trace cysteine determination is based on its inhibiting surfactant influence on instantaneous polarographic i/t waves. He derived an equation which holds for the oxygen maximum at $-0·4$ V in 2×10^{-3} M hydrochloric acid. It is not as sensitive as the catalytic cobalt method of Brdička but gives a linear curve. The time measured was that at which the current fell to a constant value.

Holland et al (1969) determined traces of cysteine (and thiourea) through hydrogen overvoltage measurements on platinum in 0·3 N sulphuric acid. They plotted a graph of overvoltage against concentration of the thiol which enabled them to estimate concentrations from 10^{-7} to 5×10^{-4} M.

Pradec et al. (1971) determined cysteine in rat organs in vivo via cyclic voltammetry yielding $I–E$ peaks because of its oxidation.

REFERENCES

Berge, H. and Jeroschewski, D. (1965). Z. Anal. Chem. **212**, 278.
Benišek, L. (1960). Z. Anal. Chem. **175**, 244; also (1961). Faserforsch. u. Textiltech. **12**, 23.
Blazsek, V. A. and Bukaresti, L. (1967). Anal. Biochem. **18**, 572.
Brdička, R. (1933). Coll. Czech. Chem. Commun. **5**, 112, 238; also (1934). Mikrochem. **15**, 167.
Čerňák, V. J., Blažej, A., Štefanec, J. and Síleš, B. (1961). Acta Chim. Acad. Sci. Hung. **27**, 87.
Charles, R. and Knevel, A. M. (1968). Anal. Biochem. **22**, 179.
Coult, D. B. (1958). Analyst (London), **83**, 422.
Dušinský, G. and Faith, L. (1967). Pharmazie, **22**, 475.
Engel, A. J., Lawson, J. and Aikens, D. A. (1965). Anal. Chem. **37**, 203.

Gerber, M. I. (1947). *Zh. Anal. Khim.* **2**, 265.

Hird, F. J. R., Croker, I. W. D. and Jones, W. L. (1968). *J. Sci. Food Agric.* **19**, 602.

Holland, P. E., Peeler, J. T. and Wehby, A. J. (1969). *Anal. Chem.* **41**, 153.

Holzapfel, H. and Stottmeister, U. (1967). *Z. Anal. Chem.* **232**, 331.

Kaláb, D. (1956). *Pharmazie*, **11**, 265.

Kashiki, M. and Ishida, K. (1965). *Rev. Polarog.* (*Kyoto*), **12**, 168; *Chem. Abs.* **65**, 18476; also (1967). *Bull. Chem. Soc. Japan*, **40**, 97.

Kolthoff, I. M. and Barnum, C. (1940). *J. Amer. Chem. Soc.* **62**, 3061.

Konupčik, M., Liška, M. and Kupčik, F. (1960). *Česk. Farm.* **9**, 502.

de Lange, P. and Hintzer, H. M. R. (1955). *Cereal Chem.* **32**, 307.

Mairesse-Ducarmois, C. A., Patriarche, G. J. and Vandenbalck, J. L. (1974). *Anal. Chim. Acta* **71**, 165.

Matsuura, N., Muroshima, K. and Takizawa, M. (1964). *Japan Analyst*, **13**, 324.

Obolentsev, R. D., Rafikova, L. G. and Baikova, A. Ya. (1964). *Khim. Seraorgan. Soedin. Soderzhashch. v Neft. i Nefteprod.*, *Akad. Nauk SSSR, Bashkirsk. Filial*, **7**, 269; *Chem. Abs.* **63**, 399.

Paleček, E. (1958). *Z. Anal. Chem.* **162**, 1.

Phillips, S. L. (1967). *Anal. Chem.* **39**, 679.

Pradac, J., Pradacová, J. and Koryta, J. (1971). *Biochim. Biophys. Acta*, **237**, 450.

Ratovskaya, A. A. and Gavrilova, L. D. (1964). *Khim. Seraorgan. Soedin. Soders-hashch. v Neft. i Nefteprod.*, *Akad. Nauk SSSR, Bashkirsk. Filial* **6**, 105; *Chem. Abs.* **61**, 8100.

Ratovskaya, A. A., Bedarev, N. G. and Kiselev, B. A. (1961). *ibid.* **4**, 75; *Chem. Abs.* **56**, 11883.

Soběslavský, C. and Kůtová, M. (1967). *Clin. Chim. Acta* **15**, 69.

Srivastave, S. C. and Prakash, S. (1973). *Vijnana Parishad Anusandhan Patrika*, **16**, 187; *Chem. Abs.* **80**, 152015.

Stricks, W. and Kolthoff, I. M. (1952). *J. Amer. Chem. Soc.* **74**, 4646.

Stricks, W., Kolthoff, I. M. and Tanaka, N. (1954). *Anal. Chem.* **26**, 299.

Tiwari, S. K., Kumar, A. and Jain, S. (1973). *J. Prakt. Chem.* **315**, 817.

Ueno, Y. (1968). *Experientia* **24**, 970.

Vacek, J. (1960). *Česk. Farm.* **9**, 126; also (1965). *ibid.* **14**, 216.

Voorhies, J. D. and Parsons, J. S. (1959). *Anal. Chem.* **31**, 516.

Wenck, H., Schwabe, E., Schneider, F. and Flohé, L. (1972). *Z. Anal. Chem.* **258**, 267.

Zakharchenko, I. P. and Antropov, L. I. (1971). *Vestn. Kiev. Politekh. Inst.*, *Ser. Khim. Mashinostr. Taknol. No. 8, 6*; *Chem. Abs.* **78**, 10860.

Zhantalai, B. P. (1964). *Biokhimiya*, **19**, 1009; *Chem. Abs.* **62**, 7590.

Zhustareva, S. S., Krunchak, V. G., Mikhailova, V. P., Lomova, M. A., Dybtsina, N. F. and Oleinik, A. T. (1972). *Bum. Prom.* **13**; *Chem Abs.* **77**, 156097; see also *Chem. Abs.* **79**, 34841.

Zuman, P. (1951). *Coll. Czech. Chem. Commun.* **16**, 510; also *Sbor. Mezinar. Polar. Sjezdu Praze*, 1st *Congr. Proc.* 603, *Chem. Abs.* **47**, 11596.

Zuman, P. (1952). *Chem. Listy*, **46**, 73.

Zumanová, R. and Zuman, P. (1954). *Pharmazie*, **9**, 554.

Zumanová, R., Zuman, P. and Teisinger, J. (1955). *Coll. Czech. Chem. Commun.* **20**, 139.

11. PURELY PHYSICAL METHODS

1. UV-SPECTROPHOTOMETRY

Since the early fifties there have been publications on the detection and determination of compound classes containing thiol or potential thiol groups with the help of absorbance measurements in the ultraviolet region.

Thus 6-mercaptopurine was determined in the presence of uric acid, xanthine and hypoxanthine by Braschi and Cerri (1957) through measurements of absorbance at 326 nm (the others had absorption maxima at 284, 265 and 248 nm, respectively). Noto *et al* (1958) determined the same compound at 325 nm in the presence of hypoxanthine in acid (pH 1) solution.

Gastovo and Pileri (1954) noted that 6-mercaptopurine and adenine migrated together on chromatograms but could be distinguished through the absorption maxima of 325 and 260 nm, respectively. Absorption in the u.v. was utilized by Teshima *et al.* (1955) for visualizing numerous mercapto–purine and pyrimidine derivatives on paper chromatograms.

Thiouracils possess absorption maxima in the 260–275 nm range, depending also on pH. Rapaport and Verzina (1966) studied the u.v. spectra of pyrimidines and determined methylthiouracil in the presence of phenobarbital by measuring absorbance at 272 nm in neutral solution (or at 304 nm at pH 11); they chose 260 nm as the best wavelength for estimation in the presence of other pyrimidines. Nekrasov (1966) also determined methylthiouracil in tablets by dissolving in 0·1 N sodium hydroxide and evaluating at 261 nm; the Beer–Lambert law held over a wide concentration range. The estimation of thiouracil derivatives in feeds was investigated by Brueggeman and Schole (1967) and they based a method on absorbance measurements on the extract with 0·1 N sodium hydroxide adjusted to pH 7·5 with tris hydrochloride. They chose 260 nm for 4-methyl- and 4-propyl-2-thiouracil, and 263 nm for 2-thioorotic acid. Hom *et al.* (1971) based their determination of *ca.* 5 μg ml^{-1} amounts of propylthiouracil on absorbance measurements at 275 nm on the solution in 0·1 N hydrochloric acid.

2-Mercaptobenzothiazole was determined by Gurvich and Kostikova

(1962) through evaluation at 309 nm of the solution containing sodium hydroxide and borax. Tsuruoka *et al.* (1968) studied the contaminating influence of vulcanization accelerators in rubber on medical injection solutions. They estimated 2-mercaptobenzothiazole through its absorbance at 322, 314 or 310 nm at pH 5, 7 and 8, respectively. Absorbance at 309 nm was utilized also by Salamatina and Anokhina (1970) to determine the same compound in an aliquot of alkaline solutions containing sodium and calcium hydroxides, diluted by borax. Absorbance at 310 nm was chosen by Jones and Woodcock (1973) to determine 2-mercaptobenzothiazole in flotation plant liquors; they worked at pH above 9 and found that the Beer–Lambert law was valid at concentrations of 12 parts/10^6 and less.

Recently Shekdar and Venkatachalam (1972) determined butanethiol in hydrocarbon solvents such as heptane through u.v. measurements at 225 nm; at higher concentrations, 238 nm was preferred.

2. IR-SPECTROPHOTOMETRY

Absorbance measurements in the infrared region have been little used analytically for thiols. The alkanethiol stretching frequency at ca. 2600 cm^{-1} is feeble. Adams *et al.* (1953) (See also Coleman *et al.* (1956)) identified numerous thiols among sulphur compounds in a petroleum distillate of b.p. up to 100°C through the IR spectrum; they were able to estimate amounts semiquantitatively. Stanescu *et al.* (1971) determined 2-mercaptobenzothiazole by absorbance measurements at 600 cm^{-1}, ascribed to the C—S bond in the ring, however.

3. MASS SPECTROMETRY

There are few examples of the use of mass spectrometry in qualitative or quantitative analytical work on thiols. Hoogendonk and Porsche (1960) extracted C_1–C_7 thiols from naphtha with alkali (see p. 187), acidifying the extract to recover the thiols. A final extract in isooctane was submitted to low voltage mass spectrometry for analysis. Levy and Stahl (1961) studied the mass spectra of thiols and sulphides exhaustively and their information permitted identification of primary, secondary and tertiary alkanethiols. Examples of analytical work on thiols are found in the recent GLC–MS coupling procedures. Thus Brink *et al.* (1971) identified in this way the compounds in kraft black liquor pyrolysis products, which included methanethiol (see also "Gas Chromatography", p. 245).

Sharkey *et al.* (1957) converted alcohols to trimethylsilyl ethers before subjecting to mass spectrometry. This enabled determinations to be carried out in the presence of compounds which would otherwise interfere. The authors indicated that this could be applied to amines and thiols also and quote data for *n*-butanethiol.

4. PHOSPHORIMETRY

This physical property is highly sensitive and two examples of its utilization may be given: Sanders *et al* (1969) studied phosphorescence characteristics of several anti-metabolites, quoting a table of compounds with analytically useful phosphorescence. Among these is 2-thiouracil, with excitation maximum at 312 nm and emission maximum at 432 nm, yielding a detection limit of 4 ng ml^{-1} and a sensibly linear concentration range over four powers of ten.

Lukasiewicz *et al.* (1972) investigated the phosphorescence of some organic molecules, including 2-thiouracil, at $+77°K$ in water–methanol (9 + 1). They used excitation wavelengths between 220 and 280 nm and obtained linear curves relating the phosphorescence signal (in ampères) and concentration of compound. Nanogram amounts/ml could be determined.

5. METHODS BASED ON RADIOACTIVITY

The use of radioactive reagents introduces a new way of detecting or determining a reaction product, namely, through its radioactive properties with the help of a counter or autoradiography. Reagents of this nature have been mentioned under the heading of the corresponding non-radioactive reagent and are therefore only tabulated below. The most used isotope has been ^{14}C, and there are examples also of the application of ^{203}Hg, ^{110}Ag and ^{35}S:

^{14}C-N-ethylmaleimide (p. 152);

^{14}C-active halides, such as bromoacetate, iodoacetate, 1-fluoro-2.4-dinitrobenzene (pp. 135, 140);

^{14}C-mercurials, such as *p*-chloromercuribenzoate (p. 88);

^{110}Ag in silver nitrate for titration (pp. 105, 113) or ion exchange (p. 115);

^{203}Hg-organic reagents, such as 3-chloromercuri-2-methoxy-1-propyl-urea or methylmercuric nitrate (p. 88);

^{35}S-tetraethylthiouram disulphide (p. 184).

In a few examples, the compound for determination was already radioactive, e.g. ^{25}S-thiols by conversion *via* the silver salt to barium sulphate (p. 115), and *via* conversion to the mercury derivative (p. 88); ^{14}C-methanethiol, also through conversion to the mercury(II) derivative (p. 88); and ^{35}S-cysteine by means of u.v. measurements on the lead derivative (p. 65).

Cysteine and glutathione have been determined by permanganate oxidation; the manganese dioxide formed was converted to [36]Cl–manganese(II)-chloride with hydrochloric acid containing [36]chlorine, and this was assayed with a counter (p. 38).

Thiophenol in coal tar acids was determined by the isotope dilution method using ^{35}S–thiophenol; the thiophenol was precipitated with mercuric chloride and assayed radioactively (p. 88).

REFERENCES

Adams, N. G., Coleman, H. J., Eccleston, B. H., Hopkins, R. L., Mikkelson, L., Rall, H. T., Richardson, D., Thompson, C. J. and Smith, H. M. (1953). *Am. Chem. Soc., Div. Petroleum Chem., General Papers* No. 30, 93; *Chem. Abs.* **50**, 10385.

Braschi, A. and Cerri, O. (1957). *Boll. Chim. Farm.* **96**, 148.

Brink, D. L., Pohlman, A. A. and Thomas, J. F. (1971). *Tappi*, **54**, 714.

Brueggeman, J. and Schole, J. (1967). *Landwirt. Forsch.* No. 21, 134.

Coleman, H. J., Adams, N. G., Eccleston, B. H., Hopkins, R. L., Mikkelson, L., Rall, H. T., Richardson, D., Thompson, C. J. and Smith, H. M. (1956). *Anal. Chem.* **28**, 1380.

Gastovo, F. and Pileri, A. (1954). *Rass. Med. Sper.* **1**, 122.

Gurvich, Ya. A. and Kostikova, V. P. (1962). *USSR Patent* No. 144,636 of Feb. 15.

Hoogendonk, W. P. and Porsche, F. W. (1960). *Anal. Chem.* **32**, 941.

Hom, F. S., Verseh, S. A. and Miskel, J. J. (1971). *J. Assoc. Offic. Anal. Chem.* **54**, 1420.

Jones, M. W. and Woodcock, J. T. (1973). *Can. Met. Quart.* **12**, 497.

Levy, E. J. and Stahl, W. A. (1961). *Anal. Chem.* **33**, 707.

Lukasiewicz, R. J., Rozynes, P. A., Sanders, L. B. and Winefordner, J. D. (1972). *Anal. Chem.* **44**, 237.

Nekrasov, V. I. (1966). *Aptech. Delo* **15**, 37; *Chem. Abs.* **65**, 8676.

Noto, T., Sawada, H., Sato, Y., Fukuda, N. and Inoui, Y. (1958). *Ann. Rep. Gohei Tanabe Co., Ltd.* **3**, 53; *Anal Abs.* **7**, 258.

Rapaport, L. I. and Verzina, G. V. (1966). *Aptech. Delo*, **15**, 30; *Chem. Abs.* **65**, 19926.

Salamatina, G. A. and Anokhina, D. I. (1970). *Fiz.-Khim. Metody Anal.* No. 1, 91, *Chem. Abs.* **77**, 42873.

Sanders, L. B., Cetorelli, J. J. and Winefordner, J. D. (1969). *Talanta*, **16**, 407.

Sharkey jr., A. G., Friedel, R. A. and Langer, S. H. (1957). *Anal. Chem.* **29**, 770.

Shekdar, A. V. and Venkatachalam, K. A. (1972). *Indian J. Tech.* **10**, 343.

Stanescu, G., Smorjevski, M. and Tatu, I. (1971). *Chim. Anal. (Bucharest)*, **1**, 213.

Teshima, I., Matsuura, S., Inukai, Y. and Ichikawa, Y. (1955). *Res. Repts., Nagoya Ind. Sci. Research Inst.* No. 8, 62; *Chem. Abs.* **50**, 15693.

Tsuruoka, M., Nakao, Y. and Horioka, M. (1968). *Kyshu Yakugakkai Kaiho* No. 22, 17; *Chem. Abs.* **71**, 6521.

12. CHROMATOGRAPHIC AND RELATED METHODS

These are best subdivided into the various techniques.

1. PAPER AND THIN-LAYER CHROMATOGRAPHY

Two subheadings are employed for these chromatography techniques: The first covers samples which are predominantly thiols or at least sulphur-containing compounds amongst which there are some thiols; the second refers to chromatography of derivatives of thiols. No information is given about the preparation of these derivatives since that is to be found under the relevant reagent heading (the cross-reference is quoted).

In the many investigations of amino acids, a thiol (usually cysteine) has often been among the examples. This chance occurrence is not considered to warrant inclusion of this vast mass of material. The same consideration applies to derivatives of amino acids, most of which are prepared by reactions of the amino or carboxylic acid group and have therefore no relation to any thiol group present.

A short section is devoted to newer reagents for chromatographic detection of thiols or sulphur-containing compounds. These are mentioned also under the particular reagent heading but it is probably useful to have them collected together (Tables XI, XII, XIII).

Newer reagents for detection in chromatography
Table XIV gives some newer reagents, most of which have been developed expressly for paper or thin-layer chromatographic visualization. Some are more general tests which would evidently be applicable in the chromatographic field.

2. GAS CHROMATOGRAPHY

Most applications of gas chromatography have been to lower alkanethiols in various samples, such as cigarette smoke, car exhaust, sour natural gases, petroleum fractions, beer, water and products from the paper industry. Other lower sulphur compounds, e.g. dimethyl sulphide and dimethyl disulphide, are often present in such samples and the method was aimed more at separation of thiols from these compounds than from one another.

Much gas chromatographic work has been carried out on derivatives of amino acids, such as the N-acylated alkyl esters or trimethylsilyl products. The derivatives from cysteine or homocysteine still contain a thiol group which generally resists the treatment of acylation. These examples have not been included here, however, since this would go unreasonably beyond the scope of this book.

As is customary, two main quantitative measuring steps are encountered: the classical method based on peak area or some equivalent measurement; and chemical estimation of the separated thiols. The latter procedures include, for instance, combustion to sulphur dioxide, then coulometrically titrated with iodine; coulometric titration with silver ion; oxidation with excess iodine, evaluating the unused amount through the change of potential of a redox half element immersed in the solution.

Some applications of GLC were preceded by chemical modification of the thiols to accomplish prior separation from other compound classes, or to yield products better adaptable to successful chromatography. These are tabulated here as well as being quoted under the particular heading of the reagent used for this conversion; examples are precipitation as heavy metal salts (Hg, Cd), followed by regeneration with acid; conversion to disulphides by oxidation by iodine; conversion to sulphides with α-bromo-2,3,4,5,6-pentafluorotoluene.

Physical data, potentially useful for analytical purposes have been obtained by some investigators. This includes retention data or the response of detector types to certain compound classes. These articles also are included in Table XV.

Gas chromatography of thiol derivatives

The problem of difficulty separable mixtures, especially of members of the same compound class, can often be solved by prior conversion into derivatives which are more easily separated. Where gas chromatographic separation is the aim, the derivatives are usually more volatile than the original compounds (e.g. with amino acids or sugars) but may also be less volatile (e.g. with lower thiols).

Probably the best known application is to the amino acids, using derivatives such as N-acylated (acetyl-, trifluoroacetyl-) or N-trimethylsilyl esters. As has been said, the chance presence of cysteine or glutathione among the amino acid examples is hardly justification to discuss here the extensive

Table XI. Paper chromatography of thiols

Sample	Conditions	Reference
S-containing amino acids (e.g. cysteine)	Phenol–ammonia system; visualized with KI/H_2PtCl_6, holding damp strip in HCl vapour; gives colourless zones on pink	Winegard et al. (1948)
Glutathione	2-dimensional; phenol for 17 h, then collidine for 42 h; Whatman No. 4; visualized with ninhydrin	Gordon (1949)
Thiouracil; 5-methyl- and 5-propylthiouracil	Butanol or amyl alcohol saturated with water; visualized with iodine or Gibbs' reagent followed by 0·1 N NaOH, giving orange-yellow on purple	Lederer and Silberman (1952)
Thiouracil derivatives (evidently for $=N-CS-N=$ group)	Benzene–ethanol (16 + 6) on Whatman No. 1; visualized with $RuCl_3$ in dilute trichloroacetic acid→pale green to blue on red-brown	Reinhardt (1954)
Purine, pyrimidine derivatives, including mercapto-compounds	2 N HCl–isopropanol (35 + 65), ascending; visualized in u.v.	Teshima et al. (1955)
Heterocyclic thiols. e.g. thiazolines, imidazoles, benzothiazoles	Whatman No. 1, also buffered; water-saturated n-butanol, descending; visualized on dried paper with 5 % Bi nitrate in 0·5 N HNO_3	Zijp (1956)
Pharmaceuticals, including thiouracils	Schleicher & Schüll paper 2034b strips, buffered to pH 5·7 with phosphate, in test tubes; butanol–ammonia ($d = 0.97$) (1 + 3), butanol–isopentanol–ammonia (6 + 5 + 3) and butanol–$CHCl_3$–ammonia (10 + 1 + 2); visualized with iodine/azide giving nitrogen bubbles and decoloration	Fischer and Otterbeck (1958)
Mercaptobenzothiazole	Whatman No. 4, impregnated with acetic acid–ammonia buffer pH 4; mobile phase of isopropanol–ammonia–CCl_4 (5 + 1 + 4); visualized with Bi nitrate, spots cut out and weighed	Stepień and Gaczyński (1961)

Thiouracils extracted from food (alcohol)	Whatman No. 1; isopentanol saturated with water for 16–18 h; air dried and visualized with iodine	Giannessi (1961)
Thiouracil (and urea)	Solvent of butanol–acetic acid–water (4 + 1 + 5)	Duro (1961)
Products from synthesis of 6-mercaptopurine	Disc technique with butanol–ethylene glycol–water (4 + 1 + 1), butanol–dimethyl sulphoxide–water (4 + 1 + 1), butanol–formamide–ethylene glycol–water (9 + 3 + 2 + 4)	Waksmundzki et al. (1963)
Mercaptoanilines, also disulphides and org. thiocyanates (authors believe —SH more or less oxidized to —S—S— during the run)	Paper impregnated with DMF–benzene (1 + 3); mobile phase of cyclohexane–benzene–ethanol (6 + 2 + 1); visualized with dimethyl-aminobenzaldehyde (reaction of amino group)	Papke and Pohloudek-Fabini (1967)
Sulphur-containing amino acids, including cysteine, substituted cysteines and homo-cysteine	7 solvents; R_f-values quoted for butanol–acetic acid–water (4 + 1 + 1); visualized with ninhydrin	Villanueva and Barbier (1967)
Ethanolic extract of thiouracils in feeds	Schleicher & Schüll No. 2040bM; overnight development with butanol saturated with 0·2 N ammonia; visualized with 0·4% Gibbs' reagent giving brown	Brueggeman and Schole (1967)
Sulphur-containing amino acids, e.g. cysteine, glutathione	Ascending with mobile phase of butanol–acetic acid–water (4 + 1 + 4); visualized with aniline, then bromine, giving mauve–brown	Bayfield and Cole (1969)

Table XII. Thin-layer chromatography of thiols.

Sample	Conditions	References
Alkane- and arene thiols, also sulphides	Alumina D, using light petroleum b.p. 40–60°C; also impregnated with cetane, using methanol–$CHCl_3$–water ($5 + 15 + 1$) or acetic acid–CH_3CN ($1 + 3$); visualized with bromophenol blue→yellow) or, in reversed phase, with alkaline nitroprusside (→red)	Prinzler et al. (1965, (1966)
Mercaptoanilines, also disulphides	Silica gel G, using cyclohexane–benzene–ethanol ($6 + 2 + 1$); visualized with dimethylaminobenzaldehyde→yellow or orange	Papke and Pohloudek-Fabini (1967)
Sulphur-containing amino acids, e.g. cysteine, substituted cysteines and homo-cysteine	Powdered cellulose; cellulose containing 0·1 M KH_2PO_4; 7 solvent systems; visualized with ninhydrin	Villanueva and Barbier (1967)
Sulphur amino acids, including cysteine, from human cerebrospinal fluid	Silica gel, using butanol–acetic acid–water ($4 + 1 + 2$), then phenol–water ($3 + 1$) in second dimension; visualized with iodine–azide and ninhydrin	Kumar and Sen (1969)
Sulphur compounds in petroleum (thiols, sulphides, disulphides)	Alumina, activity grade II, using hexane; visualized with iodine vapour	Snegotskii and Snegotskaya (1969)
Antithyroid drugs, e.g. thiouracils	Silica gel G, using $CHCl_3$–isopropanol–acetic acid ($50 + 6 + 0·1$)	Begliomini and Fravolini (1970)
2-Mercaptobenzo-thiazole in extracts from rubber goods used in the food industry	Mobile phase of acetone–benzene ($1 + 9$); visualized with bromophenol blue in 1% $AgNO_3$ after exposure to Br_2	Ganeva (1971)

Purines and pyrimidines, including mercapto derivatives	Layer of protein prepared from wood cortical cells; mobile phase of n-butanol saturated with water at 25°C; visualized in u.v.	Brady and Hoskinson (1971)
Alkane- and arene thiols	Layer of porous glass powder with 5% gypsum as binder, activated at 200°C; mobile phase of light petroleum; visualized with acidic periodate, followed by benzidine	Wolf et al. (1971)
S-(2-aminoethyl)-2-thio-pseudourea and trans-formation products, e.g. 2-mercapto-ethyl-guanidine	Layer of cellulose MN300 or Lucefol-Quick chromoplate (cellulose on Al foil); various solvents; visualized with nitroprusside-ferricyanide giving white spots on yellow	Kefurt et al. (1972)
Thiophenol; cysteine; mercaptoacetic, -propionic and benzoic acids; heterocyclic thiols	Si gel, using $CHCl_3$–methanol $(9 + 1)$; visualized with ninhydrin/conc. H_2SO_4, a Cu(I) reagent or in u.v.	Bhatia et al. (1972)
Separation of BAL from TSH (1,2,3-trimercapto-propane) and degradation products	Silica gel G, using 6 solvent systems; visualized with I_2 vapour after drying (p. 22)	White et al. (1974)
Vulcanization accelerators, including mercaptobenzothiazole, in unvulcanized rubber compounds	On silica gel G_{254}, using various solvents, notably benzene–ethyl acetate–acetone $(100 + 7 + 2)$; visualized with iodoplatinate (p. 39) and dibromobenzoquinonechloroimide (p. 143), giving yellow and orange, respectively	Millingen (1974)
Glutathione	PC with propanol–water + NEM(N-ethylmaleimide)	Hanes et al. (1950)

Table XIII. Paper and thin-layer chromatography of thiol derivatives

Sample	Reagent/Derivative	Conditions	Reference
Amino acids, glutathione	Chloroacetyl derivatives	PC with butanol–acetic acid–water (20 + 5 + 20) or pyridine–water (4 + 1); visualized with ninhydrin on in u.v.	Bheemeswar and Sreenwasaya (1952)
Amino acids, including cysteine	Thiohydantoins	PC on Whatman No. 1, using butanol saturated with water, or petrol ether, 100–120°–propionic acid (1 + 1); visualized with molybdophosphoric acid, then exposure to ammonia→blue	Edward and Nielsen (1953)
S-containing amino acids, e.g. cysteine, homocysteine	+ HCHO	PC with butanol–formic acid–water (77 + 10 + 13), separating from non-reacting —S— and —S—S—; visualized with ninhydrin	Strack et al. (1956)
Tested on e.g. cysteine, thiomalic and mercapto-acetic acids	+ N-(4-Hydroxy-1-naphthyl) iso-maleimide	PC on Whatman No. 20 with 7 solvent systems; air-dried, briefly exposed to NH_3, then dipped into tetrazotized di-o-anisidine→blue or red within secs; also→red fluorescent spots in u.v. of 366 nm	Price and Campbell (1957)
Thiols, mercapto-acids	Converted to 2.4-dinitrophenyl sulphides	PC on paper saturated with methanol, using heptane; visualized in u.v. at 360 nm. Mercaptoacids on paper saturated with aqueous phase of n-butanol–ethanol–water (4 + 1 + 5) using organic phase as mobile phase or org. phase of benzene–1% acetic acid	Day and Patton (1959)

Volatile alkane thiols to C_5	+acrolein→3-alkylthiopropionaldehydes (converted to 2,4-dinitrophenylhydrazones)	Descending PC with DMF stationary phase and mobile phases of cyclohexane saturated with DMF	Jacot-Guillarmod and Ceschini (1959)
Thiols, also hydroxy- and aminocompounds	Derivatives with 3,5-dinitrobenzoyl chloride	Tested 20 solvent systems, also on impregnated paper. Visualized via fluorescence quenching of α-naphthylamine/Rhodamine 6GBN, or by reduction with $SnCl_2$ then +N-dimethylaminobenzaldehyde	Gasparič and Borecký (1961)
E.g. cysteine, glutathione, thiomalic acid	Reaction with iodoacetamide, also with p-chloromercuribenzoate or p-mercuriphenylsulphonate (water soluble products)	PC, paper washed free of metal ions with EDTA, then mobile phase of butanol–water–acetic acid (21 + 21 + 15); mercury derivatives visualized with dithizone; also ninhydrin	White and Wolfe (1962)
Volatile thiols and disulphides, separated from cultures of marine organisms or reaction mixtures	+$Hg(CN)_2$ or $HgCl_2$, then freed from metal derivatives with acid and reacted with 1-chloro-2,4-dinitrobenzene	PC on Whatman 3MM paper, impregnated with liquid paraffin; mobile phase of $CHCl_3$-methanol–water–liquid paraffin (15 + 20 + 9 + 6)	Folkard and Joyce (1963)
	+1-fluoro-2,4-dinitrobenzene	Suggested preparation for PC or TLC (study of analytical uses of poly-nitrocompounds)	Génévois (1964)
Cysteine with ^{35}S	Pb salt	PC, using t-butanol–formic acid–water (14 + 3 + 3) or butanol–water–ethanol (2 + 1 + 2);	Khusmitdinova (1965)
Glutathione in blood and tissues	+NEM	PC, using propanol–water (81 + 19); visualized with ninhydrin–Cd, red product eluted and determined spectrophotometrically	Wernze and Koch (1965)

Table XIII (*Cont.*)

Sample	Reagent/Derivative	Conditions	Reference
Thiols in general	+ 1-fluoro-2,4-dinitrobenzene/NaHCO$_3$	TLC on silica gel, using benzene–xylene–CCl$_4$ (2 + 1 + 1); located in u.v. and quantitatively determined	Obara *et al.* (1966)
Flour thiols	+ ^{14}C-NEM, then hydrolysed→S-succinyl-L-cysteine (^{14}C:)	Descending PC with butanol–pyridine–acetic acid–water (30 + 20 + 6 + 24); visualized with ninhydrin; radioactive quantitative evaluation	Lee and Lai (1967)
Amino acids, e.g. cysteine, glutathione	2,4-dinitro-5-aminophenyl derivatives (of —NH$_2$ group)	2-dimensional PC on Whatman 3MM, using 1% pyridine in water and 1·5 M phosphate buffer, pH 6; also one-dimensional with these solvents, and 5% acetic acid	Deyl *et al.* (1967)
Thiols (with alcohols and amines)	+ pyruvic acid chloride, then converted to 2,6-dinitrophenyl-hydrazones	TLC on silica gel G, using hexane–benzene–diethylamine (3 + 2 + 5) for 20 min	Schwartz and Brewington (1968)
Aqueous solutions of thiols	+ Hg acetate; derivatives extracted with ethyl acetate	TLC on silica gel G, using ether–light petroleum (1 + 9); visualized with acidic 0·1% dithizone in ethyl acetate→orange	Howard and Baldry (1969)
β-Mercaptoethylamine in urine and stomach contents	+ NEM	PC, using *n*-propanol–N acetic acid (10 + 1); visualized with ninhydrin/Cd reagent; spots cut out, extracted with CH$_3$OH and solution evaluated at 509 nm	Titov (1969)

Cysteine and esters in presence of cystine, etc.	+iodoacetamide	TLC on silica gel GF$_{254}$, using many solvents, mostly basic or neutral; visualized with ninhydrin	Sanso and Rigoli (1970)
Glutathione in rat epidermis	+NEM		Miyagawa et al. (1971)
Amino acids, including cysteine; also mercaptopurines	2,4-dinitrophenyl derivatives (—SH and —NH$_2$ groups of cysteine reacted)	TLC on protein prepared from wood cortical cells, some esterified, some unmodified: 2-dimensional using butanol–water–acetic acid (3 + 2 + 1), then t-amyl alcohol–NH$_3$ (d = 0.88) (5 + 1)	Brady and Hoskinson (1971)

Table XIV

Sample	Procedure	Reference
Amino acids, including cysteine	PC sprayed with 1% KMnO$_4$ containing 2% Na$_2$CO$_3$ →yellow on red after a few min	Dalgleish (1950)
Thiols	PC dipped in ammoniacal methanolic nitroprusside→red; also using Pt- and Pd-containing reagents	Toennies and Kolb (1951)
Amino acids	PC sprayed with saturated KIO$_4$, then with benzidine hydrochloride in ethanol; S-containing acids→yellow, others blue. Also sprayed with KIO$_4$, heated 20–30 min at 50–60°C, then spotted with 5% NaI; cysteine→brown turning blue	Cifonelli and Smith (1955)
Homocysteine	PC sprayed with ninhydrin, heated and treated with Hg(NO$_3$)$_2$→ cherry red on deep blue; reaction catalysed by Cl$^-$, PO$_4^{3-}$, HCO$_3^-$; 80 other substances, including 28 S-compounds,→no colour	Pasieka and Morgan (1955)

Table XIV (*Cont.*)

Sample	Procedure	Reference
Sulphur(II), e.g. thiols, sulphides, disulphides	PC, TLC sprayed with 0·1% $NaIO_4$ in 8 N acetic acid; after 4 min, sprayed with 0·5% benzidine in butanol–acetic acid (4 + 1)→white on blue	Stephan and Erdman (1964)
Sulphur(II)	$C(NO_2)_4$ in $CHCl_3$ or *n*-hexane→yellow (very intense with thiols)	Kawanami (1964)
Thiols and some other compounds	Suppression of fluorescence of 1,3,6,8-tetramercuri-tetraacetatofluorescein	Havíř et al. (1965)
S-containing amino acids, e.g. cysteine, glutathione	PC (glass-fibre paper) sprayed with 1% $KMnO_4$–1% H_2SO_4 (1 + 1); MnO_2 product treated with HCl containing ^{36}Cl isotope, giving $Mn^{36}Cl_2$, assayed by Geiger counter	Beer et al. (1967)
Free radical precursors, e.g. thiols	PC or TLC sprayed with 0·3% 7, 7, 8, 8-tetracyanoquinodimethane in acetone or pyridine–acetone→colours	Sawicki et al. (1967)
Thiols, tested on cysteine, cysteamine and mercaptoacetic acid	PC treated with ammonium hexanitratocerate in dilute HNO_3→white; same reagent + $Hg(NO_3)_2$ + $KMnO_4$→yellow, turning brown in a few min	Trop et al. (1968)
Thiols, even in presence of other S-compounds	TLC exposed to iodine vapour→bleached sites on tan	Brown and Edwards (1968)
E.g. cysteine, glutathione, cysteamine	2,2'-Dithiobis(5-nitropyridine)→yellow	Grassetti and Murray, jr. (1969)
S-containing amino acids, e.g. cysteine, glutathione	PC strips drawn through 3% aniline in light petroleum, BP 40–50°; solvent evaporated, strips then exposed to Br_2 vapour→mauve to brown	Bayfield and Cole (1969)
Thiols (disulphides, after reduction)	PC, TLC sprayed with 5,5-dithiobis(2-nitrobenzoic acid)→yellow in a few secs	Glaser et al. (1970)

Sample	Conditions	Reference
Thiols	PC, TLC, solution treated with 2,2'-dithio-bis(5-nitropyridine)→ yellow	Grassetti (1971)
S-containing compounds (thiols, disulphides sulphides)	Dried TLC exposed to Br_2 vapour for 15 s to several minutes; excess reagent removed and plate sprayed with 0·1% 2,3-dichloro-5,6-dicyanoquinone in benzene, ethanol or aqueous ethanol; 37 of 44 tested compounds→blue fluorescent spots with 0·1 to 1 μg	Belliveau and Frei (1971)
S-compounds, e.g. cysteine, mercapto-succinic acid	PC, TLC sprayed with chloroplatinic acid/KI/HCl; dried, then sprayed with water→white on purple	Wong (1971)
Thiols	In TLC by first spraying with an ammoniacal copper reagent, then with 20% hydroxylammonium chloride→yellow, green or green-brown	Bhatia et al (1972)
E.g. arene thiols, benzyl mercaptan, mercapto-propionic acid	Silica gel HF_{254} thin layers sprayed with ammonium hexanitratocerate in 2 N nitric acid →colourless spots on yellow.	Grossert and Langler (1974)

Table XV. Gas chromatography of thiols

Sample	Conditions	Reference
Alkanethiols up to C_6	Dioctyl phthalate containing 5% $(C_6H_5)_2NH$ on 180–220 mesh carborundum; N_2 carrier gas; effluent gas passed into $I_2/KI/70\%$ potential change of a Pt strip as redox electrode related to the thiol amount	Sunner et al.. (1955)
Volatile organic S-compounds, e.g. C_1–C_3 thiols, also with lower hydrocarbons and alcohols	Tricresyl phosphate–Celite 545 (30 + 70); He gas; thermal conductivity detector; temperature up to 120°C	Ryce and Bryce (1957)

Table XV (*Cont.*)

Sample	Conditions	Reference
Thiols, e.g. in petroleum distillates, after extraction with KOH/ethanol and regeneration	Celite–silicon DC 550 stearic acid; issuing gases passed into 75% ethanol and titrated coulometrically with Ag^+	Liberti and Cartoni (1957)
Organic S-compounds, including thiols from C_3 to C_6	Tricresyl phosphate on 35–80 mesh Johns Manville C-22 fire brick $(2 + 5)$; N_2 gas; thermal conductivity detector; eluate examined mass spectroscopically	Amberg (1958)
Low boiling S-compounds, e.g. C_3 and C_4 thiols, in a Persian crude oil	17% Dow Corning 550 silicone oil on 40–50 mesh acid-washed fire brick; He gas; 38°C; thermal conductivity detector	Coleman et al. (1958)
Retention times for thiols and sulphides	40% dinonyl phthalate on 20–30 mesh Johns Manville C-22 firebrick; He gas; 50°C	Spencer et al. (1958)
Study of GLC of 11 S-compounds (including C_3–C_5 thiols) in BP range 58–126°C	β,β'-Iminodipropionitrile on Celite 545; He gas; 84°C; also using white oil on Celite 545	Karchmer (1959)
12 thiols and 11 sulphides in BP 35–220°C range	25% squalane on 30–60 mesh firebrick; temperature–programmed from 20 to 150°C; thermal conductivity detector	Sullivan et al. (1959)
C_1–C_4 thiols, regenerated by treating Hg derivatives with toluene-3,4-dithiol	Carbowax or Apiezon grease at 86° or 60°C, respectively; He gas	Carson et al. (1960)
Lower thiols, regenerated under reduced pressure from alkali extracts	Two columns: 30% Silicone 200 and 30% tetraisobutylene on 60–100 mesh Celite; He gas; 30°C; Perkin Elmer model 154A	LeRosen (1961)

Alkanethiols/sulphides/isopropanol	Dibutyl phthalate + tritolyl phosphate, in series, also alone, on 42–60 mesh C-22 firebrick; He gas; 50°C; also obtained retention data using Dow Corning 200 silicone fluid; Perkin Elmer 154D model	Farrugia and Jarreau (1962)
Liquid odorants, containing thiols/sulphides	28·6% dodecyl phthalate on 42–60 mesh GC 22 firebrick at 50°C; better temperature programmed (2°/min from 35–180°C) using same impregnation but on 60–80 mesh silicone-treated Chromosorb W; He gas; thermal conductivity detector	Baumann and Olund (1962)
Qualitative analysis scheme for various compound classes, e.g. thiols and sulphides	Various systems, best with 20% Carbowax 20M and 4000M on 60–80 mesh firebrick; 125°C; thermal conductivity and ionization detectors; also capillary columns coated with 5% solutions of the liquid phase; also β,β'-oxydipropionitrile and Carbowax 4000M as a pair	Merritt and Walsh (1962)
Odorant compounds in natural gas (C_2–C_4 thiols and sulphides)	Three columns at 25°C: 10% β,β'-oxydipropionitrile on Chromosorb P; 5% silicone oil 550 on the same; 5% dodecanethiol on the same; Ar gas; Ar ionization detector	Adreen et al. (1962, 1963)
Disulphides derived from oxidation of thiols with iodine	Silicone oil or silicone gum rubber on 60–80 mesh Chromosorb P; Ar gas; 60 to 150°C depending on the sample	Sporek and Danyi (1963)
RSH from determination of $R—S_x—R'$ through reduction with $LiAlH_4$	20% Dinonyl phthalate on acid-washed Celite; 65°C	Porter et al (1963)
Thiols in sour natural gases	30% tricresyl phosphate on acid-washed Chromosorb W; He gas temperature programmed from room T to 130°C; thermal conductivity detector, also coulometric silver ion titration of issuing thiols	Fredericks and Harlow (1964)
Thiols in C_4–C_{16} range, e.g. α-toluenethiol, dodecanethiol	20% Carbowax 20MM on Chromosorb W gave best results; He gas; temperature-programmed from 135 to 175°C at 13°/min; FID	Wallace and Mahon (1964)
Thiols, sulphides, di- and polysulphides; S-components in garlic	Carbowax 20M and silicone oil, columns with 10% on 60–80 mesh Chromosorb W; N_2 gas; 110 or 140°C; flame or electron capture detector	Oaks et al. (1964)

Table XV (*Cont.*)

Sample	Conditions	Reference
Thiols in petroleum samples	15% silicone rubber on acid- and caustic-washed 30–60 mesh Chromosorb W; N_2 gas; temperature programmed from 60° to as high as 400°C; products burnt to SO_2, determined by coulometric titration with I_2	Martin and Grant (1965)
Thiols in gasoline boiling range stocks	30% dinonyl phthalate on 60–80 mesh, acid-washed Chromosorb P; He gas; temp. programmed from 60 to 160°; then maintained; separated thiols titrated coulometrically with silver ion	Brand and Keyworth (1965)
Thiols in light hydrocarbons after alkali extraction and regeneration	30% tricresyl phosphate on 50–80 mesh Sil-O-Cel; N_2 gas; 150°C; thermal conductivity detector	Bloemberger and Vermaak (1965)
Thiols, separated as Hg or Cd salts and then regenerated	Tested many stationary phases, e.g. Versamide 930, Silicon-Triton, Flexol, tritolyl phosphate, Ucon, TMTS, Apiezon	Feldstein *et al.* (1965)
2-Thiouracils in feeds after methylation	3% SE-30 on Chromosorb W; N_2 gas; 180°C; FID	Fravolini and Begliomini (1965)
Methanethiol in aqueous solution from sampled head space	20% silicone oil DC 200 on 60–80 mesh acid-washed Chromosorb P; N_2 gas; 50–60°C; FID with H_2 for combustion	Field, jr. and Gilbert (1966)
S-odorants (methane-, ethane-, and propane-thiols, and sulphides)	Poly(propylene glycol sebacate) on Chromosorb and Ryosorb (2 columns); N_2 gas; FID with H_2 for combustion	Weisser *et al.* (1966); Popl and Weisser (1966)
Kraft mill S-compounds (methanethiol, methyl sulphide and disulphide)	20% tri-*m*-tolyl phosphate on 60–80 mesh Chromosorb W; N_2 gas; 90°C; FID	Rayner *et al.* (1967)

Compounds	Method	Reference
Retention volumes of 185 compounds, including allyl mercaptan, propanethiol, mercaptoacetic acid	On Apiezon and polyethylene 6000 impregnations	Maekawa et al. (1967)
Methane-, ethanethiols and dimethylsulphide in air after separation as Hg(II) salt or freezing in acetone-dry ice	30% tricresyl phosphate on Celite (3 + 7); N_2 gas; 40°C	Okita (1967)
2-Hydroxyethyl derivatives, including 2-mercaptoethanol	4% Carbowax 20H column terminated with terephthalic acid on 60–80 mesh HMDS-treated Chromosorb W; N_2 gas; 73°C; H-flame detector	Fishbein and Zielinski (1967)
Lower thiols (C_1–C_4) and sulphides up to C_5	Steel tube coated with 10% Triton X-100 and 1·5% Span 80 in CH_2Cl_2; He gas; 35°C; FID	Freedman (1968)
Thiols and other compounds in car exhaust	10% Quadrol on Porapak Q; Ar gas; 115°C; effluent burnt and gases detected via electron capture, Ar ionization, FID or thermionic detector	Funasaka et al. (1968)
Propanethiol/propanol/propylamine/dipropylamine	9% Amine 220 on 80–100 mesh Chromosorb W; N_2 gas; 95°C; FID (only qualitative for propanethiol due to vapour losses)	Hermanson et al. (1968)
Primary, secondary and tertiary thiols	Capillary columns of phenyl silicone oil DC 550 or dodecyl phthalate and trimer acid on stainless steel surface; of silicon oil DC 200 on glass; Ar gas	Leppin et al. (1969)
Volatile S-compounds in beer (thiols, sulphides, disulphides)	10% Triton X-305 on 60–80 mesh Chromosorb G; isothermal procedure at 40 and 150°C, temperature programmed from 120 to 150°C also; N_2 gas; FID	Drews et al. (1969)
Methanethiol, methyl sulphide and disulphide formed in kraft pulping	30% tricresyl phosphate on Chemisorb W; He gas	Andersson and Bergstrom (1969)

Table XV (*Cont.*)

Sample	Conditions	Reference
Traces of impurities (S-compounds) in fuel gases	2 columns: 10·1 % 2-cyanopropoxyethane + 1 % dodecanethiol on Porolith at 50° for higher-boiling compounds; 15·3 % dioctyl phthalate + 1 % dodecanethiol on Porolith at 25°C for lower; β-ray Ar ionization detector	Reinhardt *et al.* (1970)
Gaseous pulp mill effluents (H_2S, methanethiol, methyl sulphide and disulphide)	10 % Triton X-305 at 25° and 60°C; poly (phenylene ether) on Haloport F at 60°C	Smith and Tauss (1970)
n-Alkanethiols (up to C_{12}), sulphides and disulphides (retention indeces)	Various silicone oils or dinonyl phthalate or its mixture with aniline at 60°, 120° or 130°C	Golovnya and Arsen'ev (1970)
Some alkane- and arenethiols and alcohols	+ excess acrylonitrile; unused determined by GLC on 10 % polyethylene glycol 3000 on 60–80 mesh Chromosorb W; He gas; 100°C	Obtemperanskaya and Nguyen Dyk Hoc (1970)
S-compounds in air, parts/10^9 amounts of methanethiol, methyl sulphide, etc.	Poly(phenyl ether) and orthophosphoric acid on 40–60 mesh Teflon; N_2 gas; flame photometric detection at 394 nm	Stevens *et al.* (1971)
Methane- and ethanethiols and corr. sulphides and disulphides in beer head space vapour	Coiled glass columns containing 3 % Igepol CA 630 on 80–100 mesh DMCS-Chromosorb G-AW, or 10 % of each of GE-SF 96 and Embaphase on acid-washed 60–80 mesh Chromosorb W; Igepal column at 40° for thiols and sulphides; N_2 gas; FID	Jansen *et al.* (1971)
Volatile S-compounds in cigarette smoke, including C_1–C_6 thiols	25 % 1,2,3-tris(2-cyanoethoxy)propane on acid-washed 60–80 mesh Chromosorb W; N_2 gas; 75°C; flame photometric detector	Groenen and van Gemert (1971)

Trace S-compounds in C_2H_6–C_2H_4 (e.g. methane- and ethane-thiols, CS_2, COS, thiophene, methyl sulphide)	15% squalane on Spherochrom-1; Ar gas; 20°C; for EtSH and CS_2, a second stage of 20% dimethylsulpholane on Spherochrom-1	Lulova and Tsifrinovich (1971)
Tests on the response of the flame photometric detector to S-compounds, including C_2–C_4 thiols and sulphides	5% silicone oil QF-1-6500 on Porapak Q-S; response measured at 394 nm and found proportional to $[S]^2$	Stubbs (1971)
Pyrolysis products (480–1137°C) of kraft black liquor	Glass column packed with 20% DC-710 on 60–80 mesh Chromosorb W; N_2 gas; temperature programmed from 50 to 200°C at 2·5°/min; gas stream divided, one way to FID, one to microcoulometer specially for S-compounds	Brink et al. (1971)
S-compounds, e.g. methanethiol, in coke oven gas	50% liquid petrolatum or 20% silicon oil on Chromosorb W; subjected to h.f. discharge and intensity of the 921·99 nm line measured	Shapunov et al. (1971)
S-compounds in cigarette smoke	GSC on 80–100 mesh Porapak Q or 80–100 mesh Chromosorb 104; isothermal at 50° or programmed from 50° (after 5 min) to 150°C at 5°/min; N_2 gas; S-specific flame photometric detector	Guerin (1971)
Methanethiol and other S-pollutants, e.g. COS,H_2S	Teflon column packed with deactivated silica gel ("Deactigen"); FID	Hartmann (1971)
Parts/10^9 amounts of S-compounds, e.g. methanethiol, in air	Teflon tubing packed with graphitized C black with 0·5% phosphoric acid and 0·3% Dexsil; 80° for methanethiol; flame photometric detector	Bruner et al. (1972)

Table XV (*Cont.*)

Sample	Conditions	Reference
Homologous series of thiols, sulphides and disulphides (retention indices)	Best with Apiezon M and PEG 1000 at 60°C, and Apiezon M, PEG 20000 and Triton X-305 at 130°C; N_2 gas; FID	Golovnya and Arsen'ev (1972)
Methanethiol in determination of reaction products from synthesis of methyl sulphide	Combined columns of 1·5 m of PEG 600 on Polychrome and 1 m Chromosorb 102; He gas; 100°C; thermal conductivity detector	Pechernikova *et al.* (1972)
S-compounds, including methane- and ethane-thiols, H_2S, COS	"Gas flow reversal procedure" on columns of 30% tritolyl phosphate on Chromosorb P; He gas; 22°C; thermal conductivity detector; use 2 columns, one 10', one 20' with gas flow reversed to return through the longer	Kremer and Spicer (1973)
Volatile thiols in rat whole blood (methane- and ethanethiols)	20% SE-20 on Chromosorb P; He gas; 25°C; dual H flame detector	Doizaki and Zieve (1973)
Identification of volatile S-compounds, including propanethiol	Teflon column packed with Chromosorb T; N_2 carrier gas; 100°C; flame photometric detection	Bremmer and Banwart (1974)
Volatile S-compounds including alkane- and alkenethiols and dithiols	PTFE column packed with 3,3'oxydipropionitrile, bonded chemically to 80–100 mesh Porasil C; N_2 gas; 20°C, programmed to 130°C at 2°/min, then kept there; flame ionization detector	Raulin and Toupance (1974)
Thiols up to C_8	5% PEG 100 on 80–100 mesh Chromosorb W. AW or DMCS; N_2 gas; T° programmed from 40 to 130°C at 10°/min; flame ionization detector	Korolczuk *et al.* (1974)

literature on analytical methods for this class. This applies even more strongly here, since this section is entitled "gas chromatography of thiol derivatives" and the stated derivatives are based on reactions of the amino or carboxylic acid groups.

Some authors narrowed the range by working on sulphur-containing amino acids. Two such examples may be quoted: Caldwell and Tappel (1968) studied the GLC of trimethylsililated sulphur amino acids on 2% SE-20 on 90–100 mesh Anakrom SD at 100–250°C, using nitrogen carrier gas and a hydrogen-FID. Shahrokhi and Gehrke (1968) also converted such amino acids to trimethylsilyl derivatives with bis(trimethylsilyl)acetamide or bis(trimethylsilyl)trifluoroacetamide and then chromatographed on 0·5% SE-20 on acid-washed Chromosorb G, pretreated with dimethyldichlorosilane, at 75–200°C, using nitrogen carrier gas and also a hydrogen-FID.

Some examples of the preparation of thiol derivatives suitable for GLC have been given under various reagent headings; it may be convenient to list them together here with the appropriate reference:

Reaction with benzoyl chloride	(Chapter 3, Section 1)
Reaction with 1-chloro-2,4 (and 2,6)-dinitrobenzene	(Chapter 3, Section 3)
Reaction with α-bromo-2,3,4,5,6-pentafluorotoluene	(Chapter 3, Section 5)

3. COLUMN CHROMATOGRAPHY

Column chromatography is often used in general cleanup procedures especially of larger and more complex samples, such as petroleum distillation fractions, for instance. These occasionally contain thiols but individual references are not quoted here since the interest has been centred generally on the non-thiol components.

Separations of amino acids, where cysteine, glutathione or penicillamine are present, or of purine and pyrimidine compounds where the occasional mercapto-derivative is present, are not cited for analogous reasons.

The work of Rozova (1970) may be mentioned; he absorbed mercaptoacetic acid from air on silica gel, ultimately stripping it with methanol and concluding by forming the ferric chelate of the hydroxamate and estimating this colorimetrically at 470 nm.

Some examples of column chromatography of derivatives of thiols have been mentioned under the reagents concerned. For example:

Chapter 3, Section 1: pyruvic acid 2,6-dinitrophenylhydrazones, from preliminary reaction with pyruvic acid chloride.
Chapter 3, Section 3: formation of the 2,4-dinitrophenyl derivatives

I*

4. ION EXCHANGE

Little work appears to have been done on the direct separation of thiols with the help of ordinary ion exchangers. Among the many studies of ion exchange separation of amino acids, thiols such as cysteine have rarely been included; they often cause difficulties with overlapping, and oxidation to the sulphonic acid is generally recommended, e.g. cysteine to cysteic acid or penicillamine to penicillanic acid. Examples have been mentioned under "Oxidation", Sections 1.23 and 1.47.

Woodroffe and Munro (1970) separated mercaptobenzothiazole from benzotriazole on the anion exchange resin Bio-Rad AG1-X2 (chloride form, 50–100 mesh). Both compounds were retained and they eluted the benzotriazole with 0·2 M hydrochloric acid and the mercapto compound with 0·2 M hydrochloric acid in ethylene glycol.

Two techniques have been applied in ion exchange work on thiols, especially of biological origin:

(a) conversion to derivatives more easily separable and identifiable (the oxidation of thiol amino acids really belongs here).

(b) preparation of ion exchangers selective for thiols.

Examples of both these have been given under the relevant reagent headings. For convenience, they are tabulated here:

Type (a)

Reaction on a [110]silver-containing column	Chapter 2, Section 19.3.1
Alkylation with ^{14}C-bromoacetate	Chapter 3, Section 4.4.4
Reaction with acrylonitrile	Chapter 4, Section 1.2.1
Reaction with 4-vinylpyridine	Chapter 4, Section 1.2.2
Reaction with N-ethylmaleimide	Chapter 4, Section 1.5.1(a)

Type (b)

Mercury-containing exchangers	Chapter 2, Section 14.3.1(h)
Maleimide-containing polymer	Chapter 4, Section 1.5.1(b)

Ishibashi *et al.* (1967) studied the adsorption of some thiols and disulphides on macroreticular ion-exchange resins, such as Amberlyst 15, in various metal forms, e.g. copper(II), cobalt(II), nickel(II), zinc(II) and silver(I). The samples were dissolved in toluene, *n*-hexane or methanol at 25°C. They found that the thiols were most strongly adsorbed by the copper-forms, evidently yielding copper mercaptides. The compounds could be eluted with alkaline or acidic aqueous or acetone solutions. They also desorbed *n*-propanethiol by heating to 110°C. It was also adsorbed moderately on Amberlyst A-27 in the hydroxyl-form.

Hashida and Nishimura (1968) studied the adsorption in the gas phase of propane- and butanethiols and of benzyl mercaptan on some Amberlites, Duolites and other ion exchangers. The best adsorbents were basic exchangers in the hydroxyl-form. Copper- and silver-forms of cation exchangers evidently adsorbed by forming mercaptides. All the compounds could be eluted with sodium hydroxide or hydrochloric acid in acetone.

5. ELECTROPHORESIS

Comparatively few electrophoretic separations have been carried out on thiols or mixtures in which thiols were present more than occasionally. Schrauwen (1963) referred to separation of thiols by paper electrophoresis at pH 6, using acetate buffer, and under 220 V. He dried the paper in nitrogen and at low temperature to minimize oxidation of the thiol group. His work concerns primarily the visualization procedure which he performed with 1-(4-acetoxymercuriphenylazo)-2-naphthol. Wronski (1973) described the high voltage paper electrophoretic separation of thiols, disulphides and other sulphur compounds. He detected with o-hydroxymercuribenzoic acid and dithiofluorescein; here, too, the accent was on the detection.

Amino acids have often been subjected to electrophoresis but the chance presence of, say, cysteine, among the examples scarcely justifies comprehensive inclusion here. A few examples are nevertheless given: Ambert et al. (1966) separated urinary amino acids, including homocysteine, by electrophoresis at pH 5·4; they also carried out two-dimensional separations, electrophoresis/chromatography. Bondarev et al. (1970) undertook PC and electrophoretic studies of sulphur- and nitrogen-containing substances, including cysteine and cysteamine. An isotachophoretic separation of amino acids, including cysteine, was performed by Everaerts and van der Put (1970); they used an electrolyte of pH 7·1, 0·02 M in collidine, 0·01 M in formic acid and containing 17·5% formaldehyde to convert the amine groups to hydroxymethylamino groups, $-N(CH_2OH)_2$ and modify its pK value. Wronski (1971) separated glutathione and also oxidized glutathione and the sulphone from other amino acids in low molecular weight fractions of blood, using medium voltage (12 V/cm) electrophoresis; the medium was acetic acid–formic acid–water (10 + 3 + 37), after which he developed in butanol–acetic acid–water (4 + 1 + 1) at 90° to the original direction and finally visualized with a ninhydrin reagent.

Some examples of electrophoretic separation of derivatives of amino acids have been given under the headings of the reagents used to prepare the derivatives. To summarize, mention may be made here of the following. Zuber et al. (1955) determined cysteine in wool by conversion to the 2,4-dinitrophenyl derivative, hydrolysing with hydrochloric acid and submitting to paper electrophoresis at pH 1·93. The spot corresponding to the cysteine product (detected in u.v. light) was cut out, eluted and ultimately evaluated at 320–340 nm. Stratton and Frieden (1967) treated the thiol groups in proteins with 1-(3-chloromercuri-2-methoxypropyl)urea containing ^{203}mercury; estimation was carried out by electrophoresis and autoradiography. Sloan (1969) subjected the DANS(diaminonaphthalenesulphonyl) derivatives of amino acids, including cysteine, to electrophoresis (1 h at 1000 V at 12–15°C) on cellulose thin layers in a pH 4·4 buffer containing 0·4% pyridine and 0·8% acetic acid. After drying and exposing to ammonia he visualized the amino acid derivatives through their yellow fluorescence in u.v. light. States and Segal (1969) separated the N-ethyl-maleimide derivatives of non-protein thiols by TLC and electrophoresis.

The oxidation products of certain sulphur-containing amino acids are often easier to separate from others than are the unoxidized forms. Cysteine is a good example, oxidized to cysteic acid containing a sulphonic acid group. Thus Spencer and Wold (1960) treated cysteine (and cystine) groups in proteins with dimethyl sulphoxide, then hydrolysed with 6 N acid in the usual way and identified the cysteic acid by high voltage paper electrophoresis.

REFERENCES

Andreen, B. H. and Kniebes, D. V. (1962). *Proc. Operating Sect., Am. Gas Assoc.* CEP-62-13; *Chem. Abs.* **60**, 7843.
Andreen, B. H., Kniebes, D. V. and Tarman, P. B. (1963). *Inst. Gas Technol., Tech. Rept.* No. 7; *Chem. Abs.* **62**, 2645.
Amberg, C. H. (1958). *Can. J. Chem.* **36**, 590.
Ambert, J. P., Pechery, C., Charpentier, C. and Hartmann, L. (1966). *Ann. Biol. Clin. (Paris)* **24**, 17.
Andersson, K. and Bergstrom, J. G. T. (1969). *Svensk Papperstidn.* **72**, 375.
Bayfield, R. F. and Cole, E. R. (1969). *J. Chromatogr.* **40**, 470.
Baumann, F. and Olund, S. A. (1962). *J. Chromatogr.* **9**, 431.
Beer, J. Z., Budzynski, A. Z. and Malwinska, K. (1967). *Chem. Anal. (Warsaw)*, **12**, 1055.
Begliomini, A. and Fravolini, A. (1970). *Arch. Vet. Ital.* **21**, 63.
Belliveau, P. E. and Frei, R. W. (1971). *Chromatographia*, **4**, 189.
Bhatia, M. S., Bajaj, K. L., Singh, S. and Bhatia, I. S. (1972). *Analyst (London)* **97**, 890.
Bheemeswar, B. and Sreenwasaya, M. (1952). *Current Sci. (India)*, **21**, 213.
Bloembinger, R. H. and Vermaak, C. (1965). *Erdol u. Kohle*, **18**, 185.
Bondarev, G. N., Grachev, E. V., Mus, G. I. and Leonova, E. V. (1970). *Zh. Anal. Khim.* **25**, 1219.
Brady, P. R. and Hoskinson, R. M. (1971). *J. Chromatogr.* **54**, 65.
Brand, V. T. and Keyworth, D. A. (1965). *Anal. Chem.* **37**, 1424.
Bremmer, J. M. and Banwart, W. L. (1974). *Sulphur Inst. J.* **10**, 6–9, 17; *Chem. Abs.* **81**, 103715.
Brink, D. L., Pohlman, A. A. and Thomas, J. F. (1971). *Tappi*, **54**, 714.
Brown, P. R. and Edwards, J. O. (1968). *J. Chromatogr.* **38**, 543.
Brueggeman, J. and Schole, J. (1967). *Landwirt. Forsch.* No. 21, 134.

Bruner. F., Liberti, A., Passanzini, M. and Allegrini, I. (1972). *Anal Chem.* **44**, 2070.
Caldwell, K. A. and Tappel, A. L. (1968). *J. Chromatogr.* **32**, 625.
Carson, J. F., Weston, W. J. and Ralls, J. W. (1960). *Nature (London)*, **186**, 801.
Cifonelli, J. A. and Smith, F. (1955). *Anal. Chem.* **27**, 1501.
Coleman, H. J., Thompson, C. J., Ward, C. C. and Rall, H. T. (1958). *Anal. Chem.* **30**, 1592.
Dalgleish, C. E. (1950). *Nature (London)*, **166**, 1076.
Day, E. A. and Patton, S. (1959). *Microchem. J.* **3**, 137.
Deyl, Z., Schinkamannová, L. and Rosinus, J. (1967). *J. Chromatogr.* **30**, 614.
Doizaki, W. M. and Zieve, L. (1973). *J. Lab. Clin. Med.* **82**, 674.
Drews, B., Baerwald, G. and Niefind, H.-J. (1969). *Mschr. Brau,* **22**, 140; *Anal. Abs.* **19**, 1748.
Duro, F. (1961). *Boll. Sedute Accad. Groenia Sci. Nat. Catania,* **6**, 125; *Chem. Abs.* **57**, 17152.
Edward, J. T. and Nielsen, S. (1953). *Chem. Ind. (London)*, 197.
Everaerts, F. M. and van der Put, A. J. M. (1970). *J. Chromatogr.* **52**, 415.
Farrugia, V. J. and Jarreau, C. L. (1962). *Anal. Chem.* **34**, 271.
Feldstein, M., Balestrieri, S. and Levaggi, D. A. (1965). *J. Air Pollution Control Assoc.* **15**, 215.
Field jr., T. G. and Gilbert, J. B. (1966). *Anal. Chem.* **38**, 628.
Fishbein, L. and Zielinski jr., W. L. (1967). *J. Chromatogr.* **28**, 418.
Fischer, R. and Otterbeck, N. (1958). *Sci. Pharm.* **26**, 184.
Folkard, A. E. and Joyce, A. E. (1963). *J. Sci. Food Agric.* **14**, 510.
Fravolini, A. and Begliomini, A. (1965). *J. Assoc. Offic. Agric. Chem.* **48**, 908.
Fredericks, E. M. and Harlow, G. A. (1964). *Anal. Chem.* **36**, 263.
Freedman, R. W. (1968). *J. Gas Chromatogr.* **6**, 495.
Funasaka, W., Kojima, T. and Seo, Y. (1968). *Japan Analyst,* **17**, 464.
Ganeva, M. (1971). *Khig. Zdraveopazvane* **14**, 308; *Chem. Abs.* **75**, 132686.
Gasparič, J. and Borecký, J. (1961). *J. Chromatogr.* **5**, 466.
Génévois, L. (1964). *Chim. Anal. (Paris)*, **46**, 539.
Giannessi, P. (1961). *Ann. Staz. Chim.-Agrar. Sper., Roma, Ser. III Pubbl.* No. 190; *Chem. Abs.* **58**, 3830.
Glaser, C. B., Maeda, H. and Meienhofer, J. (1970). *J. Chromatogr.* **50**, 151.
Golovnya, R. V. and Arsen'ev, Yu. N. (1970). *Izv. Akad. Nauk SSSR, Ser. Khim.* 1399.
Golovnya, R. V. and Arsen'ev, Yu. N. (1972). *Izv. Akad. Nauk SSSR, Ser. Khim.* 1402; *Chem. Abs.* **78**, 37701.
Gordon, A. H. (1949). *Biochem. J.* **45**, 99.
Grassetti, D. A. (1971). *U.S. Patent* No. 3,597,160 of Aug. 3.
Grassetti, D. R. and Murray, jr., J. F. (1969). *J. Chromatogr.* **41**, 121.
Groenen, P. J. and van Gement, L. J. (1971). *J. Chromatogr.* **57**, 239.
Grossert, J. S. and Langler, R. F. (1974). *J. Chromatogr.* **97**, 83.
Guerin, M. R. (1971). *Anal. Letters* **4**, 751.
Hanes, C. S., Hird, F. R. and Isherwood, F. A. (1950). *Nature (London)* **166**, 288.
Hartmann, C. H. (1971), *Joint Conf. Sensing Environ. Pollutants, Coll. Tech. Pap.* I, 71; *Chem. Abstr.* **77**, 9308.
Hashida, I. and Nishimura, M. (1968). *Kogyo Kagaku Zasshi* **71**, 1939.
Havíř, J., Vrestal, J. and Chromý, V. (1965). *Chem. Listy,* **59**, 431.
Hermanson, H. P., Helrich, K. and Carey, W. F. (1968). *Anal. Letters,* **1**, 941.
Howard, G. E. and Baldry, J. (1969). *Analyst (London)*, **94**, 589.
Ishibashi, N., Kamata, S. and Matsuura, M. (1967). *Kogyo Kagaku Zasshi* **70**, 1036.

Jacot-Guillarmod, A. and Ceschini, P. (1959). *Helv. Chim. Acta*, **42**, 713.
Jansen, H. E., Strating, J. and Westra, W. M. (1971). *J. Inst. Brewing*, **77**, 154.
Karchmer, J. H. (1959). *Anal. Chem.* **31**, 1377.
Kawanami, J. (1964). *Mikrochim. Acta*, 106.
Kefurt, K., Kefurtová, Z. and Jarý, J. (1972). *J. Chromatogr.* **67**, 193.
Khusmitdinova, Z. S. (1965). *Metody Analiza Radioaktivn. Preparatov, Sb. Statei* 148; *Chem. Abs.* **63**, 11985.
Korolczuk, J., Daniewski, M. and Mielniczuk, Z. (1974). *J. Chromatogr.* **100**, 165.
Kremer, L. and Spicer, L. D. (1973). *Anal. Chem.* **45**, 1963.
Kumar, R. and Sen, A. K. (1969). *Indian J. Biochem.* **6**, 82.
Lederer, M. and Silberman, H. (1952). *Anal. Chim. Acta*, **6**, 133.
Lee, C. C. and Lai, Ts-S. (1967). *Can. J. Chem.* **45**, 1015.
Leppin, E., Gollnick, K. and Schomburg, G. (1969). *Chromatographia*, **2**, 535.
Le Rosen, H. D. (1961). *Anal. Chem.* **33**, 973.
Liberti, A. (1957). *Anal. Chim. Acta*, **17**, 247; also Liberti, A. and Cartoni, G. P. (1957). *Chim. e Ind. (Milan)* **39**, 821.
Lulova, N. J. and Tsifrinovich, A. N. (1971). *Zav. Lab.* **37**, 899.
Maekawa, K., Kodama, M., Kushii, M., Mitamura, T., Yada, N. and Fukino, H. (1967). *Nem. Ehime Univ.*, Sect. VI, 12, 131; *Chem. Abs.* **68**, 92772.
Martin, R. L. and Grant, J. A. (1965). *Anal. Chem.* **37**, 1644.
Merritt, C. and Walsh, J. T. (1962). *Anal. Chem.* **34**, 903
Millingen, M. B. (1974). *Anal. Chem.* **46**, 746.
Miyagawa, T., Kaji, H., Sakata, Y. and Minakawa, S. (1971). *Kurume Med. J.* **18**, 177; *Chem. Abs.* **76**, 137660
Oaks, D. M., Hartman, H. and Dimick, K. P. (1964). *Anal. Chem.* **36**, 1560.
Obara, Y., Ishikawa, Y. and Nishino, C. (1966). *Agric. Biol. Chem.* **30**, 164.
Obtemperanskaya, S. I. and Nguyen Dyk Hoc (1970). *Vestn. Mosk. Univ., Khim.* **11**, 369; *Chem. Abs.* **74**, 82828.
Okita, T. (1967). *Koshu Eiseiin Kenkyu Hokoku*, **16**, 59; *Chem. Abs.* **68**, 117009.
Papke, K. and Pohloudek-Fabini, R. (1967). *Pharmazie*, **22**, 485.
Pasieka, A. E. and Morgan, J. F. (1955). *Biochim. Biophys. Acta*, **18**, 236.
Pechernikova, Z. D., Chernyshkova, R. E. and Kruglov, E. A. (1972). *Nov. Sorb. Khrom.* No. 1889; *Chem. Abs.* **80**, 90932.
Popl. M. and Weisser, O. (1966). *Sb. Vys. Sk. Chem.-Technol v Praze, Technol. Paliv* **11**, 35; *Chem. Abs.* **67**, 50141.
Porter, M., Saville, B. and Watson, A. A. (1963). *J. Chem. Soc. (London)*, 346.
Price, C. A. and Campbell, C. W. (1957). *Biochem. J.* **65**, 512.
Prinzler, H. W., Pape, D. and Teppke, M. (1965). *J. Chromatogr.* **19**, 375.
Prinzler, H., Pape, D., Tauchmann, H., Teppke, M. and Tzscharnke, C. (1966). *Ropa Uhlie*, **8**, 13; *Chem. Abs.* **65**, 9710.
Raulin, F. and Toupance, G. (1974). *J. Chromatogr.* **90**, 218.
Rayner, H. B., Murray, F. F. and Williams, I. H. (1967). *Pulp Pap. Mag. Canada*, **68**, T301; *Anal. Abs.* **15**, 6348.
Reinhardt, F. (1954). *Mikrochim. Acta*, 219.
Reinhardt, M., Otto, R. and Koch, H. (1970). *Chem. Tech. (Berlin)*, **22**, 481.
Rozova, N. D. (1970). *Gig. Tr. Prof. Zabol.* 14, 58; *Chem. Abs.* **74**, 15543.
Ryce, S. A. and Bryce, W. A. (1957). *Anal. Chem.* **29**, 925.
Sanso, G. and Rigoli, A. (1970). *Boll. Chim. Farm.* **109**, 266.
Sawicki, E., Engel, C. R. and Elbert, W. C. (1967). *Talanta*, **14**, 1169.
Schrauwen, J. A. M. (1963). *J. Chromatogr.* **10**, 113.

Schwartz, D. P. and Brewington, C. R. (1968). *Microchem. J.* **13**, 310.
Shahrokhi, F. and Gehrke, C. W. (1968). *J. Chromatogr.* **36**, 31.
Shapunov, L. A., Parnyuk, N. I., Gosteminskaya, T. V. and Karatova, T. S. (1971). *Zav. Lab.* **37**, 1066.
Sloan, B. P. (1969). *J. Chromatogr.* **42**, 426.
Smith, O. D. and Tauss, K. H. (1970). *S. Pulp Paper Mfr.* **33**, 32, 34; *Chem. Abs.* **72**, 101987.
Snegotskii, V. I. and Snegotskaya, V. A. (1969). *Zav. Lab.* **35**, 429.
Spencer, R. I. and Wold, F. (1969). *Anal. Biochem.* **32**, 185.
Spencer, C. F., Baumann, F. and Johnson, J. F. (1958). *Anal. Chem.* **30**, 1473.
Sporek, K. F. and Danyi, M. D. (1963). *Anal. Chem.* **35**, 956.
States, B. and Segal, S. (1969). *Anal. Biochem.* **27**, 323.
Stephan, R. and Erdman, J. G. (1964). *Nature (London)*, **203**, 749.
Stepień, M. and Gaczyński, R. (1961). *Chem. Anal. (Warsaw)*, **6**, 1045.
Stevens, R. K., Mulik, J. D., O'Keefe, A. E. and Krost, K. J. (1971). *Anal. Chem.* **43**, 827.
Strack, E., Friedel, W. and Hambsch, K. (1956). *Z. Physiol. Chem.* **305**, 166.
Stratton, L. P. and Frieden, E. (1967). *Nature (London)* **216**, 932.
Stubbs, R. C. (1971). Amer. Gas. Ass., *Oper. Sect. Prec. D* 53-DJ 5.
Sullivan, J. H., Walsh, J. T. and Merritt jr., C. (1959). *Anal. Chem.* **31**, 1826.
Sunner, S., Karrman, K. J. and Sundeń, V. (1955). *Mikrochim. Acta*, 1144.
Teshima, I., Matsuura, S., Inukai, Y. and Ichikawa, Y. (1955). *Res. Repts. Nagoya Ind. Sci. Research Inst.* No. 8, 62; *Chem. Abs.* **50**, 15693.
Titov, A. V. (1969). *Vop. Med. Khim.* **15**, 295; *Chem. Abs.* **71**, 67814.
Toennies, G. and Kolb, J. J. (1951). *Anal. Chem.* **23**, 823, 1095.
Trop, M., Sprecher, M. and Pinsky, A. (1968). *J. Chromatogr.* **32**, 426.
Villanueva, V. R. and Barbier, M. (1967). *Bull. Soc. Chim. France*, 3992.
Waksmundzki, A., Wawrzynowicz, T. and Wolski, T. (1963). *Acta Polon. Pharm.* **20**, 259.
Wallace, T. J. and Mahon, J. J. (1964). *Nature (London)*, **201**, 817.
Weisser, O., Popl, M. and Landa, S. (1966). *Gas Wasserfach*, **107**, 282.
Wernze, H. and Koch, W. (1965). *Klin. Wochschr.* **43**, 459.
White, B. J. and Wolfe, R. G. (1962). *J. Chromatogr.* **7**, 516.
White, L. B., Torosian, G. and Becker, C. H. (1974). *Anal. Chem.* **46**, 143.
Winegard, H. M., Toennies, G. and Block, R. J. (1948). *Science*, **108**, 506.
Wolf, F., Kotte, G. and Hannemann, J. (1971). *Chem. Tech. (Berlin)*, **23**, 550.
Wong, F. F. (1971). *J. Chromatogr.* **59**, 448.
Woodroffe, G. L. and Munro, J. L. (1970). *Analyst (London)*, **95**, 153.
Wronski, A. (1971). *Z. Med. Labortech.* **12**, 151.
Wronski, M. (1973). *Z. Anal. Chem.* **264**, 406.
Zijp, J. W. H. (1956). *Rec. Trav. Chim.* **75**, 1060.
Zuber, H., Traumann, K. and Zahn, H. (1955). *Z. Naturforsch.* **10b**, 457.

13. ENZYMIC METHODS

Microbiological methods usually require many hours for completion but are often highly specific, sensitive and cheap. Although such methods fall outside the planned scope of this work, some information is given here:

1. GLYOXALASE METHOD FOR GLUTATHIONE

Lohmann (1932) discovered that glutathione is a specific activator of the enzyme glyoxalase which converts methylglyoxal to lactic acid. Using a mixture of yeast, methylglyoxal and sodium hydrogen carbonate, Woodward (1935) obtained carbon dioxide as reaction product and plotted the volume obtained in 20 min against concentration of (reduced) glutathione in a calibration curve to enable this thiol to be determined. Her method, or a modification, found subsequent wide application, e.g. by Ennor (1939) who obtained results on tissue extracts in good agreement with those from iodate titration; Racker (1951); Krimsky and Racker (1952); Bhattacharya *et al.* (1955); Martin and McIlwain (1959) who used yeast apoglyoxalase and obtained total glutathione values in brain.

Schroeder and Woodward (1939) modified the original method to eliminate the manometric measurement. They followed the reaction by determining residual methylglyoxal substrate using a modification of Clift and Cook's bisulphite method (1932); after 15 min reaction with excess bisulphite, unreacted reagent was removed with iodine, the bisulphite addition compound then decomposed by adding disodium hydrogen phosphate and the released sulphite was titrated with standard iodine. Krimsky and Racker (1952) followed the disappearance of methylglyoxal by converting it to its 2,4-dinitrophenylhydrazone and estimating colorimetrically the product which this yielded with alkali.

The influence of other thiols and compounds on the assay has been studied, e.g. by Fodor *et al.* (1953) who found for instance that cysteine was inactive but not cysteinylglycine (a degradation product of glutathione). Shamshikova and Ioffe (1947) stated that Woodward's method is reliable only in the absence of cysteine which otherwise catalyses oxidation of the glutathione. A nitrogen or hydrogen atmosphere prevents this.

Rausch *et al.* (1957) compared the enzymic method with iodate titration and colorimetry using nitroprusside.

Racker (1951) showed that two enzymes are involved in the change. Glyoxalase I catalyses condensation of methylglyoxal with glutathione to give an intermediate; glyoxalase II then splits this to lactic acid and glutathione. Wieland *et al.* (1955) based their enzymic procedure for glutathione on the first of these two reactions. They mixed methylglyoxal, sample, phosphate buffer of pH 6·3 and glyoxalase I at 20°C. The intermediate, S-lactoylglutathione, is formed quantitatively within a few minutes and they estimated it by absorbance measurements at 240 nm. They applied the method to determine glutathione in blood and organ extracts.

2. GLUTATHIONE AS A COENZYME

Glutathione functions also as coenzyme in the *cis–trans* isomerization of maleyl- to fumarylpyruvate by a bacterial isomerase prepared from a Pseudomonas strain (Lack and Smith, 1964). They determined the thiol by following the reaction through the decrease in absorbance at 330 nm and were able to determine amounts down to a few nanograms.

REFERENCES

Bhattacharya, S. K., Robson, J. S. and Stewart, C. P. (1955). *Biochem. J.* **60**, 696.
Clift, F. P. and Cook, R. P. (1932). *Biochem. J.* **26**, 1788.
Ennor, A. H. (1939). *Australian J. Exptl. Biol. Med. Sci.* **17**, 157.
Fodor, P. J., Miller, A., Neidle, A. and Waelsch, H. (1953). *J. Biol. Chem.* **203**, 991.
Krimsky, I. and Racker, E. (1952). *J. Biol. Chem.* **198**, 721.
Lack, L. and Smith, M. (1964). *Anal. Biochem.* **8**, 217.
Lohmann, K. (1932). *Biochem. Z.* **254**, 332.
Martin, H. and McIlwain, H. (1959). *Biochem. J.* **71**, 275.
Racker, E. (1951). *J. Biol. Chem.* **190**, 685.
Rausch, L., Nick, H. and Tegeler, R. (1957). *Arch. Klin. Exptl. Dermatol.* **205**, 245.
Schroeder, E. F. and Woodward, G. E. (1939). *J. Biol. Chem.* **129**, 283.
Shamshikova, G. A. and Ioffe, A. L. (1947). *Biokhimiya,* **12**, 437.
Wieland, T., Dose, K. and Pfleiderer, G. (1955). *Biochem. Z.* **326**, 442.
Woodward, G. E. (1935). *J. Biol. Chem.* **109**, 1.

AUTHOR INDEX

Numbers in italic refer to the pages where references are listed in full

K

SUBJECT INDEX

A hyphen joining two page numbers indicates that these and all intermediate pages contain references to the entry; it is used to save space.

The word "to" joining two page numbers indicates that this is a chapter or section heading, running from the one number to the other.